MUON PHYSICS

VOLUME III

Chemistry and Solids

CONTRIBUTORS

J. H. Brewer

K. M. Crowe

V. S. Evseev

S. S. Gershtein

F. N. Gygax

L. I. Ponomarev

A. Schenck

Muon Physics

VOLUME III

Chemistry and Solids

Edited by

VERNON W. HUGHES

Physics Department
Yale University
New Haven, Connecticut

C. S. WU

Department of Physics
Columbia University
New York, New York

ACADEMIC PRESS New York San Francisco London 1975

A Subsidiary of Harcourt Brace Jovanovich, Publishers

ACADEMIC PRESS, INC.
111 Fifth Avenue, New York, New York 10003

United Kingdom Edition published by
ACADEMIC PRESS, INC. (LONDON) LTD.
24/28 Oval Road, London NW1

Library of Congress Cataloging in Publication Data
Main entry under title:

Muon physics.

Includes bibliographies and index.
1. Muons. I. Hughes, Vernon W. II. Wu, Chien-shiung.
QC793.5.M82M86 539.7'2114 75-11829
ISBN 0-12-360603-9 (v. 3)

PRINTED IN THE UNITED STATES OF AMERICA

CONTENTS

Section 3 DEPOLARIZATION OF NEGATIVE MUONS AND INTERACTION OF MESONIC ATOMS WITH THE MEDIUM

V. S. Evseev (Translated by S. J. Amoretty)

LIST OF CONTRIBUTORS

Numbers in parentheses indicate the pages on which the authors' contributions begin.

J. H. Brewer (3), Department of Chemistry and TRIUMF, University of British Columbia, Vancouver, British Columbia, Canada

K. M. Crowe (3), Department of Physics and Lawrence Berkeley Laboratory, University of California, Berkeley, California

V. S. Evseev (235), Laboratory of Nuclear Problems, Joint Institute for Nuclear Research, Dubna, USSR

S. S. Gershtein (141), Theoretical Department, Institute of High Energy Physics, Serpukhov, USSR

F. N. Gygax (3), Lawrence Berkeley Laboratory, University of California, Berkeley, California, and Swiss Institute for Nuclear Research, Villigen, Switzerland

L. I. Ponomarev (141), Laboratory of Theoretical Physics, Joint Institute for Nuclear Research, Dubna, USSR

A. Schenck (3), Laboratory for High Energy Physics, Swiss Federal Institute of Technology, Zurich, Switzerland

PREFACE

The contents of this treatise are grouped into seven chapters:

Chapter II, on the electromagnetic interaction of muon, describes in detail the latest experimental and theoretical developments concerning the static properties of the muon and the validity tests of QED in the simple muonic system such as muonium (μ^+e^-), muonic hydrogen (μ^-p), and heavier muonic atoms (μ^-A). Possible tests of QED at much higher energy and large momentum transfers are also discussed. An explanation of the unified gauge theories of electromagnetic and weak interactions in very simple and easily understandable terms is also included.

Chapter III is on muonic atoms, a subject that has been reviewed recently in several excellent articles. Therefore, the aim is to present the field in a way that can serve as a starting point for new work on muonic atoms with the next generation of experiments. Those aspects that are relatively new and are likely to raise special interest in the future are discussed in detail.

Chapter IV on cosmic-ray muons emphasizes the character of very high-energy nucleon–nucleon interactions and the properties of the electromagnetic and weak interactions at very high energies.

Chapter V is on weak interactions. Sections 1 and 4 are on theories: Section 1 deals with elementary particle aspects of muon decay and muon capture. The conventional two- and one-neutrino-field theories are presented. The law of lepton conservation is examined in both cases. Section 4, on semileptonic weak interactions in nuclei, includes neutrino reactions, charged-lepton capture, and β decay. Because of the close analogy between the semileptonic weak processes and electron scattering, these two processes are discussed together. As is pointed out, in principle, the relationship between semileptonic weak processes and electromagnetic processes can be obtained quite directly, without going through the intermediary of any nuclear model. Many beautiful examples can be found throughout this section, particularly in the appendices, where a few selected recent developments and some discussion and speculation about future desirable experiments are presented.

The experimental results on weak interactions (low energies) are reviewed and discussed in three separate sections: Section 2 on experimental muon decay, Section 3 on rare and ultrarare muon decays, and Section 5 on muon capture. More precise determinations in some of the experiments seem to be highly desirable.

Chapter VI is on interactions of muon neutrinos, with emphasis on the high-energy type only. This is a very new experimental approach in the study of high-energy physics and weak interactions and will probably be the superstar of the next generation. The experimental setup is gigantic and the technique is difficult. The statistics are not always as good as desired. Yet the outcome of these experiments will determine the future theoretical approaches to the weak interaction and the unified theory of weak and EM interactions. The stake on this type of experiments is indeed very high.

Chapter VII examines muon chemistry and muons in matter, a very broad and rapidly expanding field. In order to include the most recent developments, three outstanding and active groups have pooled their efforts: Section 1, on positive muons and muonium in matter; Section 2, on mesomolecular processes induced by muons; and Section 3, on depolarization of negative muons and interaction of muonic atoms with the medium.

CONTENTS OF OTHER VOLUMES

CHAPTER VII

MUON CHEMISTRY AND MUONS IN SOLIDS

Section 1

POSITIVE MUONS AND MUONIUM IN MATTER

J. H. BREWER

Department of Chemistry and TRIUMF
University of British Columbia
Vancouver, British Columbia, Canada

K. M. CROWE

Department of Physics and Lawrence Berkeley Laboratory
University of California
Berkeley, California

F. N. GYGAX

Lawrence Berkeley Laboratory
University of California
Berkeley, California
and
Swiss Institute for Nuclear Research
Villigen, Switzerland

A. SCHENCK

Laboratory for High Energy Physics, ETH
Zurich, Switzerland

I. Introduction

The purpose of this paper is to provide a status report on a field of muon physics that is in a state of rapid expansion: the study of interactions of positive muons and muonium with matter. We have chosen to concentrate on recent results currently under extensive study instead of providing an historical account of progress to date. We apologize in advance to those whose pioneering work will be mentioned only briefly, with the excuse that they have led the way to so many new and exciting topics that sometimes there is only room left for a brief reference to the early publications. We shall also take generous advantage of descriptions in other texts; these will provide, from time to time, the framework necessary for the reader to become his own innovator. Finally, we will often refer to original papers, which, of course, must be the ultimate source for the serious reader. At times we will only be able to provide a brief sketch to show the way and to convey our excitement about this relatively new field.

The first step of an experimental study in this field is to bring polarized positive muons to rest in condensed or gaseous matter. The usual μ^+ beam characteristics and the involved stopping mechanism leave the thermalized muons distributed over an extended macroscopic zone of the target. They can, therefore, be used as a probe for testing the bulk properties of the target material. Unlike the μ^-, the μ^+ will not be captured in atomic or molecular orbits and will have no nuclear interactions. Until its decay, the implanted positive muon behaves much like a proton, playing the role of the nucleus of muonium, a "light isotope of atomic hydrogen." We will be mainly interested in the muon spin's magnetic interaction with the target medium. The muon will "see"—through its magnetic moment—all the magnetic field components at its location. These local fields may originate from nuclei, electrons, paramagnetic ions, or a variety of hyperfine interactions, all of which may be influenced by externally controlled parameters, such as temperature and applied magnetic field. When the μ^+ decays, the direction of positron emission is correlated with the direction of the muon magnetic moment. Thus, counting decay positrons in a given direction as a function of the time spent by the muons in the medium provides information about the evolution of their polarization. If many muons see an identical local field, the energy of interaction between that field and the μ^+ magnetic moment may be observed as a precession frequency; random or time-dependent local fields will cause relaxation of the monitored polarization.

In Section II we will discuss how the polarized muon is obtained, how it is detected, and how it decays. Section III will deal with the process by which a high-energy μ^+ comes to rest in matter, and the consequences thereof. Sections IV–VI will treat muonium—its formation, its depolarizing effect on the μ^+, and its chemical reactions. We will then discuss experimental studies of muonium chemistry in Section VII and of muonium in solids in Section VIII. Section IX will treat the interactions of quasi-free muons in matter.

The properties of the positive muon are well measured and understood; we will start by mentioning only those that are relevant to a phenomenological understanding of the complex interactions of positive muons with their environment.

II. Phenomenological Description of Production and Behavior of Polarized Positive Muons

Table 1 contains the basic information about muons relevant to this section. Apart from its mass and lifetime, the positive muon is in nearly

TABLE 1

MUON PROPERTIES

Spin	$\frac{1}{2}$
Mass	$m_\mu = 105.6595(3)$ MeV $= 206.7684(6)m_e$ $= 0.1126123(6)m_p$
Magnetic moment	$\mu_\mu = \lvert g_\mu S_z \rvert \dfrac{e\hbar}{2m_\mu c} = 3.183347(9)\mu_p$ $= 28.0272(2) \times 10^{-18}$ MeV/G
Bohr radius	$a_B{}^\mu = \dfrac{\hbar^2}{m_\mu e^2} = 255.927$ fm $= 2.55927 \times 10^{-11}$ cm
Compton wavelength	$\lambda\!\!\!^{-}_\mu = \dfrac{\hbar}{m_\mu c} = 1.86758$ fm
Lifetime	$\tau_\mu = 2.1994(6) \times 10^{-6}$ sec

every respect similar to a positron. In the following discussions, however, the proton-like aspects of its behavior, which have received relatively minor attention in the past, will emerge as more important to an understanding of its interactions with matter. The muonium atom (Mu), as a light radioactive isotope of hydrogen, promises to supplement greatly the information obtained from studies of tritium and positronium. (Negative muons play an entirely different role in matter from that of positive muons, but it is not the purpose of this paper to cover that field.)

The theory of the production and decay of the muon is well established; a few words are appropriate here about the mechanism by which nature has contrived to provide almost completely polarized muon beams. The earliest source of positive muons is, of course, cosmic rays; and even they are polarized. See, for example, Chapter IV, Volume II, on cosmic ray muons, or Bradt and Clark (1963). Artificially produced muons arising from pion decay, $\pi^+ \rightarrow \mu^+ + \nu_\mu$, are 100% polarized opposite to their momentum in the pion center of mass (c.m.) frame. In the laboratory frame, where the pions are relativistic, a muon beam arises from disintegrations at various decay angles. Preservation of a high average polarization requires only that one establish geometrical conditions that accept a reasonably narrow decay cone in the c.m. system. Fortunately, the momentum of the decay muon in the c.m. system (29 MeV/c) is large enough to separate forward decays from backward decays by simply requiring the muon to have either a higher magnetic rigidity than the pion or a longer range than the backward decay; the situation can easily be reversed. This argument assumes, of course, that the pion beam is monoenergetic. In fact, for a continuous spectrum of decaying pions, the muons are polarized only if the pion momentum spectrum has a steep slope in the region from which the observed muons originate.

The first cyclotrons operating in the meson-producing region yielded pion beams from internal targets located in the main circulation field. Positive pions usually can escape only if they are produced in the backward direction. Here the spectrum drops rapidly with momentum, producing highly polarized beams. Negative muon beams coming from pions produced in forward directions off internal targets will be less polarized.

With externally produced pions, a polarized muon beam can be made by providing a crude pion momentum selection and tuning the last elements of the channel for muons away from the mean pion momentum. Many muon channels also have sufficiently narrow momentum transmissions to provide excellent polarized beams simply by specifying the final muon range (see, for example, Culligan *et al.*, 1964).

Thus, using parity violation in π–μ decay and taking reasonable pains in the beam design, one can produce muon beams with polarizations typically between 60 and 90%. The direction of polarization can even be reversed, by tuning the muon momentum to be either above or below the mean pion momentum. The tighter the geometry and momentum resolution, the higher the polarization—at the expense, of course, of net flux.

The muon is decelerated in matter by normal collisions with electrons; the resultant multiple Coulomb scattering produces a broadened beam, but the muon's spin is unaffected by the electrostatic interactions and remains pointed in its initial direction. Let us assume that the muon has come to rest, still unaffected by local magnetic fields in the medium. This assumption will be justified in the next section.

If the muon is in a vacuum, it precesses in a magnetic field B at the classical Larmor frequency,

$$|\nu^{\mu}| = \left|\frac{2\mu_{\mu}B}{h}\right| = \left|\frac{g_{\mu}eB}{4\pi m_{\mu}c}\right| = 13.55\,\frac{\text{kHz}}{\text{G}} \times |B| \tag{1}$$

Table 2 shows the frequency and time scales corresponding to the range of fields we shall consider. Note that for future reference we have included the corresponding periods for the muonium atom.

The decay of the μ^+ lepton goes according to the reaction $\mu^+ \rightarrow$

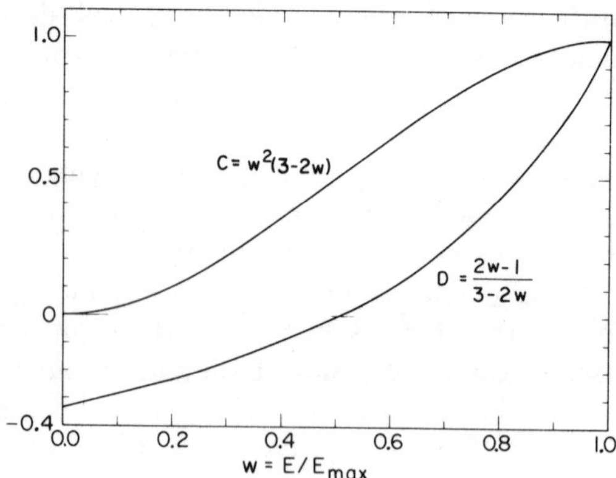

FIG. 1. μ^+-decay spectrum: isotropic contribution for the energy spectrum of the decay positron C (upper curve), and energy dependence of the asymmetry factor D, assuming $\xi = 1$ (lower curve).

$e^+ + \bar{\nu}_\mu + \nu_e$. The positron spectrum (treated in Chapter V) is described by

$$\frac{dN(w, \theta)}{dw\, d\Omega} = \frac{w^2}{2\pi}[(3 - 2w) - \xi(1 - 2w)\cos\theta] = \frac{C}{2\pi}[1 + D\cos\theta] \quad (2)$$

where $w = E/E_{\max}$ is the positron energy, measured in units of $E_{\max} = m_\mu/2$, θ is the angle between the spin of the μ^+ and the e^+ momentum, and ξ stands for the degree of polarization of the decaying muons. The spectrum is shown in Fig. 1.

In practice, the positrons are detected with an efficiency $\epsilon(w)$, which is not constant over their entire energy range. The observed probability is then

$$\left\{\int_0^1 \epsilon(w) \left[\frac{dN(w, \theta)}{dw\, d\Omega}\right] dw\right\} \frac{d\Omega}{4\pi} = \bar{\epsilon}[1 + \tilde{A}\cos\theta]\frac{d\Omega}{4\pi} \quad (3)$$

If positrons of all energies were detected with the same efficiency, the observed average asymmetry $\tilde{A}$ would be $\frac{1}{3}\xi$. As the detection efficiency for low-energy e^+ is reduced, the observed $\tilde{A}$ increases from $\frac{1}{3}\xi$ toward the limiting value of ξ. Although this effect has been exploited only in a few experiments, it is especially valuable for detecting very small muon polarizations.

In addition to reduction of the asymmetry by kinematic depolarization ($\xi < 1$) and by the average over positron energy, one observes an experimental asymmetry A smaller than $\tilde{A}$ because of the inevitable finite detector solid angle.

A related effect of detector geometry is the average effective misalignment of the muon spin with respect to the axis of symmetry of the positron telescope. To illustrate this effect, first consider the idealized case in which all the muons stop at the same site and the positron detector subtends a very small solid angle. We then consider only those decays in which the positron is coplanar with the telescope axis and the muon spin direction. The angle θ between muon spin and positron directions cannot generally be measured directly, since the muon spin direction is often the unknown we seek to define by experiment. It is, in principle, the angle θ' between the telescope axis and the e^+ direction that is directly observable. For events in the plane defined above, these two angles are related by $\theta' = \theta + \theta_0$, where θ_0 is the angle between the telescope axis and the muon polarization, as shown in Fig. 2. Since decays are only detected if $\theta' = 0$, the counting rate in the positron detector will depend on geometry as

$$R(\theta_0) \sim 1 + A \cos \theta_0 \tag{4}$$

An actual experimental apparatus, of course, has a finite stopping target and a finite positron detector, as shown in Fig. 3. The effect of each departure from the ideal is to require definition of an average effective θ_0. The typical experiment also involves application of a magnetic field to the region, which further complicates the definition of θ_0 due to the curved paths of incoming muons and decay positrons. In the end, however, Eq. (4) will always hold with an appropriate definition of A and θ_0, the latter being considered an empirical parameter.

One experimental approach to studying the muon polarization is to place two positron telescopes at approximately $\theta_0 = 0$ and $\theta_0 = \pi$. The difference between the properly normalized positron detection rates in these two directions will be proportional to the muon asymmetry A, which

TABLE 2

Observable Muon and Muonium Precession Frequencies and the Corresponding Cycle Periods for a Magnetic Field Strength B

B (kG)	0.01	0.1	1	10	100
$\lvert\nu^{\mu}\rvert$ (MHz)		1.355	13.55	135.5	1355
$T_{\mu} = 1/\lvert\nu^{\mu}\rvert$ (nsec)		740	74	7.4	0.74
$\lvert\nu^{\mathrm{Mu}}\rvert$ (MHz)	13.94	139.4	1394		
$T_{\mathrm{Mu}} = 1/\lvert\nu^{\mathrm{Mu}}\rvert$ (nsec)	72	7.2	0.72		

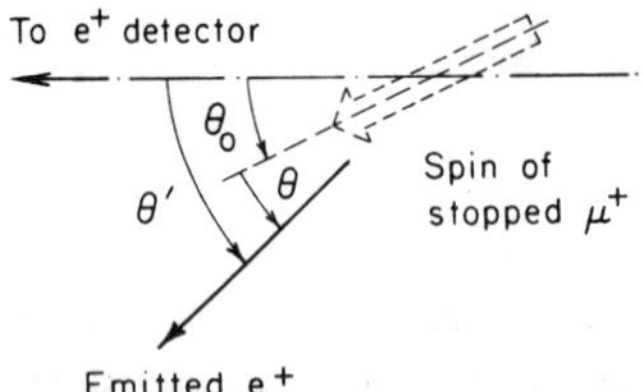

FIG. 2. Angular relations between detector axis, spin orientation of the stopped μ^+, and a coplanar positron emission direction.

may be a function of time, $A(t)$, due to spin relaxation by interactions with the medium. The application of a magnetic field parallel to the muon polarization often has dramatic effects on the muon asymmetry, as will be seen later. This "longitudinal field technique" has been used extensively in measuring the muonium hyperfine splitting (see Chapter II), and for studying certain interactions of muons with the medium. However, with somewhat greater experimental effort one can often obtain more information by using the "transverse field technique" described below.

In transverse field experiments, an external field B is applied perpendicular to the mutual plane of the muon polarization and the detector axis, causing the muon spins to precess at an angular frequency

$$\omega^\mu = \mu_\mu B/\hbar = g_\mu \mu_0{}^\mu B/\hbar = g_\mu eB/2m_\mu c \tag{5}$$

This causes θ_0 to be replaced by $\theta_0 + \omega^\mu(t_1 - t_0)$ in Eq. (4), where t_0 is the time of the muon's entry into the stopping target and t_1 is the time of the decay. The distribution of positron detections thus becomes oscillatory in time.

This distribution also decays exponentially with a lifetime $\tau_\mu = 2.2$ μsec due to the disintegration of the muon, and in some cases has a time dependence $A(t)$ of the envelope of the oscillations, which reflects a relaxation of the average polarization due to interactions with the medium and/or inhomogeneities in the external field.

The distribution can be expressed by the following formula, which incorporates all the effects mentioned above:

$$dN(t)/dt = N_0 \exp(-t/\tau_\mu)\{1 + A(t)\cos[\omega^\mu t + \phi]\} + BG \tag{6}$$

where $t = t_1 - t_0$ and $\phi = \theta_0 + \Delta\phi$, in which $\Delta\phi$ represents medium-dependent shifts of the apparent initial phase of the precession, to be discussed later. The term BG accounts for additional events, accidentals, and electronic breakthroughs, which occur at random generally and are usually rare. N_0 is a normalization factor, essentially determined by the solid angle

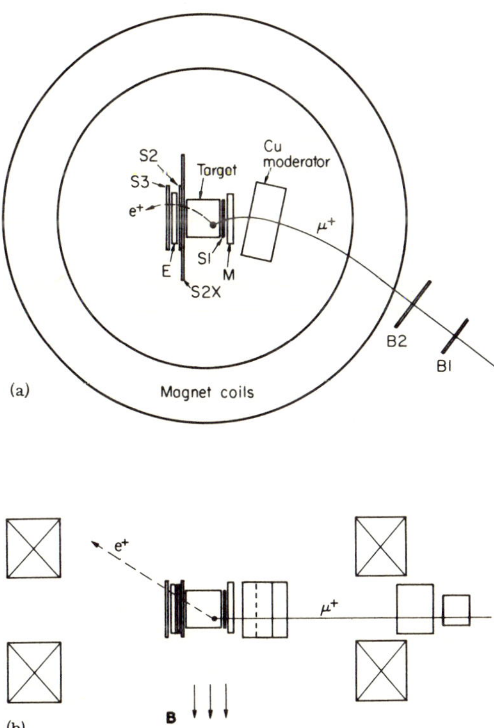

FIG. 3. Typical experimental setup. Top (a) and side (b) views of stopping target, counter arrangement, and magnet coils for transverse field. Not to scale.

of the e^+ detector, its average detection efficiency, and the total number of stopped muons.

Experimentally, one collects data by recording the arrival time t_0 of the muon and the decay time t_1, and constructing an elapsed-time histogram for $t = t_1 - t_0$, an example of which is shown in Fig. 4. In this situation, the ensemble of events is collected one at a time; the event signature is constructed so that only a single muon can be in the sample at once, and subsequently only one positron can appear. The time interval resolution Δt can be set either electronically or in the binning program, and is typically on the order of 10^{-9} sec. The expected number of counts in a histogram bin of width Δt at t_i is given by $[dN(t_i)/dt]\,\Delta t$; by varying the parameters N_0, $A(t)$, ω^μ, ϕ, and BG, the optimal fit of Eq. (6) to the entire histogram is determined, defining the best values for these parameters. Usually $A(t)$ is expressed as $A\exp(-t/T_2)$ in terms of the two independent parameters A and T_2. Of the four parameters A, T_2, ω^μ, and $\Delta\phi = \phi - \theta_0$, each repre-

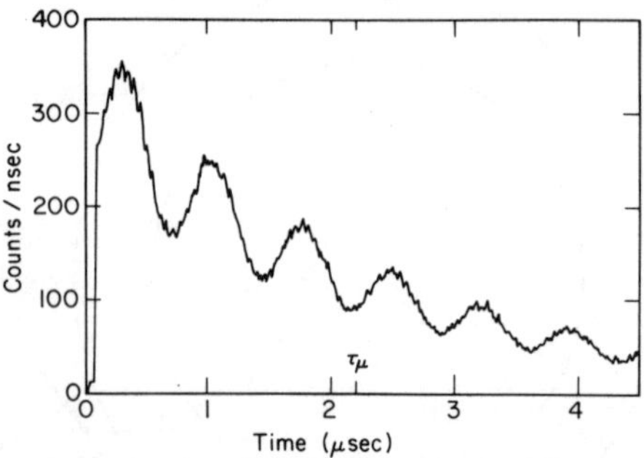

FIG. 4. A typical experimental histogram for muon precession in a transverse magnetic field. The target is carbon tetrachloride at 100 G; the data are binned into 10-nsec bins. The mean muon lifetime, $\tau_\mu = 2.20$ μsec, is indicated.

sents a piece of valuable information about the interactions of the positive muon.

The asymmetry A is proportional to the apparent initial muon polarization. As an indication of the amount of polarization one observes in practice, Table 3 shows a number of results for asymmetries measured in various materials. The data shown fall into three general groups. One group is essentially completely polarized ($A \approx \frac{1}{3}$), the second group is roughly 50% depolarized, and the third has a polarization of 10% or less.

Relaxation times in the range $10^{-7} < T_2 < 10^{-5}$ sec can easily be measured by this technique, providing information complementary to proton NMR data. These phenomena will be discussed in Section IX.

The fitted phase ϕ consists of two parts: the average effective angle θ_0 between the positron telescope axis and the mean polarization of the stopping muon, and the medium-dependent apparent phase shift $\Delta\phi$. Since in the usual situation muons of various polarizations stop over an extended target volume, θ_0 may be a poorly calculable quantity; it is usually left as an empirical parameter. Experimental uncertainties in t_0 are usually also absorbed into the definition of θ_0. The additional apparent phase shift $\Delta\phi$ is due exclusively to interactions of the muon in the target medium after it comes to rest. A nonzero value of $\Delta\phi$ generally reflects a very short-lived formation of muonium just after the μ^+ enters the target. During this brief episode, rapid precession takes place, which sometimes will leave the muons in their ultimate diamagnetic environment with somewhat rotated spins. All this happens too quickly to be directly observable, but the resultant $\Delta\phi$ provides very useful information about the fast processes involved. In Section VI we will discuss some of the recent applications of these phase data.

The precession frequency of a single muon in a magnetic field B is given by Eq. (5). Equation (6) embodies the assumption that all the muons see the same field B. Occasionally, in addition to small field inhomogeneities and local effects leading to muon spin relaxation by dephasing, there may be several distinct local fields seen by separate ensembles of muons. In this case $\cos(\omega^{\mu}t + \phi)$ must be replaced by $\sum_i P_i \cos(\omega_i^{\mu}t + \phi_i)$ to preserve the generality of Eq. (6) (P_i = fraction of the muons precessing at the frequency ω_i^{μ}). Often it is advantageous to perform a simple Fourier analysis of the time histogram (after subtracting background and dividing out the exponential decay). This treatment may reveal precession signals that otherwise might go unsuspected.

In the brief discussion above we have attempted to use only classical concepts; in our detailed discussions of these phenomena we will attempt to be more precise.

III. Deceleration and Thermalization of Positive Muons in Matter

A qualitative understanding of how a fast μ^+ slows down and stops in a target is essential to a complete picture of the behavior of muons in matter. Several questions relating to this process are of basic importance:

(1) What are the essential stages of the deceleration process?
(2) What is the time scale for each stage?
(3) Is the muon polarization affected?
(4) In what state or states does the muon finally thermalize?

In this section we will deal with these questions in some detail. Other questions, related to problems that are not yet well understood, can also be raised: What sorts of radiation damage are caused by the muon, and how might they influence the muon after it stops? In what media could there be significant exceptions to the established rules of how a μ^+ slows down? We will not attempt to answer these questions in the absence of specific experimental evidence, but in some instances interesting conjectures can be made.

The most important gross features of the energy loss process are displayed as a flow diagram in Fig. 5. The boxes represent the various stages of deceleration and contain descriptions of the mechanisms involved. On the right side are time estimates for the duration of the different stages, and the approximate kinetic energy at each stage is also indicated.

A muon entering a target with a kinetic energy of $\sim$50 MeV (momentum 115 MeV/c) will first lose energy by scattering with electrons, until its

TABLE 3

PRECESSION OF POSITIVE MUONS AND MUONIUM ATOMS IN PURE SUBSTANCES IN TRANSVERSE MAGNETIC FIELDS[a]

Target substance	Temp. (°K)	Field (G)	Asymmetry	T_2 (μsec)	Reference[b]
Metals and semimetals					
Aluminum	300	35	0.27 ± 0.01		1
Aluminum	300	50	0.209 ± 0.010		2
Beryllium	300	50	0.222 ± 0.012		2
Carbon (graphite)	300	50	0.229 ± 0.008		2
Carbon (lampblack)	300	50	0.253 ± 0.021		2
Carbon	300	800	0.219 ± 0.008		3
Carbon	300	1300	0.225 ± 0.004		3
Carbon	300	3500	0.203 ± 0.008		3
Lithium	300	50	0.201 ± 0.014		2
Magnesium	300	50	0.254 ± 0.013		2
Molybdenum	300	50	0.086 ± 0.009		5
Iron (powder, sphere)	300	50	0.015 ± 0.003	>20	LBL
Lead	300	350	0.020 ± 0.006	>20	LBL
Lead	5	350	<0.01		LBL
Copper	300			>50	4
Copper	150			8.80 ± 0.20	4
Copper	77			4.60 ± 0.30	4
Semiconductors					
Silicon	300	50	0.253 ± 0.012		2
Si crystal (P doped)	300	35	0.082 ± 0.020		1
Si crystal (B doped)	300	35	0.14 ± 0.014		1
Si crystal (B doped)	300	100	0.094 ± 0.002	>20	LBL
Si crystal (B doped)	77	50	0.012 ± 0.002	>20	LBL

Si crystal (B doped)	77	100	0.046 ± 0.005	10.26 ± 5.11	LBL
Si crystal (B doped)	77	150	0.014 ± 0.001	>20	LBL
Ge crystal (P doped)	300	35	0.25 ± 0.02		1
Ge crystal (As doped)	300	50	0.269 ± 0.068		6
Ge crystal (impure)	300	50	0.172 ± 0.012	7 ± 2	6
SiC	300	50	0.213 ± 0.011		2
		Insulators			
Carbon (diamond)	300	50	0.045 ± 0.008		2
Sulfur	300	35	<0.005		1
Sulfur	300	50	0.014 ± 0.011		2
Sulfur	300	800	0.056 ± 0.008		3
Sulfur	300	3500	0.048 ± 0.009		3
Sulfur	300	4000	0.046 ± 0.007	0.03 ± 0.005	3
SiO_2 (quartz)	300	50	0.043 ± 0.003		7
SiO_2 (fused)	300	50	0.038 ± 0.006		2
SiO_2 (fused)	300	50	0.050 ± 0.006		8
SiO_2 (fused)	300	100	0.037 ± 0.003	>20	LBL
SiO_2 (crystal)	300	50	0.056 ± 0.006		8
SiO_2 (crystal)	300	88	<0.03		2
SiO_2 (powder)	300	50	0.110 ± 0.005		8
Al_2O_3 (corundum)	300	35	<0.07		1
Al_2O_3	300	50	0.022 ± 0.009		2
Al_2O_3	300	800	0.137 ± 0.007		3
Al_2O_3	300	3500	0.108 ± 0.007		3
Al_2O_3 (powder)	300	50	0.111 ± 0.007		8
B_2O_3 (fused)	300	50	0.127 ± 0.006		8
LiH	300	50	0.166 ± 0.011	5.3 ± 1.0	7
LiH	300	11,000	0.127 ± 0.002	4.51 ± 0.16	LBL
LiF	300	35	0.14 ± 0.03		1
LiF	300	50	0.169 ± 0.009	6.7 ± 1.0	7

TABLE 3—Continued

Target substance	Temp. (°K)	Field (G)	Asymmetry	T_2 (μsec)	Reference[b]
LiF	300	1000	0.191 ± 0.004	5.30 ± 0.33	LBL
LiF	300	11,000	0.184 ± 0.009	5.24 ± 0.29	LBL
LiF	77	50	0.163 ± 0.014	1.3 ± 0.3	7
LiF (fused)	300	51	0.203 ± 0.009	6.0 ± 0.9	9
LiF (crystal)	300	50	0.148 ± 0.008	4.6 ± 0.8	9
LiF (crystal)	300	405	0.150 ± 0.006	4.4 ± 0.8	9
KCl (crystal)	300		<0.017		10
NaCl	300	50	0.041 ± 0.009		2
NaCl (powder)	300	800	0.035 ± 0.007		3
NaCl (powder)	300	3500	0.023 ± 0.007		3
NaCl (crystalline)	300		<0.073		10
NaCl (crystal)	300	2	0.099 ± 0.014	>15	11
NaCl (crystal)	300	50	0.084 ± 0.005	>15	11
MgO	300	35	0.12 ± 0.03		1
MgO	300	50	0.079 ± 0.012		2
MgF_2	300	50	0.136 ± 0.009		2
$CuSO_4$ (anhydrous)	300	4500	0.156 ± 0.003	>20	LBL
$CuSO_4$	300	11,000	0.275 ± 0.002	13.42 ± 0.64	LBL
$CuSO_4 \cdot 5(H_2O)$	300	4500	0.261 ± 0.007	5.04 ± 0.11	LBL
$CuSO_4 \cdot 5(H_2O)$	300	11,000	0.257 ± 0.002	3.46 ± 0.06	LBL
NaOH (solid)	300	11,000	0.191 ± 0.002	7.53 ± 0.25	LBL
$Ca(OH)_2$ (solid)	300	11,000	0.177 ± 0.002	7.51 ± 0.32	LBL
$CaSO_4 \cdot 2(H_2O)$ (gypsum)	300	1000	0.157 ± 0.003	5.5 ± 0.5	LBL
$CaSO_4 \cdot 2(H_2O)$ (avg)	300	4500	0.166 ± 0.002	4.0 ± 0.6	LBL
$CaSO_4 \cdot 2(H_2O)$	300	11,110	0.167 ± 0.002	5.30 ± 0.14	LBL
KH_2PO_4	300	1000	0.122 ± 0.002	17.65 ± 2.19	LBL
KH_2PO_4	300	4500	0.135 ± 0.004	2.55 ± 0.15	LBL

$FeCl_2 \cdot 4(H_2O)$	300	4500	0.159 ± 0.003	2.88 ± 0.08	LBL
$FeCl_3 \cdot 6(H_2O)$	300	4500	0.224 ± 0.003	1.99 ± 0.04	LBL
$Fe(NO_3)_3 \cdot 6(H_2O)$	300	11,000		0.04 ± 0.01	LBL
FeF_3	300	4500	0.007 ± 0.002	4.67 ± 1.02	LBL
$FeBO_3$	300	1000	0.068 ± 0.004	5.13 ± 0.86	LBL
$FeBO_3$	300	4500	0.010 ± 0.006	1.14 ± 1.02	LBL
$FeBO_3$	423	4500	0.096 ± 0.002	5.10 ± 0.34	LBL
$Gd(NO_3)_3 \cdot 6(H_2O)$	300	11,000		0.04 ± 0.01	LBL
Cr_2O_3	300	1000	0.037 ± 0.007	0.81 ± 0.17	LBL
Cr_2O_3	310	4500	0.222 ± 0.003	11.64 ± 0.70	LBL
Silicone DC-200	300	50	0.139 ± 0.010		2
Assorted elements and inorganic compounds					
Boron	300	1000	0.274 ± 0.005	3.02 ± 0.10	LBL
Boron	300	11,000	0.312 ± 0.004	3.13 ± 0.09	LBL
B_4C	300	50	0.23 ± 0.02	6.5 ± ?	2
B_4C	300	50	0.295 ± 0.009	5.3 ± 0.5	7
B_4C	300	100	0.286 ± 0.009	5.3 ± 0.5	7
B_4C	300	400	0.295 ± 0.009	5.6 ± 0.7	7
B_4C	300	800	0.225 ± 0.007		3
B_4C	300	1000	0.286 ± 0.02	4.5 ± 0.6	LBL
B_4C	300	3500	0.196 ± 0.007		3
B_4C	300	4500	0.264 ± 0.003	3.94 ± 0.10	LBL
B_4C	373	11,000	0.274 ± 0.003	4.25 ± 0.10	LBL
B_4C	173	11,000	0.266 ± 0.005	3.13 ± 0.11	LBL
B_4C	77	50	0.286 ± 0.009	3.6 ± 0.4	7
Phosphorus	300	50	0.025 ± 0.017		2
Phosphorus (red)	300	35	0.019 ± 0.038		1
Phosphorus (red)	300	800	0.022 ± 0.008		3
Phosphorus (black)	300	35	0.19 ± 0.04		1
N_2 (liquid)	77	50	0.037 ± 0.006		8

TABLE 3—Continued

Target substance	Temp. (°K)	Field (G)	Asymmetry	T_2 (μsec)	Reference[b]
N_2 (liquid)	77	50	0.028 ± 0.002	>20	LBL
N_2 (liquid)	77	100	0.114 ± 0.003	9.01 ± 1.20	LBL
H_2O (water)	300	50	0.146 ± 0.003		5
H_2O (water)	300	100	0.155 ± 0.001	>20	LBL
H_2O (water)	300	150	0.154 ± 0.005	>20	LBL
H_2O (water)	300	1000	0.160 ± 0.002	>20	LBL
H_2O (water)	300	3500	0.136 ± 0.011		3
H_2O (water)	300		0.176 ± 0.005		12
H_2O (water)	300	4400	0.147 ± 0.001	>20	LBL
H_2O (water)	300	4500	0.160 ± 0.001	>20	LBL
H_2O (water)	300	11,000	0.165 ± 0.001	>20	LBL
H_2O (ice)	261	11,000	0.130 ± 0.003	5.04 ± 0.33	LBL
H_2O (ice)	195	3500	0.046 ± 0.009		13
H_2O (ice)	77	50	0.066 ± 0.004		8
H_2O (ice)	77	50	0.060 ± 0.006	>5	7
D_2O (heavy water)	300	100	0.139 ± 0.004	>20	LBL
CO_2 (dry ice)	195	50	0.038 ± 0.013		8
CO_2 (dry ice)	195	800	0.058 ± 0.009		3
CO_2 (dry ice)	195	3500	0.046 ± 0.009		3
CO_2 (dry ice)	77	100	0.051 ± 0.001	>20	LBL
$Fe_2(SO_4)_3$	300	4500	0.253 ± 0.003	4.47 ± 0.13	LBL
CsI	300	50	0.031 ± 0.013		2
$AgNO_3$	300	4500	0.237 ± 0.003	>20	LBL
TiH_2 (powder)	300	100	0.316 ± 0.006	5.09 ± 0.26	LBL
NiO (powder)	300	4500	0.004 ± 0.001	>20	LBL

Saturated organic compounds					
CCl_4	300	50	0.237 ± 0.012		5
CCl_4	300	100	0.271 ± 0.001	>20	LBL
$CHBr_3$ (bromoform)	300	50	0.223 ± 0.008		5
$CHBr_3$	300	271	0.219 ± 0.009		5
$CHBr_3$	300		0.280 ± 0.006		12
$CHBr_3$	300		0.286 ± 0.004		14
$CHCl_3$ (chloroform)	300	50	0.184 ± 0.015		2
$CHCl_3$	300	50	0.190 ± 0.009		5
$CHCl_3$	300	100	0.229 ± 0.010	>20	LBL
CH_2I_2	300	50	0.227 ± 0.009		5
CH_2Cl_2	300	100	0.176 ± 0.003	>20	LBL
CH_3NO_2 (nitromethane)	300	100	0.144 ± 0.002	>20	LBL
CH_3OH (methanol)	300	50	0.137 ± 0.009		5
CH_3OH	300	50	(0.60 ± 0.02)*A*		15
CH_3OH	300	50	0.133 ± 0.002	>20	LBL
CH_3OH	300	90	(0.61 ± 0.03)*A*		15
CH_3OH	300	100	0.140 ± 0.002	>20	LBL
CH_3OH	300	1000	0.136 ± 0.001	>20	LBL
CH_3OH	300	3400	(0.62 ± 0.05)*A*		15
CH_3OH	300	4500	0.140 ± 0.003	>20	LBL
CH_3OH (liquid)	175	50	0.203 ± 0.006	>20	7
CH_3OH (solid)	77	50	0.154 ± 0.009	4.5 ± 1.4	7
CH_3CH_2OH (ethanol)	300	100	0.148 ± 0.002	>20	LBL
n-propyl alcohol	300	100	0.151 ± 0.001	>20	LBL
$C_2H_4Cl_2$	300	50	0.152 ± 0.010		5
$C_2H_5OC_2H_5$ (liquid)	157	50	0.180 ± 0.009	>20	7
$C_2H_5OC_2H_5$ (solid)	77	50	0.111 ± 0.011	6.7 ± 3.3	7
C_3H_8 (liquid propane)	193	50	0.170 ± 0.020		2
$C_3H_8O_3$ (glycerol)	300	100	0.179 ± 0.003	>20	LBL

TABLE 3—Continued

Target substance	Temp. (°K)	Field (G)	Asymmetry	T_2 (μsec)	Reference[b]
C_6H_{14} (hexane)	300	50	$(0.62 \pm 0.03)A$		15
C_6H_{14}	300	100	$(0.57 \pm 0.06)A$		15
C_6H_{14} (mixed hexanes)	300	100	0.146 ± 0.002	>20	LBL
C_6H_{14}	300	3400	$(0.67 \pm 0.08)A$		15
C_6H_{14}	300	11,000	0.170 ± 0.001	>20	LBL
C_7H_{16} (heptane)	300	50	$(0.57 \pm 0.06)A$		15
C_8H_{18} (octane)	300	7	0.142 ± 0.054		5
C_8H_{18}	300	50	0.147 ± 0.008		5
C_6H_{12} (cyclohexane)	300	100	0.160 ± 0.004	>20	LBL
C_6H_{12}	300		0.196 ± 0.007		16
C_6H_{12}	300		0.20 ± 0.005		14
C_6H_{12} (solid)	77		0.080 ± 0.005		16
$C_6H_{11}F$	300		0.197 ± 0.005		14
$C_6H_{11}Cl$	300		0.203 ± 0.005		14
$C_6H_{11}Br$	300		0.248 ± 0.005		14
$C_6H_{11}I$	300		0.275 ± 0.005		14
$C_6H_{11}OH$ (cyclohexanol)	300		0.200 ± 0.004		14
$C_6H_{10}O$ (cyclohexanone)	300		0.181 ± 0.004		14
Unsaturated organic compounds					
CS_2	300	50	0.029 ± 0.003	>20	LBL
C_6H_{12} (hexenes)	300	100	0.119 ± 0.001	>20	LBL
C_6H_{12} (2-hexene)	300	11,000	0.140 ± 0.001	>20	LBL
C_6H_{10} (hexynes)	300	100	0.103 ± 0.002	>20	LBL
C_6H_{10} (1-hexyne)	300	11,000	0.119 ± 0.001	>20	LBL
C_6H_6 (benzene)	300	50	0.046 ± 0.012		2
C_6H_6 (benzene)	300	50	0.036 ± 0.006		5

C_6H_6 (benzene)	300	50	0.036 ± 0.002	>20	LBL
C_6H_6 (benzene)	300	100	0.042 ± 0.001	>20	LBL
C_6H_6 (benzene)	300	200	0.038 ± 0.001	>20	LBL
C_6H_6 (benzene)	300	271	0.034 ± 0.007		2
C_6H_6 (benzene)	300		0.052 ± 0.005		14
C_6H_6 (liquid benzene)	300	100	0.033 ± 0.003	>20	LBL
C_6H_6 (solid benzene)	77	100	0.018 ± 0.002	>20	LBL
C_6H_5F	300		0.074 ± 0.007		16
C_6H_5Cl	300	50	0.063 ± 0.007		5
C_6H_5Cl	300		0.098 ± 0.007		16
C_6H_5Br	300	50	0.106 ± 0.010		5
C_6H_5Br	300		0.142 ± 0.007		16
C_6H_5I	300		0.164 ± 0.007		16
$C_6H_5CH_2Cl$	300		0.118 ± 0.005		14
$C_6H_5CHCl_2$	300		0.158 ± 0.005		14
$C_6H_5CCl_3$	300		0.193 ± 0.005		14
$C_6H_5CH_3$ (toluene)	300	50	0.052 ± 0.003	>20	LBL
$C_6H_5CH_3$ (toluene)	300	100	0.052 ± 0.001	>20	LBL
$C_6H_5CH_3$ (toluene)	300		0.076 ± 0.005		14
C_6H_5OH (phenol)	300	50	0.089 ± 0.006	>20	LBL
C_6H_5OH (phenol)	300	100	0.089 ± 0.001	>20	LBL
$C_6H_5NH_2$ (aniline)	300	50	0.088 ± 0.004	>20	LBL
$C_6H_5NH_2$ (aniline)	300	100	0.084 ± 0.002	>20	LBL
$C_6H_5NO_2$ (nitrobenzene)	300	50	0.090 ± 0.001	>20	LBL
$C_6H_5NO_2$ (nitrobenzene)	300	100	0.082 ± 0.001	>20	LBL
$C_6H_5C_2H_5$	300		0.090 ± 0.005		14
$C_6H_5C_4H_9$	300		0.115 ± 0.005		14
$C_6H_5CH(C_2H_5)C_3H_7$	300		0.112 ± 0.005		14
$C_6H_4(CH_3)_2$ (xylene)	300		0.082 ± 0.007		14
$C_6H_3(CH_3)_3$	300		0.090 ± 0.005		14
$C_6H_2(CH_3)_4$ (durol)	300		0.087 ± 0.006		14

TABLE 3—Continued

Target substance	Temp. (°K)	Field (G)	Asymmetry	T_2 (μsec)	Reference[b]
$C_6(CH_3)_6$	300		0.13 (approx.)		14
C_6H_{10}	300	50	0.113 ± 0.009		5
C_6H_{10}	300	271	0.116 ± 0.008		5
C_6H_{10} (cyclohexene)	300		0.16 ± 0.005		14
1–4 cyclohexadiene	300		0.135 ± 0.005		14
1–3 cyclohexadiene	300		0.105 ± 0.005		14
$C_6H_5C{=}CC_6H_5$ (tolan)	300	267	0.031 ± 0.013		5
$C_{14}H_{10}$ (anthracene)	300	267	0.025 ± 0.011		5
Phenylcyclohexane	300	50	0.084 ± 0.011		2
Polystyrene	300	50	0.070 ± 0.010		2
Polystyrene	300	50	0.044 ± 0.007		5
Polystyrene	300	271	0.034 ± 0.007		5
Plexiglas	300	50	0.093 ± 0.011		5
Nuclear emulsion	300	50	0.087 ± 0.009		2
Nuclear emulsion	300	800	0.080 ± 0.010		3
Nuclear emulsion	300	1700	0.092 ± 0.010		3
Nuclear emulsion	300	3500	0.097 ± 0.012		3
Nuclear emulsion	300	4000	0.082 ± 0.009		3
Teflon	300	1300	0.175 ± 0.015		3
Teflon	300	4000	0.174 ± 0.013		3
Gelatin	300	800	0.156 ± 0.011		3
Gelatin	300	3500	0.139 ± 0.011		3
Scintillator	300	800	0.056 ± 0.009		3
Scintillator	300	3500	0.049 ± 0.009		3
Polyethylene	300	50	0.146 ± 0.012		2
Polyethylene	300	800	0.179 ± 0.009		3
Polyethylene	300	3500	0.153 ± 0.009		3

DPPH	300	4500	0.149 ± 0.03	7.7 ± 2.0	LBL
DPPH	310	4500	0.151 ± 0.004	10.19 ± 1.29	LBL
			Muonium signals in inert materials		
Argon (40 atm)	300	2	0.04 ± 0.005	5.0 ± 4	17
N_2 gas (40 atm)	300	2	<0.01		17
SF_6 gas (40 atm)	300	2	<0.01		17
N_2 (liquid)	77	7.2	<0.03		8
SiO_2 (fused)	1800	0.5	0.039 ± 0.011	5.1 ± 0.8	18
SiO_2 (fused)	300	0.5	0.040 ± 0.008	6.2 ± 1.1	18
SiO_2 (fused)	300	7.2	0.161 ± 0.012	1.3 ± 0.2	8
SiO_2 (fused)	77	7.2	0.148 ± 0.014	0.5. ± 0.1	8
SiO_2 (fused)	300	10	0.104 ± 0.008		3
SiO_2 (fused)	300	50	0.046 ± 0.030	>1	LBL
SiO_2 (fused)	300	95	0.097 ± 0.005	~ 1.5	19
SiO_2 (fused)	300	100	0.072 ± 0.005	~ 3.0	LBL
SiO_2 (fused)	300	150	0.122 ± 0.05	>0.5	LBL
SiO_2 (crystal)	300	7.2	0.167 ± 0.013	0.4 ± 0.1	8
SiO_2 (powder)	300	7.2	<0.04		8
CO_2 (solid)	195	7.2	0.070 ± 0.015	0.4 ± 0.1	8
CO_2 (solid)	77	7.2	0.075 ± 0.015	0.6 ± 0.1	8
H_2O (solid)	77	7.2	~ 0.16	~ 0.08	8
Al_2O_3 (powder)	300	7.2	<0.09		8
Si crystal (B doped)	77	50	0.053 ± 0.010	0.45 ± 0.10	LBL
Si crystal (B doped)	77	100	0.042 ± 0.005	0.315 ± 0.05	LBL
Si crystal (hi B doped)	77	150	0.022 ± 0.010	1.09 ± 0.52	LBL
Ge crystal	77	7	0.123 ± 0.026	0.16 ± 0.07	6
Ge crystal	77	98	0.082 ± 0.012	~ 0.1	19
NaCl (crystalline)	300		<0.023		10
LiF (fused)	300	7.2	<0.021		9
KCl crystal	300		<0.013		10

TABLE 3—Continued

Target substance	Temp. (°K)	Field (G)	Asymmetry	T_2 (μsec)	Reference[b]
C_6H_6 (liquid benzene)	300	2.7	<0.011		5
C_6H_6 (solid benzene)	77	7.2	<0.05		8
B_2O_3 (fused)	300	7.2	<0.05		8
B_2O_3 (fused)	77	7.2	<0.04		8
Polyethylene	77	7.2	<0.03		8
Paraffin	77	7.2	<0.03		8
Si crystal (p-type)	77	100 (M′)	0.046 ± 0.005	0.765 ± 0.25	LBL

[a] Measured asymmetries and relaxation times are arranged loosely according to the type of medium in which the muons are stopped. In each case, the asymmetry is the product of the muon's residual polarization and the maximum effective asymmetry A_0, which is measured when no depolarization occurs in the target. The value of A_0 is determined by various experimental characteristics, and generally varies by ~20% from experiment to experiment. Care should therefore be taken in comparing asymmetries, especially when measured by different groups. Convenient "reference asymmetries" are carbon, various metals, and bromoform or carbon tetrachloride, in which very little depolarization occurs, or water at 300°K, in which the residual polarization is known to be 0.55 ± 0.03. In some references, residual polarizations are quoted without specifying the experimental value of A_0. In these cases the data are listed as (polarization) A. When no relaxation time is given, or when the field strength is not specified, it is because the value was not mentioned in the reference. In the case of muonium precession at fields of more than 10 G, two frequencies are observed. The asymmetry quoted here is the sum of the asymmetries in the two "beating" precession patterns. The final entry, labeled (M′), represents the asymmetry and relaxation time for "anomalous μ^+ precession" in silicon, thought to be shallow-donor muonium precession.

[b] Reference LBL corresponds to published or unpublished results of the Lawrence Berkeley Laboratory group. For other references, the key is as follows:

1. Eisenstein *et al.* (1966)
2. Swanson (1958)
3. Gurevich *et al.* (1968)
4. Gurevich *et al.* (1972)
5. Babaev *et al.* (1966)
6. Andrianov *et al.* (1969)
7. Minaichev *et al.* (1970)
8. Myasishcheva *et al.* (1968)
9. Myasishcheva *et al.* (1969a)
10. Myasishcheva *et al.* (1967a)
11. Ivanter *et al.* (1972)
12. Myasishcheva *et al.* (1967c)
13. Gurevich *et al.* (1971b)
14. Myasishcheva *et al.* (1969b)
15. Gurevich *et al.* (1971)
16. Myasishcheva *et al.* (1967b)
17. Mobley (1967)
18. Bowen *et al.* (1972)
19. Gurevich *et al.* (1971a)

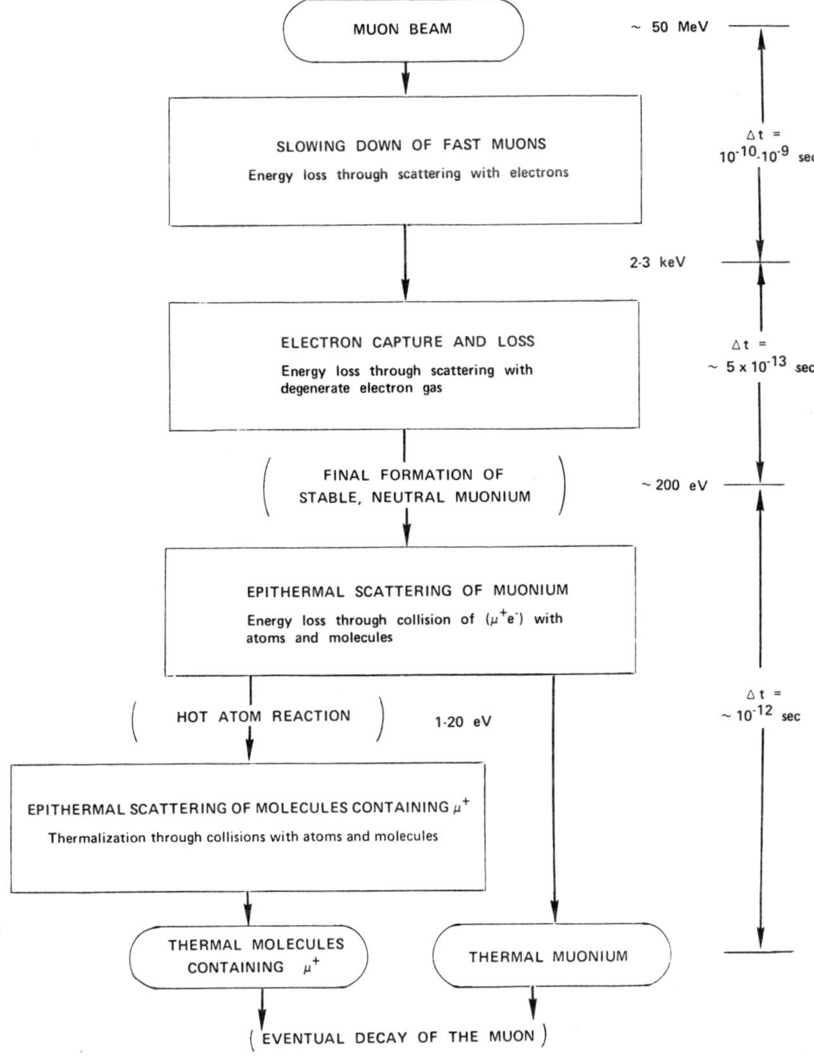

FIG. 5. Flow diagram showing the processes involved for positive muons slowing down and thermalizing in matter.

velocity approaches that of the valence electrons of the target atoms (corresponding to an energy of 2 to 3 keV). During this stage, the energy loss per unit time is given by the Bethe formula. The total time it takes the μ^+ to slow down to 2 to 3 keV in condensed matter is thus estimated to be about 10^{-10} to 10^{-9} sec; this time is probably somewhat longer in gases.

Depolarization during this stage could only be due to spin-dependent forces in the scattering processes with electrons or nuclei. In both cases, as Ford and Mullin (1957) and Wentzel (1949) have shown, such depolarizing effects are extremely small and can be neglected.

For muon energies of less than 2 keV, or for muon velocities smaller than valence electron velocities, the Bethe formula no longer represents a useful approximation. The deceleration of negative muons from this energy range down to thermal velocity has been treated by Fermi and Teller (1947). They estimate a typical slowing-down time in condensed matter of several times 10^{-14} sec. Energy loss is again due to collisions with electrons, where the electrons are now treated as a degenerate gas. In principle, positive muons would slow down through this energy range in similar times if the same mechanisms were in effect. However, there are drastic qualitative differences between the behavior of negative and positive muons at these energies. The most important one is that positive muons capture electrons from the medium to form neutral atoms. This is known to happen for protons, from studies of proton beams traversing condensed or gaseous matter, in which the positive beam is partially neutralized (Allison and Garcia-Munoz, 1962). As the velocity of the positive particle drops below that of an electron in the ground state of hydrogen ($\sim\alpha c$), corresponding to a kinetic energy of 3 keV for the muon, the proton (or muon) begins to capture and lose electrons in rapid succession. As the velocity decreases further, the neutral state begins to dominate; at a velocity of about 10^8 cm/sec the beam is more than 80% neutralized. This is shown in Fig. 6 for protons traversing aluminum or O_2 gas. It can be inferred that a positive muon beam will become almost completely neutralized at an energy of about 200 eV, forming stable atoms of muonium (Massey and Burhop, 1952). Direct confirmation that positive muon beams end up as muonium atoms was obtained from the muonium studies that have been performed by Hughes *et al.* (1960), in gaseous argon targets.

To our knowledge, no calculations have been made of the time the positive muon takes to decelerate from 3 keV to 200 eV. Allison (1958) has estimated that a μ^+ slowing down through this energy range in hydrogen gas at 1 atm will spend about 2×10^{-10} sec as muonium, passing through about 100 cycles of muonium formation and ionization. The number of capture–loss cycles should not be too dependent on the medium, but the time scale might be expected to be a factor of 1000 shorter ($\sim 2 \times 10^{-13}$ sec) in solid or liquid targets. According to Fermi and Teller, a free μ^- would slow down through this energy range in several 10^{-14} sec. One can thus safely assume that the total time the positive muon spends in the energy range 3 keV to 200 eV is less than 5×10^{-13} sec.

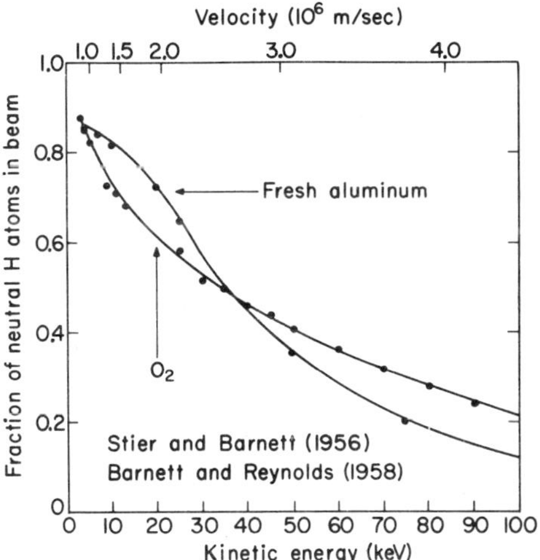

FIG. 6. Evidence that protons approach the end of their range as H atoms. The critical velocity is $\alpha c = 2.2 \times 10^6$ m/sec (from Crowe *et al.*, 1972a).

This has important consequences with regard to the depolarization that will occur in this stage. As will be seen in the next section, the spins of only half of the muons in the ground state of muonium will show a changing orientation in time. They will reverse their directions periodically at a frequency of $\sim$4.6 GHz, due to their contact interaction with the magnetic moment of the electron. This modulation is often described qualitatively as the precession of the μ^+ spin in the effective hyperfine field (about 160 kG) produced by the electron's magnetic moment. The spins of these muons will have reversed direction in about 2×10^{-10} sec. Since 5×10^{-13} sec is only a tiny fraction of this period, the depolarization in condensed media during deceleration from 3 keV to 200 eV is totally negligible, even if a single muonium atom remains intact for the entire period; if we take into account the repeated breakup and reformation of muonium, we would expect depolarization during this phase to be negligible even in gases.

At a kinetic energy of 200 eV, practically all muons are in the muonium state. Subsequent deceleration to thermal energies proceeds by collisions of muonium atoms with target atoms or molecules. Mobley (1967) has treated thermalization of muonium in argon gas as a purely kinematic process of elastic collisions, in which the average fractional energy loss per collision is constant,

$$\overline{\Delta E}/E = -2m/M$$

where m is the mass of muonium and M the mass of a target atom or molecule. In this picture, the time T_{el} to decelerate from E_1 to E_2 in a gas of density n molecules per unit volume is given by

$$T_{\mathrm{el}} = \frac{M}{2^{5/2} n\sigma \sqrt{m}} \left(\frac{1}{E_2^{1/2}} - \frac{1}{E_1^{1/2}} \right) \tag{7}$$

where σ is the cross section for elastic collisions. Note that muonium slows down faster in collisions with lighter atoms. Mobley calculates a time of 1.5×10^{-11} sec for slowing down from 200 to 0.04 eV in argon gas at 35 atm. The number of collisions in this process is about 2.5×10^3, independent of pressure.

In gases of large molecules or in condensed media, this model is of questionable validity. Many inelastic energy-loss processes doubtless exist, whose net effect can only be to speed up the deceleration process. Thus the thermalization time in condensed media with molecular weights similar to argon will generally be shorter than $\sim 10^{-12}$ sec, the estimate obtained by simply scaling Mobley's result with density.

As in the initial muonium formation stage, this time is too short to cause any appreciable depolarization. Thus the muons will still be fully polarized immediately following their thermalization in muonium atoms.

Another process is often quite significant during this thermalization stage: while a muonium atom still has an epithermal energy in the range of roughly 20 to 1 eV, it may participate in so-called "hot-atom" reactions. These are chemical reactions that are usually forbidden in the thermal region for lack of energy to overcome potential barriers (high activation energies). Hot-atom reactions of tritium (like muonium, a radioactive isotope of hydrogen) have been widely studied in gases (Rowland, 1970). There are also a number of studies of hot-tritium reactions in aqueous solutions, which may be used as a guide for conjecture regarding the outcome of hot-muonium reactions in liquids. It should be mentioned that hot-atom reactions of positronium in aqueous solution have also been studied in a number of cases (Bartal *et al.*, 1972a,b). However, the analogy between positronium and the various isotopes of hydrogen is much weaker.

The most common hot-atom reactions of muonium can be classified as follows:

1. Substitution reactions of the type

$$\mathrm{XH} + \mathrm{Mu}^* \rightarrow \mathrm{XMu} + \mathrm{H}$$

where X is some radical species;

2. Abstraction reactions such as

$$\mathrm{XH} + \mathrm{Mu}^* \rightarrow \mathrm{X} + \mathrm{MuH}$$

3. Addition reactions such as

$$X + Mu^* \rightarrow XMu$$

The third type of reaction requires a deexcitation mechanism to carry away the excess energy; it is very likely to produce a chemical radical if X is a molecule with saturated bonds. The other types may also lead to paramagnetic molecules incorporating muons, since there is a lot of energy available and the final products may be in excited triplet states.

So far, as will be shown later, only the first type of reaction has been experimentally verified in the hot-atom chemistry of muonium. In this type, muonium is built into a molecule with saturated bonds, and the total spin density of electrons at the muon site is zero. With respect to magnetic interactions, the muon can be considered quasi-free. These muons have, of course, lost none of their polarization up to the time they become part of a molecule. This is true, in fact, for all the muons that have proceeded through the hot-atom channel, independent of their final chemical status.

The situation just after thermalization can be summarized as follows:

1. The total elapsed time since the muon entered the target is between 10^{-10} sec (in condensed media) and 10^{-9} sec (in dilute gases).
2. Regardless of the muon's chemical state, its polarization is completely conserved.
3. The muon may be in any of the following states, listed in approximate order of decreasing likelihood: (a) A thermalized muonium atom; (b) Part of a molecule with saturated bonds. In this diamagnetic environment the muon can be considered quasi-free; (c) Part of a paramagnetic molecule (e.g., a radical or an excited triplet state). Here the environment is magnetically similar to muonium; (d) A free muon. This state can only occur in the rare cases where the ionization potential of the medium is greater than that of muonium (e.g., in the case of helium).

While there is much evidence to support the validity of this general picture of how a μ^+ slows down in matter, very little is known about the details of the mechanism involved and the possible implications. For instance, a better understanding of the dynamics of hot-atom reactions of muonium might lead to further elucidation of isotopic effects in hot-atom chemistry in general. However, in spite of the important role played by hot-atom reactions in the chemistry of muonium (as described in Section VII), these phenomena are not yet well understood. The rest of this chapter deals with the behavior of muons in their various states in the few microseconds of their remaining lifetime after thermalization.

IV. Qualitative Behavior of the Muon Spin in Muonium

Before we embark upon a rigorous derivation of the evolution of a μ^+ spin in a muonium atom, it is conceptually helpful to describe this behavior in semiclassical terms. To this end, we recall from Chapter II the Hamiltonian for the hyperfine interaction between μ^+ and e^- magnetic moments in the presence of an external magnetic field B; this operator can be written

$$H^{\mathrm{Mu}} = \frac{\hbar}{4}\,\omega_0(\boldsymbol{\sigma}^\mu \cdot \boldsymbol{\sigma}^e) + \frac{\hbar}{2}\,\boldsymbol{\omega}^\mu \cdot \boldsymbol{\sigma}^\mu + \frac{\hbar}{2}\,\boldsymbol{\omega}^e \cdot \boldsymbol{\sigma}^e \tag{8}$$

where $\hbar\omega_0$ is the hyperfine energy splitting, $\boldsymbol{\sigma}^\mu$ and $\boldsymbol{\sigma}^e$ muon and electron Pauli spin operators, respectively, and $\boldsymbol{\omega}^\mu$ and $\boldsymbol{\omega}^e$ muon and electron Larmor frequencies, respectively, given by $\hbar\boldsymbol{\omega}^\mu = g_\mu \mu_0{}^\mu \mathbf{B}$ and $\hbar\boldsymbol{\omega}^e = g_e \mu_0{}^e \mathbf{B}$. Here $g_\mu \approx -2$ and $g_e \approx +2$ are, respectively, the muon and electron g-factors, and $\mu_0{}^\mu$ and $\mu_0{}^e$ are their respective Bohr magnetons ($\mu_0 = e\hbar/2mc$). Note that $\boldsymbol{\omega}^\mu$ and $\boldsymbol{\omega}^e$ are antiparallel.

The eigenvalues of this Hamiltonian are given by the familiar Breit–Rabi formula. To avoid the sometimes misleading practice of labeling the energy levels by total angular momentum eigenvalues (which are good quantum numbers only in zero field), we write out the energy eigenvalues individually:

$$\begin{aligned} \omega_1 &= \frac{E_1}{\hbar} = \frac{\omega_0}{4} + \omega_-, & \omega_3 &= \frac{E_3}{\hbar} = \frac{\omega_0}{4} - \omega_- \\ \omega_2 &= \frac{E_2}{\hbar} = -\frac{\omega_0}{4} + \left(\frac{\omega_0{}^2}{4} + \omega_+{}^2\right)^{1/2}, & \omega_4 &= \frac{E_4}{\hbar} = -\frac{\omega_0}{4} - \left(\frac{\omega_0{}^2}{4} + \omega_+{}^2\right)^{1/2} \end{aligned} \tag{9}$$

where $\omega_\pm = \frac{1}{2}(|\,\boldsymbol{\omega}^e\,| \pm |\,\boldsymbol{\omega}^\mu\,|)$. These levels are shown in Fig. 7 as functions of the "natural" specific field $x = 2\omega_+/\omega_0 = |\,\mathbf{B}\,|/B_0$, where $B_0 = 1585$ G. We will later refer to the transition frequencies $\omega_{ij} = \omega_i - \omega_j$.

The energy eigenstates can be expressed most simply in terms of the "natural" specific field x and the "natural" basis $|\,m_\mu m_e\rangle_{||}$, where the subscript $||$ indicates explicitly that the axis of quantization is along the magnetic field:

$$\begin{aligned} |\,E_1\rangle &= |\,{+}{+}\rangle_{||} = |\,F = 1, m_F = +1\rangle \\ |\,E_2\rangle &= s\,|\,{+}{-}\rangle_{||} + c\,|\,{-}{+}\rangle_{||} \xrightarrow[x\to 0]{} |\,F = 1, m_F = 0\rangle \\ |\,E_3\rangle &= |\,{-}{-}\rangle_{||} = |\,F = 1, m_F = -1\rangle \\ |\,E_4\rangle &= c\,|\,{+}{-}\rangle_{||} - s\,|\,{-}{+}\rangle_{||} \xrightarrow[x\to 0]{} |\,F = 0, m_F = 0\rangle \end{aligned} \tag{10}$$

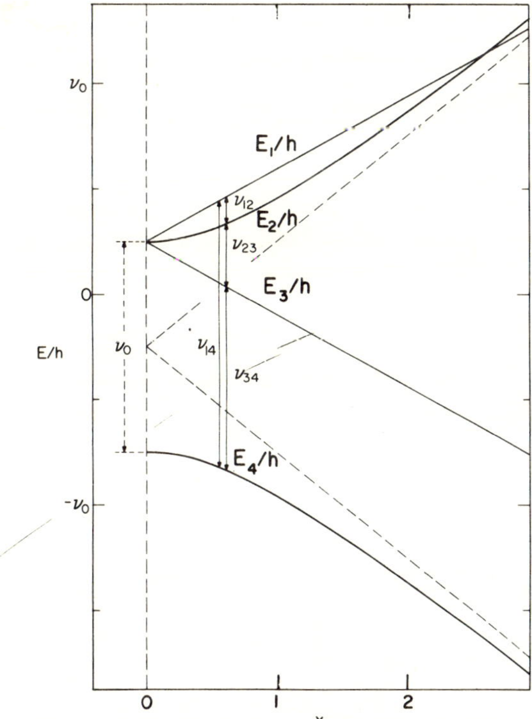

FIG. 7. Energy eigenstates of $l = 0$ muonium in an external magnetic field, as a function of the dimensionless "specific field"

$$x = 2\omega_+/\omega_0 = (g_e\mu_0^e - g_\mu\mu_0^\mu)\,|\mathbf{B}|/(\hbar\omega_0)$$

For graphical clarity, an unphysical value of $|\mu_0^e/\mu_0^\mu|$ is used to generate the plot. The four allowed transitions ($\Delta m = \pm 1$) are indicated.

where

$$c = \frac{1}{\sqrt{2}}\left(1 + \frac{x}{(1+x^2)^{1/2}}\right)^{1/2}, \quad \text{and} \quad s = \frac{1}{\sqrt{2}}\left(1 - \frac{x}{(1+x^2)^{1/2}}\right)^{1/2} \tag{11}$$

We have used the compact notation $|+-\rangle \equiv |m_\mu = +\frac{1}{2}, m_e = -\frac{1}{2}\rangle$, etc.

The muon is always assumed (without loss of generality) to be initially completely polarized, while the electron it captures is generally unpolarized. This is assumed throughout, although one can imagine situations in which a certain fraction of the electrons might have a significant polarization [e.g., in ferromagnetic media (Ivanter, 1973)]. Always choosing the quantization axis along the initial muon polarization, we can thus express the initial conditions as follows: half of the muonium ensemble is formed in the state $|a_0\rangle = |++\rangle$ and half in the state $|b_0\rangle = |+-\rangle$. Note that we have not specified by a subscript whether the quantization axis is along the field. We shall treat the two orthogonal possibilities separately.

Now, working in the Schrödinger picture, we are prepared to describe the evolution of the muonium spin state.

A. Longitudinal Field

If the magnetic field is parallel to the initial muon polarization, the first half of the ensemble is in an eigenstate, $|a_0\rangle = |++\rangle_{\parallel} = |E_1\rangle$, and is therefore stationary; the second half, however, is in a superposition of two eigenstates, $|b_0\rangle = |+-\rangle_{\parallel} = s\,|E_2\rangle + c\,|E_4\rangle$. The time dependence of this state will be

$$\begin{aligned} |b(t)\rangle &= s\exp(-i\omega_2 t)\,|E_2\rangle + c\exp(-i\omega_4 t)\,|E_4\rangle \\ &= \exp(-i\omega_2 t)[s\,|E_2\rangle + c\exp(i\omega_{24}t)\,|E_4\rangle] \\ &= \exp(-i\omega_2 t)\{[s^2 + c^2\exp(i\omega_{24}t)]\,|+-\rangle_{\parallel} \\ &\quad + sc[1-\exp(i\omega_{24}t)]\,|-+\rangle_{\parallel}\} \end{aligned} \tag{12}$$

where $\omega_{24} = \omega_2 - \omega_4$.

In zero field, $c = s = 1/\sqrt{2}$, and $\omega_{24} = \omega_0$, so that (12) becomes

$$\begin{aligned} |b(t)\rangle &= \tfrac{1}{2}\exp(-i\omega_2 t) \\ &\quad \times \{[1+\exp(i\omega_0 t)]\,|+-\rangle_{\parallel} + [1-\exp(i\omega_0 t)]\,|-+\rangle_{\parallel}\} \end{aligned}$$

That is, neglecting the physically insignificant overall phase, the state $|b(t)\rangle$ oscillates at frequency ω_0 between the original state $|+-\rangle_{\parallel}$ and the state $|-+\rangle_{\parallel}$, in which the spins are reversed. The muon polarization in this half of the ensemble thus oscillates between $+1$ and -1, averaging to zero. Combined with the constant polarization $+1$ for the first half of the ensemble, this produces a net muon polarization that oscillates between 0 and $+1$ at frequency ω_0, averaging to $\bar{P}^{\mu}\,(x = 0) = \frac{1}{2}$. In vacuum, the period of these oscillations is 0.224 nsec, too short to be resolved with any existing experimental apparatus, so that the apparent effect of muonium formation is reduction of the muon polarization by a factor of 2.

In nonzero longitudinal field, this apparent loss of muon polarization is alleviated. The muon polarization for the first half of the muonium ensemble is still a constant $+1$, while the polarization of the second half now oscillates between $P_{\min}$ and $+1$, where $P_{\min} = (x^2-1)/(x^2+1)$ can be derived from Eq. (12) with Eq. (11). Thus the net muon polarization oscillates at angular frequency ω_{24} between $x^2/(x^2+1)$ and $+1$, resulting in an average of

$$\bar{P}_{||}^{\mu}(x) = \frac{1}{2} + \frac{1}{2}\frac{x^2}{1+x^2} \tag{13}$$

The apparent depolarization of muons in muonium is thus "quenched" by strong magnetic fields ($B \gg B_0$). This phenomenon has been the subject of many experimental studies. It is important to remember that the μ^+ polarization is not actually lost in muonium, but only shared with the electron to an extent that depends on the external field. If quantum irreversible processes cause relaxation of the electron spin by interaction with the medium, all the polarization eventually disappears; but in the simple case described above, the muon polarization always returns to $+1$ periodically. Equation (13) is exact for free muonium, but refers to the average of the muon polarization over times long compared to the hyperfine period. Normally the time resolution of the apparatus forces such an average, but we will later encounter situations in which the strength of the equivalent hyperfine interaction is greatly reduced, increasing the hyperfine period to the point where these oscillations can be observed directly by the techniques described here.

B. Transverse Field

When the magnetic field is perpendicular to the initial muon polarization, neither of the states $|a_0\rangle = |++\rangle_\perp$ and $|b_0\rangle = |+-\rangle_\perp$ is an eigenstate, since the axis of quantization is no longer along the field. This situation is formally more complicated, and a thorough treatment will have to wait until the next section. However, a semiclassical model can give us a qualitative picture of the behavior.

In weak transverse fields ($B \ll B_0$), we may treat the effect of the external field as a perturbation on the zero-field eigenstates. In this approximation, the two halves of the muonium ensemble in states $|a\rangle$ and $|b\rangle$ represent, respectively, a polarized spin-1 (triplet) system in which muon and electron spins are "locked" together by the hyperfine interaction, and a spin-0 (mixed) system in which the muon polarization oscillates about zero at high frequency, somewhat as in the field-free case. The triplet state, with a magnetic moment dominated by the electron, will precess in the external field at its Larmor frequency, in a sense opposite to that of the free muon in the same field:

$$\omega^{\mathrm{Mu}} = \tfrac{1}{2}(\omega^{\mathrm{e}} + \omega^{\mu}) \approx -103\,\omega^{\mu} \tag{14}$$

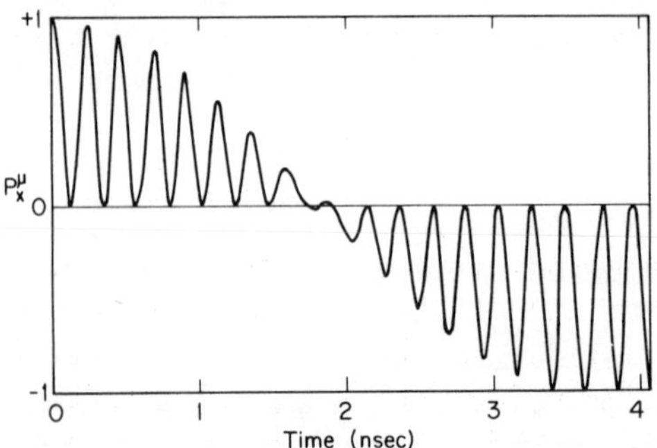

FIG. 8. Evolution of muon polarization in free muonium in 100-G transverse field. P_x^μ is the projection of μ^+ polarization along original polarization direction.

Note that $|\boldsymbol{\omega}^{\mathrm{Mu}}| = \omega_-$. Since the muon and electron spins are "locked" together, this precession frequency can be observed in the muon's decay pattern. The motion of the muon polarization in muonium in a weak transverse field thus consists approximately of rapid hyperfine oscillations superimposed upon a slower muonium precession. The actual time dependence of the projection of the μ^+ polarization along its original axis is shown in Fig. 8, for a transverse field of 100 G. The qualitative features described above are evident. Note that the mean amplitude of the muonium precession, averaged over the hyperfine oscillations, is $\frac{1}{2}$. In low field ($B \lesssim 100$ G) this precession can be observed directly (Gurevich *et al.*, 1971a).

When the transverse field is no longer very small compared to $B_0 = 1585$ G, the naive picture used above is no longer adequate. The splitting of both the muonium precession and the hyperfine oscillation frequencies can no longer be ignored. As is generally the case, these frequencies correspond to $\Delta m = \pm 1$ transitions between Zeeman eigenstates, in this case ω_{12}, ω_{23}, ω_{14}, and ω_{34}. The selection rules that govern which frequencies actually appear are a function of the field. The limiting cases are

$$\Delta m_{\mathrm{F}} = \pm 1, \quad \text{for} \quad B \ll B_0, \qquad \Delta m_\mu = \pm 1, \quad \text{for} \quad B \gg B_0 \tag{15}$$

Thus for $B \ll B_0$ all the four frequencies appear with the same amplitude, whereas only ω_{12} and ω_{34} remain for $B \gg B_0$. The frequency splittings have been observed and analyzed to extract the hyperfine frequency of muonium in various media. This new and rapidly expanding branch of μ^+ spin physics will be discussed in more detail in Section VIII.

In this section we have defined a number of useful quantities and described the behavior of the μ^+ polarization in a free muonium atom, taking advantage of a few simplifying approximations. In the next section we will construct the formalism necessary for an exact description of the evolution of muonium in an external field, including interactions of the electron with the medium. Section VI will be devoted to a description of the effect on the muon polarization of formation of muonium for very short times, and the implications for the study of muonium chemistry.

V. Muon Spin Evolution in Quasi-Free Muonium: An Advanced Treatment

In the previous section we described the time dependence of the spin state of muonium in semiclassical terms; for a few simple cases, that description was complete, but in transverse field or in cases where the muonium electron interacts with the medium, a somewhat more elaborate formalism must be developed. This will be the task of this section.

A. Free Muonium in Longitudinal Field

For the time dependence of the muon polarization in this simplest system, the treatment of the previous section is exact. The polarization of the muon in the second half of the ensemble (state $|b\rangle$) is given by $\mathbf{P}^{\mu}_{b,||}(x,t) = P^{\mu}_{b,||}(x,t)\hat{z} = \langle b(t) \mid \sigma_3{}^{\mu} \mid b(t)\rangle\hat{z}$, where $\mathbf{B} = B\hat{z}$, $\boldsymbol{\sigma}^{\mu}$ is the Pauli spin operator for the muon, and $|b(t)\rangle$ is given by Eq. (12). The muon polarization for the entire ensemble is $\mathbf{P}_{||}{}^{\mu}(x,t) = P_{||}{}^{\mu}(x,t)\hat{z} = \frac{1}{2}[1 + P^{\mu}_{b,||}(x,t)]\hat{z}$. The result is

$$P_{||}{}^{\mu}(x,t) = \frac{1}{2} + \frac{1}{2}\left[\frac{x^2 + \cos(\omega_{24}t)}{1 + x^2}\right] \tag{16}$$

B. Free Muonium in Transverse Field

The exact time dependence of the muon polarization in muonium in a transverse magnetic field can be obtained in similar fashion as long as the muonium atom does not interact with its material environment. For this we must construct a few 4×4 matrices for changes of basis.

We can most easily describe the initial state in the "transverse" basis $|m_{\mu}m_{e}\rangle_{\perp}$ (formally written $|T\rangle$), in which the axis of quantization is along the initial muon polarization, as in Section IV. A matrix of Clebsch–Gordan coefficients J transforms this basis into the basis $|F, m_{f}\rangle_{\perp}$ in which the total angular momentum along the same quantization axis is diagonal. If we write $| F, m_F\rangle_\perp$ formally as $| F_\perp\rangle$, then $J_{ij} = \langle F_\perp{}^i | T_j\rangle$. It is easy to find the rotation matrix R that transforms vectors from the $| F_\perp\rangle$ basis into the basis $| F, m_F\rangle_{||}$, or $| F_{||}\rangle$, with the new quantization axis along the magnetic field. The inverse matrix of Clebsch–Gordan coefficients, J^{-1}, then transforms the $| F_{||}\rangle$ basis into $| m_\mu m_e\rangle_{||}$, or $| L\rangle$, in terms of which the energy eigenstates $| E\rangle$ can be described by the matrix $\epsilon_{jk} = \langle E_j | L_k\rangle$ implicit in Eqs. (10). We can thus expand the initial states in energy eigenstates, allow the stationary components to evolve in time as $\exp(i\omega_j t)$, and then reexpand the result in a basis in which the muon polarization can easily be expressed. A convenient final-state basis will be $| L\rangle$, or $| m_\mu m_e\rangle_{||}$, where the effect of the operator $\boldsymbol{\sigma}^\mu$ is obvious. Thus the equations of motion for the first half of the muonium ensemble can be written

$$| a(t)\rangle = | L_i\rangle\langle L_i | E_j\rangle \exp(i\omega_j t)\langle E_j | L_k\rangle\langle L_k | F_{||}{}^l\rangle \times \langle F_{||}{}^l | F_\perp{}^m\rangle\langle F_\perp{}^m | T_n\rangle\langle T_n | a(0)\rangle \tag{17}$$

or

$$| a(t)\rangle = | L_i\rangle\epsilon_{ij}{}^+ \exp(i\omega_j t)\epsilon_{jk} J_{kl}{}^+ R_{lm} J_{mn}\langle T_n | a(0)\rangle$$

where summation over repeated indices is understood. The time evolution of the second half of the muonium ensemble, $| b(t)\rangle$, is obtained by following the same steps.

The two transverse components of muon polarization can be expressed simultaneously in terms of the complex quantity $\tilde{P}_\perp{}^\mu(x, t)$, whose real part is the μ^+ polarization along the initial direction $\hat{x}$ and whose imaginary part is the μ^+ polarization along the direction $\hat{y}$ perpendicular to both $\hat{x}$ and $\hat{z}$, the field direction, chosen so that $\hat{x} \times \hat{y} = \hat{z}$. The time dependence of this complex polarization is given by

$$\tilde{P}_\perp{}^\mu(x, t) = \tfrac{1}{2}\langle a(t) | (\sigma_1{}^\mu + i\sigma_2{}^\mu) | a(t)\rangle + \tfrac{1}{2}\langle b(t) | (\sigma_1{}^\mu + i\sigma_2{}^\mu) | b(t)\rangle \tag{18}$$

The explicit expressions of the matrices J, R, and ϵ allow solution of Eq. (18) in terms of (17). The final result takes the form

$$\begin{aligned}\tilde{P}_\perp{}^\mu(x, t) &= P_x{}^\mu(x, t) + iP_y{}^\mu(x, t)\\ &= \tfrac{1}{4}[(1 + \delta)\exp(i\omega_{12}t) + (1 - \delta)\exp(i\omega_{23}t)\\ &\quad + (1 + \delta)\exp(-i\omega_{34}t) + (1 - \delta)\exp(i\omega_{14}t)]\\ &= \exp(i\omega_- t)\cos\frac{\omega_0}{2}t\left[\cos\left(\frac{\omega_0}{2} + \Omega\right)t - i\delta\sin\left(\frac{\omega_0}{2} + \Omega\right)t\right]\end{aligned} \tag{19}$$

where

$$\delta = c^2 - s^2 = x/(1 + x^2)^{1/2}$$

$$\omega_- = \tfrac{1}{2}(\omega_{12} + \omega_{23}) = \tfrac{1}{2}(|\,\boldsymbol{\omega}^e\,| - |\,\boldsymbol{\omega}^\mu\,|) \qquad \text{(see Section IV)}$$

$$\Omega = \tfrac{1}{2}(\omega_{23} - \omega_{12}) = \frac{\omega_0}{2}[(1 + x^2)^{1/2} - 1]$$

In weak fields ($x \ll 1$), Eq. (19) can be approximated by

$$\tilde{P}_\perp{}^\mu(x, t) \approx \tfrac{1}{2}\exp(i\omega_- t)[\cos \Omega t + \cos(\omega_0 + \Omega)t] \tag{20}$$

and then its real part takes the form

$$P_x{}^\mu(x, t) \approx \tfrac{1}{2}\cos \omega_- t[\cos \Omega t + \cos(\omega_0 + \Omega)t] \tag{20a}$$

which is shown in Fig. 8. Since the frequency ($\omega_0 + \Omega$) is too high to observe experimentally, this appears as "muonium precession" at frequency ω_-, modulated at the "beat frequency" Ω, with half the initial μ^+ polarization amplitude. This pattern is clearly shown in a measurement by Gurevich *et al.* (1971a), who observed muonium precession in quartz and germanium at 98 G (Fig. 9). From the precession frequency ω_- and the

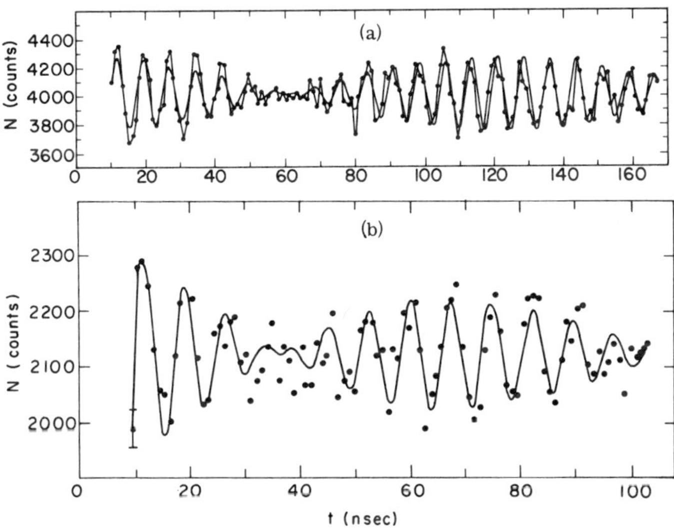

FIG. 9. "Two-frequency precession" of the muon (a) in fused quartz with a transverse field of 95 G and (b) in germanium at 98 G. The smooth curves represent the best fits of the theoretical dependence to the data. The theoretical function $N(t)$ and the data are corrected for the decay exponential $\exp(-t/\tau_\mu)$ (from Gurevich *et al.*, 1971a).

beat frequency Ω, the hyperfine frequency ω_0 can be calculated:

$$\omega_0 \approx \omega_-^2/\Omega$$

At early times or in very weak fields, the condition $\Omega t \ll \pi/2$ is fullfilled. Equation (20) can then be written

$$\tilde{P}_{\perp}{}^{\mu}(x, t) \doteqdot \tfrac{1}{2} \exp(i\omega_- t)\,(1 + \cos \omega_0 t) \tag{21}$$

That is, the motion consists of a comparatively slow "muonium precession" at frequency ω_-, superimposed upon a "hyperfine modulation" at frequency ω_0. This is just the behavior described semiclassically in the previous section.

C. Quasi-Free Muonium

A muonium atom embedded in matter can rarely be considered "free"; usually we will have to account for the perturbing effects of its material environment.

The first such effect we must consider is the possible collective action of charge carriers or of an ionic crystal field upon the Coulomb attraction in the electron–muon charge system. A large local dielectric polarizability of the medium, due to collective shielding effects, can change the size of the muonium atom, thus drastically modifying the hyperfine coupling constant. Effects of this sort, observed mainly in semiconductors, will be treated in Section VIII. Since these interactions change only the value of ω_0, we will consider them only implicitly in this section. It should be kept in mind, however, that ω_0 can be a function of the medium; since this hyperfine frequency determines the time scale in muonium, such effects influence the μ^+ depolarization rate.

An interaction with the medium that we must treat explicitly is the coupling of the muonium electron's magnetic moment with those of charge carriers or with the crystal field. (We can neglect direct coupling of the μ^+ magnetic moment to such fields in the medium, at least on the time scale of muonium precession, since these interactions are at least $\sim$200 times smaller than those of the electron.) In order to take that interaction into account, we must turn to the elegant and general formalism used by Ivanter and Smilga (1968), also illustrated in a paper by Fischer (1973). Their descriptions are based on earlier work by Nosov and Yakovleva (1963, 1965), who in turn used an approach introduced by Wangsness and Bloch (1953) for treatment of depolarization by quantum irreversible processes. Here we briefly summarize that formalism and its predictions; the results will be used in the next section in connection with μ^+ depolarization related to chemical reactions of muonium.

We start by defining the 4×4 density matrix for the muonium spin system in terms of the Pauli spin matrices $\boldsymbol{\sigma}^\mu$ and $\boldsymbol{\sigma}^e$ for the muon and electron:

$$\rho^{\mathrm{Mu}} = \frac{1}{4}\left[1 + \mathbf{P}^\mu \cdot \boldsymbol{\sigma}^\mu + \mathbf{P}^e \cdot \boldsymbol{\sigma}^e + \sum_{i,j=1}^{3} p_{ij}\sigma_i^\mu \sigma_j^e\right] = \frac{1}{4} \sum_{i,j=0}^{3} p_{ij}\sigma_i^\mu \sigma_j^e \quad (22)$$

where $\mathbf{P}^\mu$ is the muon polarization vector, with components p_{10}, p_{20}, p_{30} (x, y, z), and $\mathbf{P}^e$ is the electron polarization vector, with components p_{01}, p_{02}, p_{03}. The quantities σ_0^μ and σ_0^e are 2×2 unit matrices. Thus the 16 matrices $\sigma_i^\mu \sigma_j^e$ $(i, j = 0$ to $3)$ form a complete orthonormal basis for the space subtended by the spin-$\frac{1}{2}$ $\otimes$ spin-$\frac{1}{2}$ system.

For free muonium, the equation of motion of this density matrix is

$$\partial \rho^{\mathrm{Mu}}/\partial t = -(i/\hbar)[H^{\mathrm{Mu}}, \rho^{\mathrm{Mu}}] \quad (23)$$

where H^{Mu} is given by Eq. (8). In the quasi-free case, however, the Hamiltonian of the whole system has to be considered. One must add to H^{Mu} a term V representing an "electron–lattice" interaction, whereby polarization is transferred by the electron to the immediate environment, and a term H^{L} for the "lattice–lattice" interaction, which distributes the transferred polarization throughout the medium:

$$H = H^{\mathrm{Mu}} + V + H^{\mathrm{L}} \quad (24)$$

We must now also consider the density matrix ρ^{tot} for the combined system, muonium and medium. Following the development used by Wangsness and Bloch (1953) and by Fano (1957) we can write the time evolution of ρ^{Mu} as follows:

$$\frac{d\rho^{\mathrm{Mu}}}{dt} = -\frac{i}{\hbar}[H^{\mathrm{Mu}}, \rho^{\mathrm{Mu}}] - \frac{\pi}{\hbar}\mathrm{Tr_L}[\bar{V}, [\bar{V}, \rho^{\mathrm{Mu}}\rho_0^{\mathrm{L}}]] \quad (25)$$

where $\mathrm{Tr_L}$ is the trace over the part of the total density matrix that describes the surrounding medium "L," ρ_0^{L} the equilibrium density matrix of the medium, and $\bar{V}$ the portion of V that is diagonal in the total energy.

Following Nosov and Yakovleva (1963), we can express the effect of the double commutator phenomenologically in terms of a relaxation rate ν imposed upon the electronic components of the polarization. Writing out the commutator of Eq. (25) and equating coefficients of orthogonal operators, we are led to the system of 15 Wangsness–Bloch equations:

$$
\begin{aligned}
\dot{p}_{k0} &= -\frac{\omega_0}{2}\sum_{i,j=1}^{3}\epsilon_{ijk}p_{ij} + \sum_{i,j=1}^{3}\epsilon_{ijk}\omega_i{}^{\mu}p_{j0} \\
\dot{p}_{0k} &= \frac{\omega_0}{2}\sum_{i,j=1}^{3}\epsilon_{ijk}p_{ij} + \sum_{i,j=1}^{3}\epsilon_{ijk}\omega_i{}^{e}p_{0j} - 2\nu p_{0k} \\
\dot{p}_{ij} &= \frac{\omega_0}{2}\left(\sum_{n=1}^{3}\epsilon_{nij}p_{n0} - \sum_{n=1}^{3}\epsilon_{nij}p_{0n}\right) \\
&\quad - \sum_{m,n=1}^{3}\epsilon_{mni}\omega_n{}^{\mu}p_{mj} + \sum_{m,n=1}^{3}\epsilon_{mnj}\omega_m{}^{e}p_{in} - 2\nu p_{ij}
\end{aligned}
\tag{26}
$$

where $\boldsymbol{\omega}^e$ and $\boldsymbol{\omega}^\mu$ were defined in the previous section, and ϵ_{ijk} is the antisymmetric unit tensor. If $\nu = 0$, so that the two electron-damping terms can be neglected, Eqs. (26) are the same as we obtain from Eq. (23) for free muonium.

The system of Eqs. (26) can be separated into two irreducible subsystems, one involving only the components of muon and electron polarization along the magnetic field direction $\hat{z}$ (longitudinal subsystem), and the other involving only the components of $\mathbf{P}^\mu$ and $\mathbf{P}^e$ perpendicular to $\mathbf{B} = B\hat{z}$ (transverse subsystem). We will now treat each of these subsystems in more detail.

1. *Longitudinal Subsystem*

The equations of motion for those components of polarization coupled to p_{30}, the muon polarization along the field, are

$$
\begin{aligned}
\dot{p}_{30} &= -\frac{\omega_0}{2}(p_{12} - p_{21}) \\
\dot{p}_{03} &= \frac{\omega_0}{2}(p_{12} - p_{21}) - 2\nu p_{03} \\
\dot{p}_{11} &= -\omega_e p_{12} + \omega_\mu p_{21} - 2\nu p_{11} \\
\dot{p}_{22} &= \omega_e p_{21} - \omega_\mu p_{12} - 2\nu p_{22} \\
\dot{p}_{12} &= \frac{\omega_0}{2}(p_{30} - p_{03}) + \omega_e p_{11} + \omega_\mu p_{22} - 2\nu p_{12} \\
\dot{p}_{21} &= -\frac{\omega_0}{2}(p_{30} - p_{03}) - \omega_e p_{22} - \omega_\mu p_{11} - 2\nu p_{21}
\end{aligned}
\tag{27}
$$

where we have used $\omega_e \equiv |\boldsymbol{\omega}^e| = \omega_3{}^e$ and $\omega_\mu \equiv |\boldsymbol{\omega}^\mu| = -\omega_3{}^\mu$.

Solution of these equations in the general case is rather involved. We will mention here only two limiting cases, where good approximate solutions can be found. These cases were described by Nosov and Yakovleva (1963, 1965) and by Ivanter and Smilga (1971).

For very fast "spin-flipping" of the electron,

$$\nu \gg \omega_0(1 + x^2)^{1/2}$$

the time dependence of the μ^+ polarization is

$$P_{\|}^{\mu}(x, t) \approx \exp(-t/\tau_1) \tag{28}$$

where

$$\tau_1 = 4\nu/\omega_0^2 \tag{29}$$

That is, the muon spin is exponentially damped at a rate that does not depend on the field, but that decreases with increasing ν. This can be explained qualitatively as a weakening of the μ–e coupling by excessive electron relaxation (Nosov and Yakovleva, 1963, 1965). The limiting case $\nu \to \infty$ corresponds to the situation in metals, where the μ^+ behaves as if free.

At the other extreme is the case of very mild electron relaxation,

$$\nu \ll \omega_0(1 + x^2)^{1/2}$$

Here we average the time dependence over the hyperfine oscillations—that is, over a time interval Δt satisfying

$$\nu \ll 1/\Delta t \ll \omega_0(1 + x^2)^{1/2}$$

The experimentally observable result is

$$\bar{P}_{\|}^{\mu}(x, t) \approx P_0 \exp(-t/\tau_2) \tag{30}$$

where P_0 is given by Eq. (13), and

$$\tau_2 = (1 + x^2)/\nu \tag{31}$$

Not surprisingly Eq. (30) depicts an exponential decay of the μ^+ polarization, whose initial (average) value is the same as for free muonium in longitudinal field. The decay time τ_2 can, however, be lengthened by increasing the magnetic field. Experimental observations of such field dependence of τ_2 constitute evidence for the presence of this sort of depolarization mechanism.

2. *Transverse Subsystem*

Following Ivanter and Smilga (1968), we introduce the complex four-component vector:

$$\tilde{P} \equiv \begin{pmatrix} p_{10} + ip_{20} \\ p_{01} + ip_{02} \\ p_{13} + ip_{23} \\ p_{31} + ip_{32} \end{pmatrix} \tag{32}$$

whose first component is the complex transverse muon polarization $\tilde{P}_{\perp}{}^{\mu}(x, t)$ defined earlier. The time dependence of the transverse components in Eq. (26) can then be written

$$d\tilde{P}/dt = iA\tilde{P} \tag{33}$$

where A is the 4×4 complex matrix

$$A = \frac{\omega_0}{2} \begin{pmatrix} -2\zeta X & 0 & 1 & -1 \\ 0 & (i\gamma + 2X) & -1 & 1 \\ 1 & -1 & (i\gamma - 2\zeta X) & 0 \\ -1 & 1 & 0 & (i\gamma + 2X) \end{pmatrix} \tag{34}$$

where

$$\zeta \equiv |\,\omega^{\mu}\,|/|\,\omega^{e}\,| = |\,g_{\mu}\mu_0{}^{\mu}/g_{e}\mu_0{}^{\mu}\,| = 1/206.77 \approx m_e/m_{\mu} \tag{35}$$

$$\gamma \equiv 4\nu/\omega_0 \tag{36}$$

and

$$X = |\,\omega^{e}\,|/\omega_0 = (g_e\mu_0{}^{e}/\hbar\omega_0)\,|\,\mathbf{B}\,| = |\,\mathbf{B}\,|/B_0{}^{*} \tag{37}$$

an alternative version of the "specific field" discussed in Section IV. The "effective hyperfine field" $B_0{}^{*}$ has a value of 1593 G for free muonium. The two versions of specific field are very simply related:†

$$X = x/(1 + \zeta)$$

Since the polarization components parallel to the field evolve independently of the transverse components considered here, we may assume

† Ivanter and Smilga (1968, 1969a,b, 1971) refer exclusively to X in their works.

without loss of generality the initial condition

$$\tilde{P}(0) = \begin{pmatrix} 1 \\ 0 \\ 0 \\ 0 \end{pmatrix} \tag{38}$$

That is, the muon is initially fully polarized in the $\hat{x}$ direction and there is no initial electron polarization or correlation between electron and muon spins.

The equation of motion (33) with the initial condition (38) is solved by diagonalizing the matrix A. The orthogonal matrix M that diagonalizes A,

$$M^{-1}AM = \Lambda \tag{39}$$

and the resulting eigenvalues $\Lambda_{kk} = \lambda_k^m$ can be found by standard but tedious manipulations. We will not describe this process in detail, but simply proceed to the results.

The simplest case, of course, is that in which $\nu = 0$, so that A is a real symmetric matrix. Here, as expected, the result contains Eq. (19) for $P_\perp^\mu(x, t)$.

For the general case ($\nu \neq 0$), several limiting cases have been calculated and discussed by a number of authors.

As for longitudinal fields, extremely fast spin-flipping of the electron $[\nu \gg \omega_0(1 + x^2)^{1/2}]$ serves to weaken the coupling between the muon and the electron, allowing the μ^+ to precess almost as if free. Nosov and Yakovleva (1963, 1965) showed that the muon polarization in this case evolves as

$$\tilde{P}_\perp^\mu(x, t) \approx \exp\left[\left(-i\omega_\mu - \frac{1}{\tau_1}\right)t\right] \tag{40}$$

where τ_1 is given by Eq. (29). That is, as in longitudinal field, the muon polarization relaxes at a rate inversely proportional to ν.

When $\nu \sim \omega_0$, neither μ^+ nor muonium precession is experimentally observable; the muon polarization is lost to the medium through the electron in times shorter than the apparatus can resolve. It may prove necessary to treat this most difficult case in detail if electron spin-flipping is to be studied in cases where ω_0 is drastically reduced, lengthening the hyperfine time scale to observable intervals.

In the case of very mild relaxation $[\nu^2 \ll (\omega_0/2)^2x^4]$, Gurevich *et al.* (1971a) have calculated the time dependence of the x-component of the μ^+ polarization, averaged over the unobservable hyperfine oscillations as for

Eq. (30). Their result is

$$\bar{P}_x^\mu(x, t) \approx \frac{1}{2} \exp(-t/\tau_3) \left[\left(\cos \Omega_\gamma t + \frac{\Omega \sin \Omega_\gamma t}{3\tau_3 \Omega_\gamma^2} \right) \cos \omega_- t + \left(\frac{2\omega_+ \Omega^2}{\omega_0 \Omega_\gamma^2} \right) \sin \Omega_\gamma t \sin \omega_- t \right] \tag{41}$$

where the expected exponential damping factor is related to ν as

$$\tau_3 \equiv 2/3\nu \tag{42}$$

and the beat frequency is also relaxation dependent:

$$\Omega_\gamma \equiv \Omega[1 - (\nu^2/4\Omega^2)]^{1/2} \tag{43}$$

As will be discussed in the next section, if chemical reactions gradually eliminate free muonium atoms from the ensemble, the observed muonium precession in low fields will be exponentially damped as above, but there will be no shift in the beat frequency. In principle, an observation of this phenomenon, though difficult, would serve to identify the cause of such relaxation of muonium precession. Such a clue may prove vital to studies of muonium chemistry in the gas phase, where one technique is to add a reactive gas to an inert target gas in weak transverse field and watch the resultant decay of the muonium precession signal.

D. *Muonium in an rf Field*

We will briefly mention the effect of one artificially induced perturbation upon the evolution of the μ^+ spin in muonium: the application of a strong rf field at a frequency near one of the $\Delta m = \pm 1$ transition frequencies of muonium, usually ω_{12}. This technique has been used extensively for measuring the muonium hyperfine interval to extreme precision, and also for studying the chemical and spin-exchange reactions of muonium in gases (Mobley, 1967). The analogy with electron spin resonance experiments is nearly perfect, but several distinguishing features should be noticed.

The rf excitation is almost always superimposed upon a constant longitudinal field. Thus an excitation of the transition $|E_1\rangle \rightarrow |E_2\rangle$ destroys the longitudinal μ^+ polarization in the half of the ensemble that forms in the polarized triplet state. The average polarization is thereby reduced as rf pumping brings states $|E_1\rangle$ and $|E_2\rangle$ into equilibrium population. Since the population of state $|E_2\rangle$ in the original muonium formation is lower in stronger longitudinal fields [see Eq. (12)], the effect of the rf depolarization increases with field strength. This depolarization is reflected in the ratio of decay positrons detected in forward and backward directions, so that the "resonance signal" is

$$S = \frac{N_f(\text{rf on}) - N_f(\text{rf off})}{N_f(\text{rf on}) + N_f(\text{rf off})} - \frac{N_b(\text{rf on}) - N_b(\text{rf off})}{N_b(\text{rf on}) + N_b(\text{rf off})} \tag{44}$$

By contrast, the resonance signal in ESR experiments is a bulk absorption of rf power by the sample.

When the muonium electron is strongly relaxed by the medium, the depolarizing effect of an rf field is negligible in comparison, leading to a "quenching" of the resonance signal defined above. A fast chemical reaction, placing the μ^+ in a diamagnetic environment before the rf perturbation can have any effect, will also "quench" the resonant depolarization signal. Measurements of the height and linewidth of $S(\omega_{rf})$ as a function of impurity concentration can be analyzed to yield cross sections for spin exchange and chemical reaction of muonium (Mobley, 1967).

VI. Chemical Reactions of Muonium and Residual Muon Polarization: Theory

If all thermalized muonium atoms were to preserve their chemical state, most experiments on interactions of the μ^+ spin with matter would involve observation of the sort of phenomena described in the previous section. However, that situation is rather exceptional, since muonium is a radical and therefore has a strong propensity to react chemically. For the time being, we shall assume that all such reactions are of the form

$$\text{Mu} + \text{X} \rightarrow \text{D} \tag{45}$$

where X is some reagent and D is an unspecified final state in which the muon is incorporated into a diamagnetic molecule. In most media these reactions occur during the μ^+ lifetime, leaving that particle in a diamagnetic environment, where the rapid spin evolution described in the previous section stops abruptly. Other types of thermalized muonium reactions also occur, and we will describe later the effects of different concurrent reaction channels. In addition, as mentioned in Section III, a substantial fraction (h) of muons usually reaches a diamagnetic environment through hot-atom reaction channels, bypassing the muonium stage of spin evolution completely.

Considering for the moment only that fraction $(1 - h)$ of the muons that thermalize as free muonium, there are two extreme situations in which chemically meaningful measurements can be made in transverse field. If the Mu atoms remain uncombined for observable times (~100 nsec or more), the "muonium precession" phenomena described in the previous section can be studied; chemical reactions of Mu atoms are manifested in an exponential decay of that precession. This technique is well suited to gas-phase studies (Mobley, 1967), where reaction times are often many nsec. However, since μ^+ stopping density has historically been a severe experimental limitation, most studies of Mu chemistry have been in condensed matter, especially in liquids, where reaction times are usually much too short for direct observation of muonium precession. In this situation, one looks for μ^+ precession as described in Eq. (6). Muons still precessing in muonium (~103 times faster) appear completely depolarized on this time scale, so that muonium is considered a depolarizing influence upon the muon. In fact, since each muon "emerges" from the muonium stage of spin evolution at a different time (following a probability distribution), the μ^+ ensemble is depolarized (in a thermodynamic sense) by the resultant "dephasing." If τ_m, the mean chemical lifetime of free muonium, is much longer than the period of muonium precession ($2\pi/\omega_-$), this "fast depolarization" is complete, and the muon asymmetry in Eq. (6) is zero. However, if $\tau_m \ll 1/\omega_0$, the μ^+ spin has little opportunity to evolve in muonium before entering a diamagnetic state, and the "residual polarization" after its brief stay in muonium approaches unity. In addition, when $\tau_m \sim 1/\omega_-$, the short-lived muonium precession rotates the polarization through an average angle $|\Delta\phi| \sim \omega_-\tau_m$, which is observed as a shift of the apparent initial phase of the μ^+ precession, as described in Section II.

When muonium lasts long enough to destroy the μ^+ precession, but not long enough for Mu precession to be observed directly, no useful information can be obtained in transverse field about the fate of the fraction $(1 - h)$ of muons that thermalize as muonium. This limitation does not apply to experiments in longitudinal field, and in general the two techniques are complementary. However, the residual polarizations and apparent phase shifts in transverse field, when observable, provide the most revealing experimental test of very fast chemical reactions. The emphasis in this section will therefore be upon a precise theoretical prediction of these quantities in terms of the properties of the medium.

A. Proper Muonium Mechanism

Muonium reacts with the reagent X at a constant rate Λ. The probability $m(t)$ that a given muonium atom is still free at time t therefore obeys the rate equation

$$dm(t)/dt = -\Lambda m(t) \tag{46}$$

so that the probability $m(t)$ decays exponentially with a mean lifetime

$$\tau_m = 1/\Lambda \tag{47}$$

If the origin of time is chosen to be the instant of thermalization (approximately when the μ^+ enters the target, since the stopping time can be neglected in comparison with the time scale of μ^+ polarization evolution in muonium), then $m(t)$ is given by

$$m(t) = \exp(-t/\tau_m) \tag{48}$$

The probability dn that reaction (45) occurs within an interval dt' at time t' is given by

$$dn = (1/\tau_m) \exp(-t'/\tau_m)\, dt' \tag{49}$$

At the time of the reaction, the complex μ^+ polarization in the transverse subsystem (defined as in the previous sections) is just $\tilde{P}_\perp^\mu(x, t)$, the first component of $\tilde{P}$ in Eq. (32). Following reaction (45), the μ^+ will precess at essentially the free Larmor frequency ω_μ. Thus the polarization of such a muon at any time $t > t'$ is

$$\tilde{P}_\perp^\mu(x, t') \exp[-i\omega_\mu(t - t')]$$

Such a fate befalls a fraction dn of all the muons in the ensemble. These muons thus contribute

$$dP(t) = dn\tilde{P}_\perp^\mu(x, t') \exp[-i\omega_\mu(t - t')] \tag{50}$$

to the net ensemble muon polarization $P(t)$. Summing all such contributions leads to an integral over dt'. By the same token, we must include in $P(t)$ the contribution from Mu atoms that have not yet reacted; this is given by the product of the probability $\exp(-t/\tau_m)$ of Mu surviving reaction for a time t, and the polarization of such muons, $\tilde{P}_\perp^\mu(x, t)$. The global μ^+ polarization at time t is thus given by

$$P(t) = \int_0^t \tilde{P}_\perp^\mu(x, t') \exp[-i\omega_\mu(t - t')] \frac{\exp(-t'/\tau_m)}{\tau_m}\, dt' + \tilde{P}_\perp^\mu(x, t) \exp(-t/\tau_m) \tag{51}$$

For $t \gg \tau_m$, the second term vanishes and (51) can be written in the form

$$P(t \gg \tau_m) \approx \exp(-i\omega_\mu t) R_\perp \tag{52}$$

where

$$R_\perp = \int_0^\infty \tilde{P}_\perp^\mu(x, t') \exp(i\omega_\mu t') \left(\frac{\exp(-t'/\tau_m)}{\tau_m}\right) dt' \tag{53}$$

That is, long after all muonium atoms have reacted, the muons precess just as if they had begun at $t = 0$ with an initial polarization $R_\perp$. This apparent initial μ^+ polarization is called the "residual polarization"; it is generally reduced and rotated with respect to the actual initial polarization. $R_\perp$ can be calculated in several ways. If $\tilde{P}_\perp^\mu(x, t)$ is given by a manageable explicit formula such as (19), the integral in (53) can be performed directly. Alternatively, if the diagonalization of the matrix A as in Eq. (39) is known, the time dependence of $\tilde{P}_\perp^\mu(x, t)$ can be written formally as

$$\tilde{P}_\perp^\mu(x, t) = \sum_k F_k \exp(i\lambda_k^m t) \tag{54}$$

where

$$F_k = M_{1k}(M^{-1})_{k1} \tag{55}$$

This formula can be substituted into Eq. (53) and integrated easily. The result is

$$R_\perp = \frac{1}{\tau_m} \sum_k \frac{F_k}{\alpha_k} \tag{56}$$

where

$$\alpha_k = -\frac{1}{\tau_m} + i(\lambda_k^m + \omega_\mu) \tag{57}$$

This approach is useful in more complicated situations, as we shall see later. However, the most elegant approach to this problem is that used by Ivanter and Smilga (1968), who noted that Eq. (53) simply expresses $R_\perp$ in terms of the Laplace transform of the first component of $\tilde{P}$. This allows solution for $R_\perp$ by transforming the differential equations (33) into a system of linear equations:

$$(A - uI)\mathcal{L}(u) = -\tilde{P}(0) \tag{58}$$

where

$$\mathcal{L}(u) = \int_0^\infty e^{-ut}\tilde{P}(t)\, dt$$

with $u = 1/\tau_m - i\omega_\mu$, and I the identity matrix. Thus $R_\perp$ is just $1/\tau_m$ times the first component of the four-component transformed vector $\mathcal{L}(u)$.

The linear system is easily solved by matrix methods, giving the general result

$$R_\perp = \frac{1}{1 - i\tau(A + B)B/(AB^2 \quad A - B)} \tag{59}$$

where

$$\tau = \tfrac{1}{2}\omega_0\tau_{\rm m} \tag{60}$$

$$A = -i\alpha \tag{61}$$

$$\alpha = \gamma + 1/\tau \tag{62}$$

and

$$B = A - 2x \tag{63}$$

The longitudinal subsystem of polarization components parallel to the magnetic field is unaffected by μ^+ or Mu precession. Here the time dependence of the μ^+ polarization in muonium generally consists of an exponential decay, either in $P_{||}{}^{\mu}(x, t)$ [as in Eq. (28)] or in its average over hyperfine oscillations [as in Eq. (30)]. As muons react chemically, they are spared from this depolarization; the ensemble time dependence thus has a form analogous to Eq. (51). Such an integral is easily performed in the simple case

$$1/\tau_{\rm m} \ll \omega_0(1 + x^2)^{1/2} \ll \nu \tag{64}$$

where $P_{||}{}^{\mu}(x, t)$ is given by Eq. (28). The result is

$$P(t) = \frac{\tau_1}{\tau_{\rm m} + \tau_1} + \frac{\tau_{\rm m}}{\tau_{\rm m} + \tau_1} \exp\left[-\left(\frac{1}{\tau_{\rm m}} + \frac{1}{\tau_1}\right)t\right] \tag{65}$$

where τ_1 is given by Eq. (29). The effective depolarization time

$$\tau_{\rm eff} = \left(\frac{1}{\tau_{\rm m}} + \frac{1}{\tau_1}\right)^{-1}$$

is sometimes long enough to be observed directly (Gurevich *et al.*, 1968).

In the case of slow reactions and mild relaxation,

$$1/\tau_{\rm m} \ll \omega_0(1 + x^2)^{1/2} \qquad \text{and} \qquad \nu \ll \omega_0(1 + x^2)^{1/2} \tag{66}$$

integration of $\bar{P}_{||}{}^{\mu}(x, t)$ from Eq. (30) yields

$$\bar{P}(t) = \frac{\tau_2 P_0}{\tau_{\rm m} + \tau_2} + \frac{\tau_{\rm m} P_0}{\tau_{\rm m} + \tau_2} \exp\left[-\left(\frac{1}{\tau_{\rm m}} + \frac{1}{\tau_2}\right)t\right] \tag{67}$$

with P_0 from Eq. (13) and τ_2 from Eq. (31). Conditions (66) are often fulfilled in gas-phase reactions.

In general, however, we may only observe the residual μ^+ polarization $R_{||}$ at times $t \gg \tau_m$, long after all Mu atoms have reacted. The expression for $R_{||}$ is analogous to that for $R_\perp$:

$$R_{||} = \int_0^\infty P_{||}^\mu(x, t') \frac{\exp(-t'/\tau_m)}{\tau_m} dt' \tag{68}$$

where $P_{||}^\mu(x, t)$ is just $p_{30}(t)$ as in Eq. (27). After some manipulations with the Fourier transform (Ivanter and Smilga, 1968), one obtains the general result

$$R_{||} = \frac{x^2 + \frac{1}{2} + (\alpha^2/4)}{x^2 + 1 + \nu\tau_m + (\alpha^2/4)} \tag{69}$$

where α is given by Eq. (62).

In the limiting case (64), the result is $R_{||} = \tau_1/(\tau_m + \tau_1)$, as evident from Eq. (65).

In another simple limiting case

$$\nu \ll 1/\tau_m \ll \omega_0(1 + x^2)^{1/2} \tag{70}$$

we have $\alpha \to 0$ and (69) reduces to

$$\lim_{\alpha\to 0} R_{||} = \frac{x^2 + \frac{1}{2}}{x^2 + 1 + \nu\tau_m} \tag{71}$$

This result can also be obtained by integrating $\bar{P}_{||}^\mu(x, t)$ from Eq. (30) (Nosov and Yakovleva, 1965b).

The oblique field case can easily be handled as a superposition of longitudinal and transverse subsystems: if the original μ^+ polarization is in the xz-plane at an angle ψ to the z-axis, the net residual polarization is

$$\mathbf{R} = \hat{x}\,\mathrm{Re}(p_{10}(0)R_\perp) + \hat{y}\,\mathrm{Im}(p_{10}(0)R_\perp) + \hat{z}p_{30}(0)R_{||} \tag{72}$$

where $p_{10}(0) = \sin\psi$ and $p_{30}(0) = \cos\psi$.

The above description pertains to that fraction $(1 - h)$ of muons that thermalize as free muonium. In general there is also a fraction h that react through epithermal channels, essentially at $t = 0$, and propagate throughout their existence as quasi-free muons. To obtain the experimentally observed residual polarization, $\mathbf{P}_{\text{res}}$, one must combine these two components:

$$\mathbf{P}_{\text{res}} = h[p_{10}(0)\hat{x} + p_{30}(0)\hat{z}] + (1 - h)\mathbf{R} \tag{73}$$

The pure transverse field case, where $p_{30}(0) = 0$ and $p_{10}(0) = 1$, can thus be expressed as

$$P_{\perp\,\text{res}} = h + (1 - h)R_\perp \tag{74}$$

If we then observe the positrons from μ^+ decay long after all Mu atoms have reacted, the experimental asymmetry A and the phase shift $\Delta\phi$ in Eq. (6) will be given by

$$A = A_0 \,|\, P_{\perp \text{ res}} \,| \tag{75}$$

and

$$\tan \Delta\phi = \text{Im}(P_{\perp \text{ res}})/\text{Re}(P_{\perp \text{ res}}) \tag{76}$$

where A_0 is a constant factor equivalent to the effective asymmetry when no polarization is lost in the target. This factor, like θ_0 in Eq. (6), depends on beam polarization, counter geometry, and the details of the decay. Both are usually fitted as empirical parameters. Figure 10 shows the dependence of $|\, P_{\perp \text{ res}} \,|$ and $\Delta\phi$ on the average chemical lifetime τ_m for a perpendicular field of 100 G, assuming $\nu = 0$. The dashed curves are for the case without hot-atom chemistry, where $P_{\perp \text{ res}} = R_\perp$, and the solid curves show the results for $h = 0.5$. At very short chemical lifetime ($\tau_m \ll 1/\omega_0$) the average muon's stay in muonium is so brief as to have no effect on the polarization, leaving $P_{\perp \text{ res}} = 1$ and $\Delta\phi = 0$. For $\tau_m \sim 1/\omega_0$, muons "drop out" of muonium over one or two hyperfine periods, causing $R_\perp$ to drop to ~ 0.5; this situation persists for τ_m intermediate between $1/\omega_0$ and $1/\omega_-$, giving the "plateau" effect seen in $|\, P_{\perp \text{ res}} \,|$ in Fig. 10. When $\tau_m \sim 1/\omega_-$, muons leave muonium during approximately the first quarter cycle of

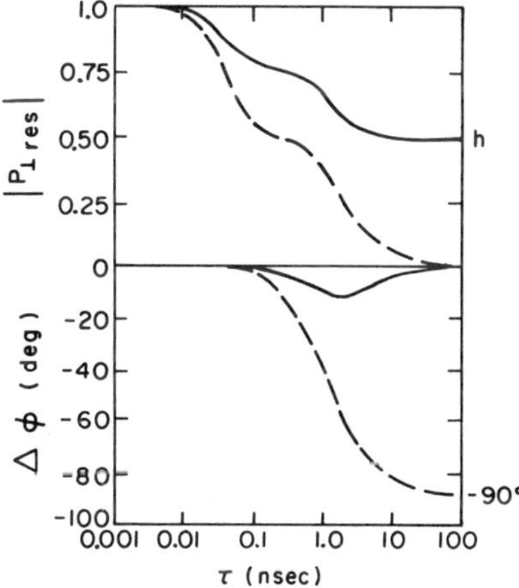

FIG. 10. Proper muonium mechanism in 100-G transverse field: dependence of magnitude and phase of residual polarization on chemical lifetime τ of free muonium. Positive phase is defined as being in the direction of μ^+ precession. Dashed curves: no hot-atom chemistry; solid curves: hot fraction $h = 0.5$.

muonium precession, leading to a rotation of the residual polarization in the opposite sense to μ^+ precession as well as a net attenuation that is essentially complete for $\tau_m \gg 1/\omega_-$. In this final limit, the phase of $R_\perp$ goes to $-90°$ (positive $\Delta\phi$ being defined here in the sense of free μ^+ precession) while $|R_\perp| \to 0$; $P_{\perp\,\text{res}}$, however, contains the constant unrotated vector h, and so its phase returns to zero as the magnitude of the rotated component vanishes.

As we have seen, irreversible depolarization of the muons is brought about by the "dephasing" of their spins due to chemical reaction of Mu atoms at random times. This is referred to as the "proper muonium mechanism" for fast μ^+ depolarization. The "plateau" in $|P_{\perp\,\text{res}}|$ and the "phase dip" in $\Delta\phi$ are particularly characteristic of this mechanism; these phenomena are not predicted in oversimplified models such as that used by Firsov and Byakov (1965).

A convenient experimental method for testing these predictions can be used in dilute solutions of a highly reactive scavenger compound (X) in a solvent (S) that is inert to thermal reactions with Mu, but in which epithermal reactions have a finite probability h. The reaction rate $\Lambda = 1/\tau_m$ is assumed to be proportional to the concentration [X] of the reagent:

$$\Lambda = k[\text{X}] \tag{77}$$

where k, the constant of proportionality, is called the chemical rate constant. Thus, by varying the reagent concentration [X], one can change the chemical lifetime τ_m to produce curves of $P_{\perp\,\text{res}}$ vs [X] such as those shown in Fig. 11, where a rate constant $k = 10^{10}$ liter/mole-sec and a "hot fraction" $h = 0.5$ are assumed. The solid curves in Fig. 11 are for an external field of 100 G, corresponding to the solid curves in Fig. 10. The dotted curves are for $B = 10$ G, and show that slowing down the muonium precession broadens the plateau. The dashed curves are for $B = B_0 = 1585$ G, where ω_- becomes comparable to ω_0 and the plateau disappears. Here the amplitude of the phase dip is reduced and the positive excursion for very short reaction times is much more noticeable than in lower fields. The reversal of the sign of $\Delta\phi$ can be understood in terms of Eq. (19) for $\tilde{P}_\perp{}^\mu(x, t)$. For high fields, $\delta \approx 1$ and the first motion of the μ^+ spin includes a rotation in the normal sense of μ^+ precession. The contrary precession of the electron moment then carries the μ^+ spin along with it. For fields stronger than B_0, the dip in the phase of $P_{\perp\,\text{res}}$ becomes too small to measure conveniently, especially when h is large. The positive excursion increases somewhat with field, but is never large enough to observe easily.

In Fig. 11 it is assumed that $\nu = 0$; any significant relaxation rate for the muonium electron would disrupt the muonium precession, as in Eq. (40),

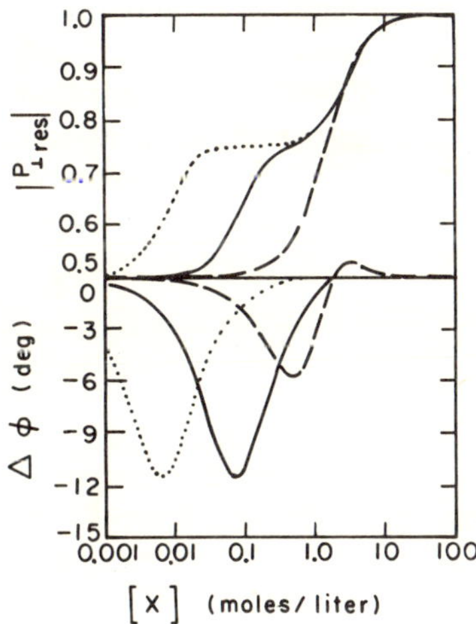

FIG. 11. Proper muonium mechanism in several transverse fields: dependence of $P_{\perp \text{ res}}$ on reagent concentration [X] when $\tau_m^{-1} = k[X]$ with $k = 10^{10}$ liter/mole-sec; hot fraction $h = 0.5$. Solid curves: $B = 100$ G; dotted curves: $B = 10$ G; dashed curves: $B = B_0 = 1593$ G.

causing the plateau effect and the phase dip, both results of coherent Mu precession, to be more or less spoiled. Equation (59) gives the dependence of $P_{\perp \text{ res}}$ on $\gamma = 4\nu/\omega_0$; one can envision a situation in which τ_m is constant while ν depends upon the concentration of a paramagnetic species (Schenck, 1970). More generally, both τ_m and ν would vary with [X]. However, there is no evidence for nonnegligible spin-flipping in even strongly paramagnetic solutions (Schenck, 1970), and so we can assume that $\nu \ll \omega_0$ in liquids.

B. Radicals and Two-Stage Mechanisms

It was assumed in the preceding description that all chemical reactions of muonium leave the μ^+ in a diamagnetic environment. This is not generally true. Since muonium is itself a paramagnetic atom, at least one of the products of its reaction with an ordinary diamagnetic molecule will be a paramagnetic molecule, or radical. If the radical product incorporates the muon, then a hyperfine interaction between the μ^+ and any unpaired electron will lead to motions of the coupled spins quite like those in muonium. The simplest sort of radical contains one unpaired electron in an s-orbital;

in this case the spin system is identical to muonium except that the strength of the hyperfine coupling is greatly reduced. That is, the hyperfine frequency in the radical, ω_r, is generally much less than that in muonium, ω_0.

Radicals created in epithermal reactions, if all of the same species, could be observed directly by virtue of their muoniumlike behavior, described most generally by Eq. (33), in which A is now A_r, the radical version of the matrix A_m defined by Eq. (34); A_r differs from A_m in the substitution of ω_r for ω_0 everywhere (including the definition of X). If a fraction r of incoming muons enter identical radicals in this way, essentially at $t = 0$, they can be detected only if they remain chemically free for an observable time ($\sim$100 nsec or more); such "radical precession" has never yet been observed.

Fast thermal reaction of an epithermally formed muon-bearing radical R leads to a μ^+ depolarization mechanism perfectly analogous to the proper muonium mechanism described earlier, provided that the thermal reaction leaves the muon in a final diamagnetic environment D:

$$\mathrm{R} + \mathrm{X} \rightarrow \mathrm{D} \tag{78}$$

However, if the radical itself arises from thermal reaction of muonium, as in

$$\mathrm{Mu} + \mathrm{X} \rightarrow \mathrm{R} \tag{79}$$

then the μ^+ spin has already evolved in muonium for an arbitrary time before entering the radical environment. If the radical subsequently reacts as in (78), and if both reactions occur swiftly enough, some μ^+ polarization still remains; but the simple Laplace transform technique described above will not suffice to solve for the residual polarization, which here involves an integral over two exponentially distributed reaction times.

A flow diagram of this "two-component mechanism" along with other competing mechanisms of μ^+ depolarization, is shown in Fig. 12. The possibilities of reactions of Mu with the solvent S are also included in this picture; these are assumed to be of the form

$$\mathrm{Mu} + \mathrm{S} \rightarrow \mathrm{D} \tag{80}$$

or

$$\mathrm{Mu} + \mathrm{S} \rightarrow \mathrm{R} \tag{81}$$

With the variety of reactions given by (45) and (78)–(81), an assortment of chemical rate constants must be defined. Table 4 lists rate constants and chemical lifetimes for the various reactions in Fig. 12, using an obvious mnemonic prescription for subscript labels. To maintain a semblance of simplicity, the possibilities of more than one reagent or of multiple radical species are not mentioned. In fact, Fig. 12 does not include more than a representative selection of reaction channels. For instance, we have not mentioned the possibility of formation of unstable diamagnetic compounds that decompose spontaneously, releasing free Mu atoms again. Ivanter and Smilga (1971) showed that the process of multiple formations of unstable diamagnetic compounds has the same effect upon the muon as the process of repeated ionization and electron capture known as charge exchange. The effect of either process is generally the same as for continuous depolarization

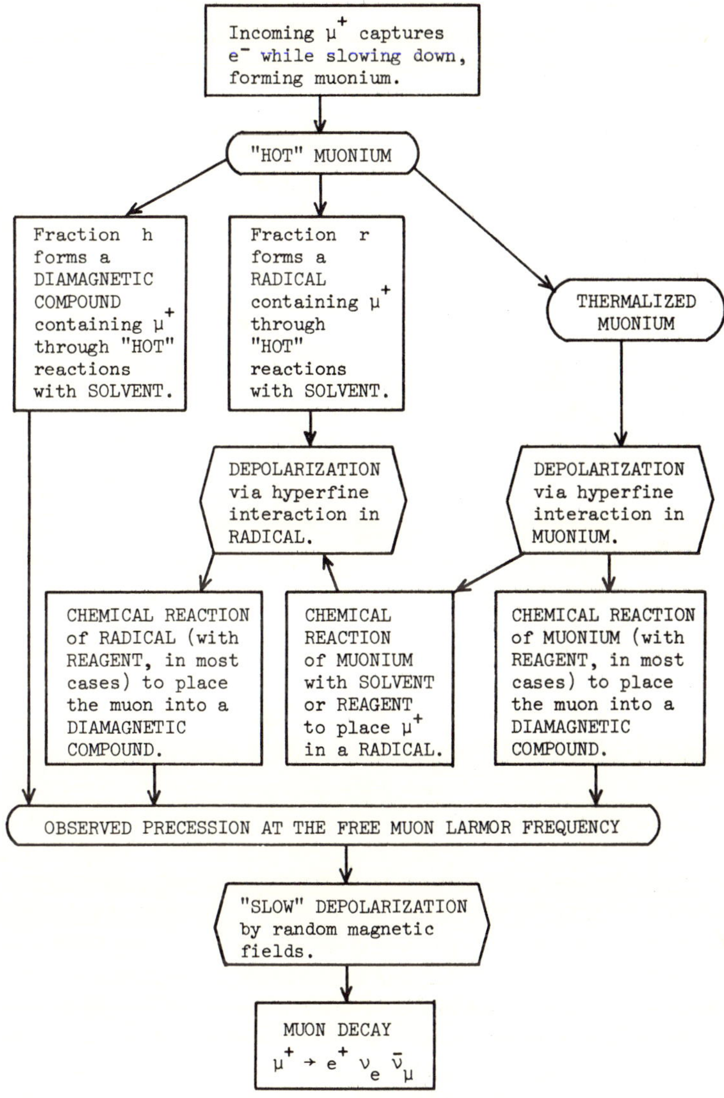

FIG. 12. Flow-chart model of depolarization mechanism in liquids.

TABLE 4

PARAMETERS OF THE THEORY

[X] = concentration of reagent X
[S] = concentration of pure solvent S (density/molecular weight)
h = fraction of muonium reacting epithermally to form a diamagnetic compound (D)
r = fraction of muonium reacting epithermally to form a radical (R)
k_{mxd} = chemical rate constant for the reaction Mu + X → D
k_{msd} = chemical rate constant for the reaction Mu + S → D
k_{mxr} = chemical rate constant for the reaction Mu + X → R
k_{msr} = chemical rate constant for the reaction Mu + S → R
k_{rxd} = chemical rate constant for the reaction R + X → D
k_{rsd} = chemical rate constant for the reaction R + S → D

In each case, the rate (in $\sec^{-1}$) at which reaction i occurs is given by $\Lambda_{\mathrm{i}} = k_{\mathrm{i}}[\mathrm{Z}]$, where Z is either X or S, whichever is the appropriate reactant.

The "lifetime" with respect to reaction i is given by $\tau_{\mathrm{i}} = 1/\Lambda_{\mathrm{i}}$. The following lifetimes are then defined:

$\tau_{\mathrm{md}} = [1/\tau_{\mathrm{mxd}} + 1/\tau_{\mathrm{msd}}]^{-1}$
$\tau_{\mathrm{mr}} = [1/\tau_{\mathrm{mxr}} + 1/\tau_{\mathrm{msr}}]^{-1}$
$\tau_{\mathrm{rd}} = [1/\tau_{\mathrm{rxd}} + 1/\tau_{\mathrm{rsd}}]^{-1}$ = "chemical lifetime" of the radical
$\tau_{\mathrm{m}} = [1/\tau_{\mathrm{md}} + 1/\tau_{\mathrm{mr}}]^{-1}$ = "chemical lifetime" of the muonium
ν_{m} = rate of depolarization of muonium electron ($\sec^{-1}$)
ν_{r} = rate of depolarization of unpaired electron in the radical ($\sec^{-1}$)
$\omega_{\mathrm{r}}/\omega_0$ = ratio of hyperfine frequencies in radical and muonium

of the muonium electron by the medium, with an appropriate definition of ν. We have also neglected such three-stage depolarization mechanisms as Mu → R → R′ → D.

The residual polarization following the two-compound depolarization process Mu → R → D can be obtained by solving a set of coupled integro-differential equations that reduce to an equivalent set of Volterra integral equations. Fischer (1973) found the solution for the case in which the unpaired electron in the radical is initially unpolarized. He also pointed out that the problem could be solved similarly for the case in which the radical simply adopts the muonium electron as its unpaired electron, so that the electron polarization [components $p_{0i}(t)$] is transferred smoothly from Mu to R. The advantage of the integrodifferential method is that the residual polarization emerges as an explicit (if somewhat unwieldy) function of the applied field, the hyperfine coupling in the radical, the electron depolarization rates ν_{m} and ν_{r} in muonium and the radical, and the chemical parameters listed in Table 4.

Another calculational technique is sometimes convenient when several complicated competitive reaction schemes are to be treated, or when A_{m}

and A_r can be diagonalized easily (e.g., by numerical techniques using a computer). This approach, described in detail in Brewer *et al.* (1973a), is an extension of the derivation described in Eqs. (54)–(57). The residual polarization is expressed as a generalized sum of contributions from all possible histories of μ^+ spin evolution, each contribution consisting of a product of the probability of that history and the final μ^+ polarization resulting from that history, as in Eq. (50). This leads to integrals over reaction times as in Eq. (51). When formal solutions for time dependence such as Eq. (54) are substituted into these integrals, they can be performed easily. The residual polarization is then expressed in terms of eigenvalues, reaction rates, and diagonalizing matrices.

Assuming that A_r is diagonalized by the matrix R [in analogy to Eq. (39)], so that

$$R^{-1}A_rR = \Lambda^r \tag{82}$$

with $\Lambda_{kk}{}^r = \lambda_k{}^r$, the residual polarization in the transverse subsystem for the model described by Fig. 12 and Table 4 can be expressed as

$$P_{\perp\,\mathrm{res}} = h - \frac{(1-h-r)}{\tau_{\mathrm{md}}}\sum_k \frac{F_k}{\alpha_k} - \frac{r}{\tau_{\mathrm{rd}}}\sum_k \frac{G_k}{\beta_k} + \frac{(1-h-r)}{\tau_{\mathrm{mr}}\tau_{\mathrm{rd}}}\sum_{ik}\frac{W_{ik}}{\alpha_k\beta_i} \tag{83}$$

where F_k is given by Eq. (55), α_k by Eq. (57),

$$G_k = R_{1k}(R^{-1})_{k1} \tag{84}$$

$$\beta_k = -1/\tau_{\mathrm{rd}} + i(\lambda_k{}^r + \omega_\mu) \tag{85}$$

and W_{ik} is given by different expressions, depending upon whether the electron polarization is lost or transmitted as Mu → R; if the radical starts over with a new unpolarized electron, then

$$W_{ik} = G_iF_k \tag{86}$$

If the radical picks up the muonium electron with no loss of polarization, then

$$W_{ik} = \sum_j R_{1i}(R^{-1})_{ij}M_{jk}(M^{-1})_{k1} \tag{87}$$

This latter formulation is less elegant, in that the dependence of $P_{\perp\,\mathrm{res}}$ on ν_m, ν_r, ω_r, and B remains implicit in the diagonalization of A_m and A_r. However, this approach has certain practical advantages. To construct a diagram such as that in Fig. 12 that includes all conceivable reaction schemes would be sheer folly, since the results are totally insensitive to processes that only involve a minute fraction of the μ^+ ensemble. How-

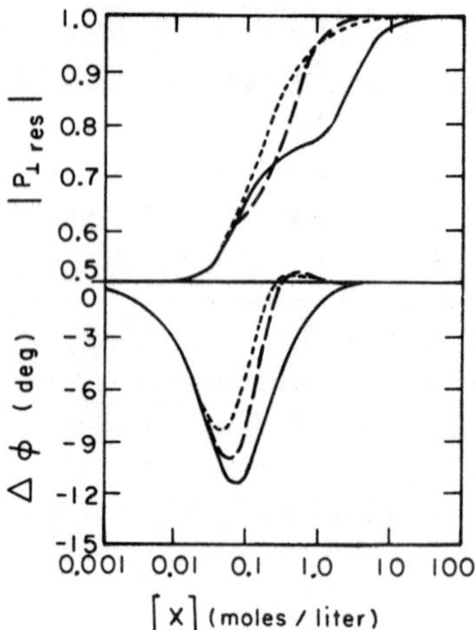

FIG. 13. Various mechanisms in 100-G transverse field. Effect of radicals on $P_{\perp\,\mathrm{res}}$ ([X]). Hot fraction $h = 0.5$, $\omega_r = 0.1\omega_0$, $r = \nu_m = \nu_r = 0$, and $k_{\mathrm{mxd}} = k_{\mathrm{rxd}} = 10^{10}$ liter/mole-sec in each case. Solid curves: proper muonium mechanism (no radical formation); dotted curves: $[\mathrm{S}]k_{\mathrm{msr}} = 10^{11}\ \mathrm{sec}^{-1}$; dashed curves: $k_{\mathrm{mxr}} = 10^{11}$ liter/mole-sec.

ever, different situations call for different treatments; for instance, depolarization in solids encompasses a wide range of mechanisms that are unimportant in liquids and completely absent in gases. Thus a formalism that can be easily modified to include new or exotic channels is sometimes more convenient in constructing "made-to-order" models.

To visualize the observable effects of radical formation, we recall the dependence of $|P_{\perp\,\mathrm{res}}|$ and $\Delta\phi$ on reagent concentration for the proper muonium mechanism (without radicals). In Fig. 11 we showed the characteristic "plateau" and phase dip for several magnetic fields. Radicals, with substantially smaller hyperfine frequencies than muonium, produce these effects at proportionally lower fields, which are usually experimentally impractical. Thus a field that is "low" for muonium may be "high" for the radical, facilitating an experimental test for the presence of radicals. Figure 13 shows several examples of physical interest. All are for $B = 100$ G, $h = 0.5$, and $r = \nu_m = \nu_r = 0$. The solid curves show the proper muonium mechanism (no radicals) with $k_{\mathrm{mxd}} = 10^{10}$ liter/mole-sec and all other rate constants zero. These are the same as the solid curves in Fig. 11. The other sets of curves show situations involving a radical with $\omega_r = 0.1\omega_0$. For the dashed curves, the radical is formed from reaction of Mu with X, and the only nonzero rate constants are $k_{\mathrm{mxd}} = 10^{10}$ liter/mole-sec, $k_{\mathrm{mxr}} = 10^{11}$ liter/mole-sec, and $k_{\mathrm{rxd}} = 10^{10}$ liter/mole-sec. For the dotted curves the radical is formed from reaction of Mu with the solvent, and the nonzero rate constants are $k_{mxd} = 10^{10}$ liter/mole-sec, $[S]k_{msr} = 10^{11}$ sec^{-1}, and $k_{rxd} = 10^{10}$ liter/mole-sec. In all examples the muonium electron is assumed to be transferred "gently" to the radical with no loss of polarization [see Eq. (87)]. The loss of the plateau effect with the addition of radicals is apparent in both situations. This is due to the shifting of the upper, "hyperfine" part of the depolarization curve towards lower concentrations (longer times).

It should be noted that, while other phenomena could cause a curve of $|P_{\perp\,res}|$ with no plateau in low field, they would also destroy the phase deviation, which persists in the situations depicted in Fig. 13. For instance, a large value for ν_m would disrupt the coherent precession of the muonium system, thereby destroying the plateau, but necessarily eliminating the phase deviation at the same time. Similarly, a mechanism consisting only of a dependence of the hot fraction h on [X] could cause a plateau-free "repolarization" but could not result in any phase deviation. Observation of phase deviations without concomitant plateaus in $|P_{\perp\,res}|$ thus serves as a specific experimental test for the presence of radicals in the depolarization mechanism.

VII. Measurements of Reactions of Muonium

The Mu atom, being paramagnetic and highly reactive, interacts chemically and/or magnetically in most media, as discussed in the preceding section. The quantities τ_m (mean chemical lifetime of free Mu) and ν (relaxation rate of the Mu electron) generally characterize these interactions; for certain ranges of τ_m and ν, phenomena can be observed that provide information about the reactions of Mu and the properties of the medium. In this section we will give a brief survey of such experimental results.

In transverse field, Mu precession cannot be observed unless both τ_m and $1/\nu$ are larger than $\sim 10^{-8}$ sec, the smallest time interval over which measurements are practical. These conditions have thus far been satisfied only for muonium in noble gases, insulators, and some semiconductors at low temperature. When Mu precession is observable, its amplitude generally relaxes; measurements of the relaxation rate are reliable in the region from about 10^5 to 10^8 sec^{-1}. This method is thus convenient for measuring rates of moderately slow chemical reaction and electron relaxation processes such as spin exchange.

Free μ^+ precession at the muon Larmor frequency, at the other extreme, is only visible if Mu reacts in very short times ($\tau_m \lesssim 1/\omega_-$) via epithermal or fast thermal processes, or if the electron's relaxation rate is so high ($\nu \gg \omega_0$) that it is effectively decoupled from the muon. The former situation occurs in many liquids, while the latter is characteristic of conductors. As mentioned in the preceding section, τ_m can be manipulated by varying reagent concentration in solutions, and the residual polarization fitted to determine chemical rate constants. Determinations of ν by transverse field techniques are apparently limited to gases in which $\nu \ll \omega_0$ and solids with $\nu \gg \omega_0$, since Mu precession has not been observed in any liquid, and even the strongest paramagnetic solutions do not show fast enough relaxation of muonium to effect the residual μ^+ polarization significantly (Brewer *et al.*, 1974).

In longitudinal field, the dephasing effect of Mu precession is absent, and the presence of muonium is signaled by a reduced value of the apparent initial polarization $R_{||}$ and/or an exponential decay of the time-averaged polarization, as in (28), (30), (65), or (67). Muonium formation alone can never "destroy" more than 50% of the muon polarization in longitudinal field [see Eq. (13)]. However, in gases or other media where muonium is long-lived, relaxation processes (e.g., spin exchange collisions) can bring about complete depolarization of the μ^+. Chemical reactions can only serve to reduce the amount of depolarization, but can speed up the apparent rate of the depolarization that does occur, as is evident from Eq. (65). These phenomena can also be studied by virtue of their quenching effect upon the resonant depolarization produced by an rf field, as discussed in Section V.

We turn now to a survey of experimental results.

A. Reactions of Mu in Gases

Mobley *et al.* (1966, 1967; Mobley, 1967) have conducted a series of experiments on chemical and spin exchange reactions of muonium with impurity gases in high-pressure (40 atm) argon. To our knowledge, these are the only studies of muonium chemistry in gases to date. They are of three types: relaxation of Mu precession in transverse field, directly observed relaxation in longitudinal field, and quenching of resonant rf depolarization in longitudinal field.

1. *Transverse Field*

Mu precession was observed in a very weak transverse field (~2 G). The rate λ of relaxation of the precession signal was measured and compared with that observed in pure argon, $\lambda_0 \approx 0.2\ \mu\text{sec}^{-1}$, which was probably due

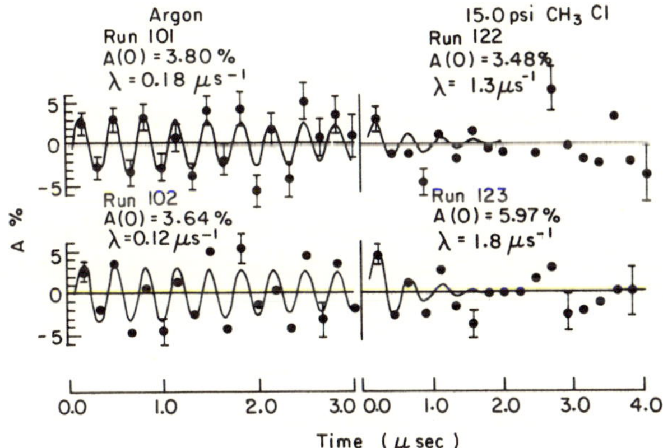

FIG. 14. Time-dependent variation (in %) of the ratio of the positron detection rate to the muon decay probability in a transverse field set up for different runs. The solid curves are fits to the data (from Mobley, 1967).

to minute O_2 impurities. Examples of Mu precession signals are shown in Fig. 14. The most reliable results were:

Added gas	Partial pressure (Torr)	λ (μsec^{-1})
O_2	0.255	4.4 ± 1.8
C_2H_4	10.4	5.8 ± 1.7
CH_3Cl	705	1.5 ± 0.5

The relaxation rate due to interactions with reagent X is proportional to the concentration [X], usually expressed in moles/liter (at 300°K, 1 Torr partial pressure is equivalent to a concentration of 5.87×10^{-5} mole/liter):

$$\lambda = k[X] \tag{88}$$

The constant of proportionality k is a second-order rate constant. Again following the chemist's convention, we express k in liter/mole-sec.

Interpretation of Eq. (88) is not unambiguous. The quantity λ is the rate of loss of polarized muonium; this may result from loss of muonium itself by chemical reaction ($k = k_{ch}$) or from depolarization of free muonium ($k = k_d$). In general $k = k_{ch} + k_d$. The mean chemical lifetime and an effective spin-flip frequency in transverse field can be expressed in terms of these rate constants as $1/\tau_m = k_{ch}[X]$ and $\nu_{eff} = \frac{2}{3}k_d[X]$ [see Eq. (42)].

Depolarization of free muonium in gases is believed to proceed by spin exchange: in a glancing collision between a Mu atom and a paramagnetic molecule, the wave function of the muonium electron briefly overlaps that of the unpaired molecular electron(s), making possible an exchange of electrons with opposite spins. The average depolarizing effect of a spin exchange process depends on the available molecular electron spin states (exchange of electrons with like spins has no effect). In general, we can write $k_d = fk_{se}$, with $f < 1$. For spin-$\frac{1}{2}$ molecules, $f = \frac{1}{2}$; for spin-1 molecules, $f = \frac{1}{2}(32/27)$ (Mobley *et al.*, 1967). By contrast, a "charge exchange" process ($Mu \rightarrow \mu^+ \rightarrow Mu$) always results in a 50% polarization loss (unless its rate competes with ω_0): $k_d = \frac{1}{2}k_{ce}$. These relations are somewhat modified in longitudinal field, as will be seen.

Relaxation in C_2H_4 and in CH_3Cl is taken to be due only to chemical reactions, $k = k_{ch}$. From Eq. (88) we obtain

$$k_{ch}(Mu + C_2H_4) = (0.95 \pm 0.28) \times 10^{10} \text{ liter/mole-sec} \tag{89}$$

and

$$k_{ch}(Mu + CH_3Cl) = (3.6 \pm 1.2) \times 10^{7} \text{ liter/mole-sec} \tag{90}$$

In O_2, both chemical reactions and spin exchange collisions are important. The experimental result cannot be separated into chemical and spin exchange parts on the basis of this measurement. Since O_2 is a spin-1 molecule, the result can be written

$$\begin{aligned} k(Mu + O_2) &= k_{ch}(Mu + O_2) + \tfrac{1}{2}(32/27)k_{se}(Mu + O_2) \\ &= (2.9 \pm 1.2) \times 10^{11} \text{ liter/mole-sec} \end{aligned} \tag{91}$$

2. *Longitudinal Field*

In longitudinal field, chemical reactions of Mu do not cause relaxation in any sense; it is easy, then, to make the mistake of equating the observed relaxation rate with a depolarization rate. However, as can be seen clearly from Eq. (65), the removal of muons from relaxing muonium causes an apparent increase in the relaxation rate for the part of the μ^+ ensemble that does relax. Thus Eq. (88) is valid for longitudinal as well as transverse field, with the more general prescription

$$k = k_{ch} + k_D \tag{92}$$

The longitudinal field depolarization rate constant k_D is related to the transverse or zero-field depolarization rate constant k_d as follows [recall Eq. (31)]:

$$k_D = k_d/(1 + x^2) \tag{93}$$

Mobley analyzed his longitudinal field results with the assumption $k = k_D$. In the most important cases, O_2 and NO, chemical reactions are significant and this assumption is not valid.

The field dependence of the measured rate constants is not described well by Eq. (93) in most cases, as can be seen from Fig. 15. The fit is best for O_2 and NO, where spin exchange is probably dominant. Even in these cases, however, the data are fitted to the dependence (93), neglecting the constant term proportional to k_{ch}. Judging from the high-field values of $\lambda/n = k$ (cm^3/sec), this term is relatively small. Its effect would be to reduce the fitted values of k_d slightly. Keeping in mind this questionable extrapolation, we can express Mobley's results as follows:

$$\lim_{x\to 0} k(\text{Mu} + \text{O}_2) = k_{ch}(\text{Mu} + \text{O}_2) + \tfrac{1}{2}(32/27)k_{se}(\text{Mu} + \text{O}_2)$$
$$= (1.7 \pm 0.2) \times 10^{11} \text{ liter/mole-sec} \tag{94}$$

$$\lim_{x\to 0} k(\text{Mu} + \text{NO}) = k_{ch}(\text{Mu} + \text{NO}) + \tfrac{1}{2}k_{se}(\text{Mu} + \text{NO})$$
$$= (1.7 \pm 0.25) \times 10^{11} \text{ liter/mole-sec} \tag{95}$$

Result (94) agrees with the transverse field result (91) within the errors.

Relaxation was also observed with NO_2, C_2H_4, C_2H_6, CH_3Cl, CO_2, and Cl_2 impurities. As can be seen from Fig. 15, the field dependence of λ/n does not fit the form (93) for NO_2 or for C_2H_4. Radical formation is likely in these cases, probably via $\text{Mu} + \text{NO}_2 \to \text{MuO}^{\cdot} + \text{NO}$ and $\text{Mu} + \text{C}_2\text{H}_4 \to \text{MuC}_2\text{H}_4^{\cdot}$; however, since some depolarization mechanism is necessary if

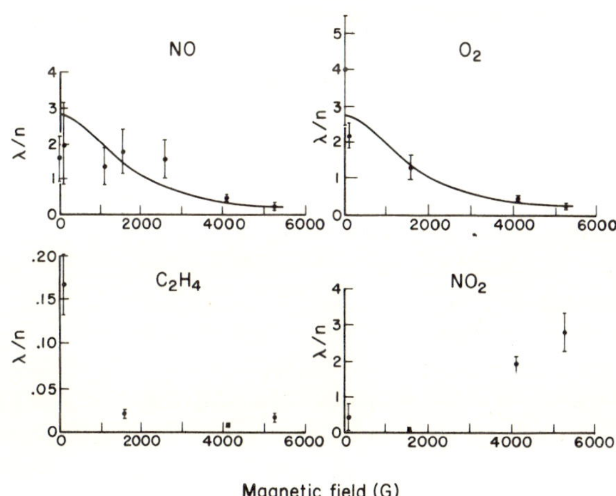

Fig. 15. Observed values of λ/n [$\propto k$ (liter/mole-sec)] versus magnetic field for NO, O_2, C_2H_4, and NO_2 in units of 10^{-16} (μsec molecule/cm^3)$^{-1}$. The solid curves for NO and O_2 are fits of Eq. (93) to the data (from Mobley, 1967).

relaxation is to be observed [Eq. (65) shows this explicitly], the overall mechanism is unclear in all these cases.

3. *Signal Quenching*

The quenching effect of impurities upon the resonant rf depolarization in longitudinal field was analyzed by Mobley to extract rate constants for chemical reaction (Mobley, 1967). We will not discuss the method in detail, but simply list the more interesting results:

$$k_{ch}(\mathrm{Mu} + \mathrm{O_2}) = (2.6 \pm 0.6) \times 10^{11} \text{ liter/mole-sec} \qquad (96)$$

$$k_{ch}(\mathrm{Mu} + \mathrm{NO}) = (1.15^{+1.1}_{-0.4}) \times 10^{11} \text{ liter/mole-sec} \qquad (97)$$

$$k_{ch}(\mathrm{Mu} + \mathrm{C_2H_4}) = (1.4 \pm 0.4) \times 10^{10} \text{ liter/mole-sec} \qquad (98)$$

$$k_{ch}(\mathrm{Mu} + \mathrm{NO_2}) > 1.1 \times 10^{12} \text{ liter/mole-sec} \qquad (99)$$

and

$$k_{ch}(\mathrm{Mu} + \mathrm{H_2}, \mathrm{N_2}, \text{or } \mathrm{SF_6}) < 5 \times 10^{8} \text{ liter/mole-sec} \qquad (100)$$

The result (96) for $k_{ch}(\mathrm{Mu} + \mathrm{O_2})$ is larger than the most precise value (94) for the total rate constant for O_2, which was expected to be dominated by the spin exchange rate. There is not yet a satisfactory explanation for this discrepancy. The results (97) and (95) for NO are consistent and can be combined to yield a spin exchange rate constant of

$$k_{se}(\mathrm{Mu} + \mathrm{NO}) = (1.1^{+2.3}_{-0.9}) \times 10^{11} \text{ liter/mole-sec} \qquad (101)$$

but the uncertainties in this result are too large for it to be very meaningful.

4. *Comparison with Hydrogen Atoms*

Unlike positronium, which has no nucleus, muonium fits neatly into the theoretical structures of physical chemistry as a light isotope of the hydrogen atom ($m_{\mathrm{Mu}} = 0.113 m_{\mathrm{H}}$). Predictions of isotopic differences in reaction rates, already tested with the trio (H, D, T), can thus be extended in a new direction with Mu. This may lead to a deeper understanding of the most basic types of reactions in physical chemistry.

For these studies the gas phase is ideal, since atoms and molecules can legitimately be thought to "collide" in gases. The purely kinetic isotope effect arising from the $m^{-1/2}$ mass dependence of the mean thermal velocity can thus be separated easily from more interesting dynamic isotope effects. In liquids, comparison of Mu and H reaction rates is generally more qualitative, as we shall see later. Even in gases, however, some care must be taken in reducing rate constants to more fundamental quantities.

For quasi-elastic two-body collisions such as the spin exchange process

$$\mathrm{Mu} + \mathrm{O_2} \xrightarrow{\text{spin exchange}} \mathrm{Mu} + \mathrm{O_2} \tag{102}$$

or for two-body abstraction reactions such as

$$\mathrm{Mu} + \mathrm{NO_2} \rightarrow \mathrm{MuO^{\cdot}} + \mathrm{NO} \tag{103}$$

the rate constant can be reduced to a cross section by using the definition

$$k\ (\text{liter/mole-sec}) = N_0 \times 10^{-3}\, \bar{v}\ (\text{cm/sec})\sigma\ (\text{cm}^2) \tag{104}$$

where $N_0 = 6.022 \times 10^{23}$ mole^{-1} and $\bar{v}$ is the mean thermal velocity of Mu atoms relative to impurity molecules, given by

$$\bar{v} = (8k_B T/\pi\mu)^{1/2} \tag{105}$$

where $\mu = [(1/m_{\mathrm{Mu}}) + (1/m_X)]^{-1}$ is the reduced mass. In most cases $m_{\mathrm{Mu}} \ll m_X$ and $\mu \approx m_{\mathrm{Mu}}$, so that $\bar{v} \propto m_{\mathrm{Mu}}^{-1/2}$. This gives the aforementioned kinetic isotope difference between $k(\mathrm{Mu})$ and $k(\mathrm{H})$ attributable to

$$\bar{v}(\mathrm{Mu}) \approx 2.98\bar{v}(\mathrm{H}) \tag{106}$$

At room temperature, $\bar{v}(\mathrm{Mu}) \approx 0.75 \times 10^6$ cm/sec. Further differences must come from σ in the form of "dynamic" isotope effects.

In a formation reaction such as

$$\mathrm{Mu} + \mathrm{C_2H_4} \rightarrow \mathrm{C_2H_4Mu^*} \tag{107}$$

the excited complex $C_2H_4Mu^*$ is long-lived enough to be considered a stable reaction product (Thrush, 1965), allowing definition of a cross section as in (104). However in highly exothermic formation reactions such as the chemical reaction of Mu with NO, the intermediate excited complex MuNO* is so unstable that a third body must actually participate in the collision to absorb the kinetic energy released. Otherwise there is simply a resonant scattering. Rate constants for these reactions cannot legitimately be expressed in terms of cross sections. For such 3-body processes as

$$\mathrm{Mu} + \mathrm{NO} + \mathrm{Ar} \rightarrow \mathrm{MuNO} + \mathrm{Ar} \tag{108}$$

the more basic quantity is the third-order rate constant κ (given in liter^2/mole^2-sec), defined by

$$k_{\mathrm{ch}}(\mathrm{Mu} + \mathrm{NO}) = \kappa(\mathrm{Mu} + \mathrm{NO} + \mathrm{Ar})[\mathrm{Ar}] \tag{109}$$

At 40 atm and room temperature, [Ar] = 1.7 mole/liter. Although the pressure dependence (109) was not checked experimentally, the H atom reaction analogous to (108) is known to be a 3-body process (Thrush, 1965), so we must assume that (108) is the correct reaction. Naturally,

TABLE 5

COMPARISON OF MU AND H REACTION RATES IN GASES AT 300°K[a]

A. Two-body collision cross sections

Reaction	Cross section (10^{-16} cm^2) Muonium	Hydrogen
(Mu, H) + O_2 spin exchange	~1	21 ± 2.1[b]
(Mu, H) + NO spin exchange	$2.4^{+5.1}_{-2.0}$	25 ± 2.5[b]
(Mu, H) + H_2 → (MuH, H_2) + H	<0.01	~10^{-5}[c]
(Mu, H) + NO_2 → (MuO, HO) + NO	>23	1.8 ± 0.2[d]
(Mu, H) + C_2H_4 → ($C_2H_4Mu^*$, $C_2H_5^*$)	0.27 ± 0.06	~0.02

B. Three-body third-order rate constants

Reaction	κ (10^{10} liter^2/mole^2-sec) Muonium	Hydrogen
(Mu, H) + O_2 + Ar → (MuO_2, HO_2) + Ar	15.3 ± 3.5	1.1
(Mu, H) + NO + Ar → (MuNO, HNO) + Ar	$6.8^{+6.5}_{-2.4}$	0.9

[a] H atom cross sections and rate constants taken or derived from the review by Thrush (1965), unless otherwise specified.
[b] From Berg (1965).
[c] From Shavitt (1959).
[d] From Phillips and Schiff (1962).

κ will still be proportional to $\bar{v}$, so we expect the same kinetic isotope effect in 3-body reactions as in the binary collisions.

Those of Mobley's results that can be compared with measured H atom cross sections and rate constants are summarized in Table 5. From these limited data we can only remark on two qualitative trends: the cross sections for spin exchange reactions seem to be significantly smaller for Mu than for H, while chemical reaction cross sections and 3-body rate constants are larger for Mu than for H. This information shows clearly that one can practically measure reaction rates of Mu in gases and compare them with analogous reaction rates of H, D, and T, illuminating the role of isotope effects in the theory of absolute reaction rates. However, Mobley's results were derived from data with very low statistics, making the uncertainties too large to allow critical quantitative comparison of Mu and H rates.

With the advent of μ^+ beams of higher stopping density (Kendall, 1972), gas-phase studies of muonium chemistry are bound to become more popular, and high-statistics determinations of rate constants, cross sections, and their pressure and temperature dependences will probably be made. Reactions such as Mu + $H_2 \rightarrow$ MuH + H or Mu + $Cl_2 \rightarrow$ MuCl + Cl, perhaps the most interesting for comparison of (Mu, H, D, T), may soon be observed and studied. These reactions were too slow at room temperature to have been detectable in Mobley's apparatus.

B. Reactions of Mu in Liquids

Most studies of muonium chemistry have been made in the liquid phase. Here, as explained in Section VI, one varies accessible parameters such as the magnetic field and the reagent concentration [X], and compares the behavior of the residual μ^+ polarization $\mathbf{P}_{res}$ with theory to determine rate constants and other physical quantities. For these studies, longitudinal field measurements are often very useful; for instance, when radical formation is important, the field dependence of $P_{\parallel\ res}$ can be fitted to a form similar to (71) to determine the hyperfine frequency ω_r (Goldanskii and Firsov, 1971). However, rate constants for diverse processes are best determined from the concentration dependence of A and $\Delta\phi$ in transverse field [see Eqs. (75) and (76)].

In a series of experiments performed at the 184-in. cyclotron in Berkeley, muons were stopped in solutions in a uniform transverse field, and time histograms of the μ^+ stop $-e^+$ emission interval fitted to the form (6), as described in Section II. The resultant values for A and $\Delta\phi$ as functions of [X] were fitted to the theory [Eqs. (75) and (76)], where $P_{\perp\ res}$ is given by Eq. (83).

Several simplifying restrictions were imposed upon the general theory (83) in these fits. First, "hot" reactions were presumed to lead only to diamagnetic compounds incorporating muons ($r = 0$); radicals were assumed to form only in thermal chemical reactions. Second, relaxation of the spin of the muonium electron was assumed to be slow by comparison with the electron Larmor frequency, and was therefore neglected ($\nu_m \approx 0$). This assumption is supported by ESR data on hydrogen atoms in solution (Neta *et al.*, 1971). Relaxation of the unpaired electron in radicals was likewise presumed to be negligible ($\nu_r \approx 0$). As noted by Fischer (1973), this assumption is questionable in the case of radicals whose unpaired electron may have hyperfine couplings with several nuclei; however, in most cases the effects of $\nu_r < \omega_r$ would be difficult to detect except in very weak fields. Third, it was assumed that only one species of radical incorporating the muon is present in a given type of solution and that it is formed by chemical reaction with either the reagent ($k_{msr} = 0$) or the solvent ($k_{mxr} = 0$), but never both in the same solution. Fourth, the form

(87) was chosen for W_{ik}, assuming "gentle" transfer of the Mu electron to the unpaired electron in the radical. The results are actually rather insensitive to the choice of (86) or (87). Finally, neither muonium nor the radical was presumed to react thermally with the solvent to form a diamagnetic species incorporating the muon, except at negligible rates ($<10^7$ liter/mole-sec). These assumptions gave the simplest form of the theory that permitted a good fit to all the data. Justifications and possible exceptions will be discussed below.

When the radical species R was assumed to be known, the hyperfine frequency ω_r in the radical was obtained by multiplying the ratio $\mu_\mu/\mu_p = 3.18$ of muon and proton magnetic moments into the measured value of the hyperfine frequency for the analogous radical in which the muon is replaced by a proton; these values were obtained from Landolt and Bornstein (1965). When the radical species was unknown, ω_r was fitted by trial and error, and the optimal value used to make the best determinations of the parameters described above. In such cases, the choice of ω_r dramatically affects the fitted values of the rate constants. This is because ω_r sets the time scale for depolarization in the radical, while the rate constants determine the duration of that chemical state. In general, when fitting data in which the radical stage plays an important role, if one changes ω_r by some factor, the resultant change in the fitted value of a given rate constant is never more than a similar multiplicative value. In the cases where ω_r was chosen by trial and error, the specified uncertainties in the rate constants shown below represent only the tolerance of the fits for the specified choice of ω_r. An additional uncertainty, which could in some cases be as large as an order of magnitude, is implicit in the crude determination of ω_r. However, this degree of accuracy is still sufficient to allow many interesting comparisons with H atom chemistry, as will be seen later. Most importantly, the qualitative conclusion that a radical depolarization mechanism is involved remains unaffected by these uncertainties. Further experiments may allow positive identification of the radical species or direct measurement of ω_r, thereby eliminating these ambiguities.

1. *Results*

A. EXAMPLE OF THE PROPER MUONIUM MECHANISM: I_2 in CH_3OH. Figure 16 shows the observed dependence of $|P_{\perp\,\text{res}}|$ and $\Delta\phi$ on the concentration of iodine in methanol solution in a field of 102 G. The leftmost point in this and all such graphs corresponds to the result for the pure solvent; due to the log scale of the concentration, the point is actually infinitely far off scale to the left. The curve through the points is the best fit to Eq. (74), where $R_\perp$ is given by (56), the case of (83) in which the muons are depolarized by the proper muonium mechanism (i.e., when no radical formation is involved). The chemical reaction involved is presumed to be

$$\mathrm{Mu} + \mathrm{I}_2 \xrightarrow{(k_{\mathrm{mxd}})} \mathrm{MuI} + \mathrm{I} \tag{110}$$

The fraction of muonium atoms reacting epithermally with CH_3OH is $h \approx \frac{1}{2}$. The phase variation is striking, and the "plateau" in $|P_{\perp\,\mathrm{res}}([\mathrm{I}_2])|$ is noticeable. Both of these phenomena are due to the coherent precession of free muonium atoms in the magnetic field, as explained in Section VI, and constitute proof of the central role of muonium in the depolarization mechanism. If a substantial number of muons were placed in radicals, the effect [see Fig. 13] would be to decrease the amplitude of the phase dip and to destroy the plateau. There does in fact seem to be a slight lessening of the plateau effect, and this may be due to a small but finite probability of reaction of muonium with CH_3OH to form a radical containing the muon, probably in epithermal collisions. This would constitute an exception to the assumption that $r = 0$. The quality of the fit is improved slightly by allowing some radical formation, but the correction is so small that the result is insensitive to the source, type, and fate of the radicals involved. Thus, since the mechanism is clearly dominated by reaction (110), this case may be practically considered to be an example of the proper muonium mechanism.

Similar results were obtained (Brewer *et al.*, 1971) for I_2 in CH_3OH at fields of 1000 and 4500 G. The 100-G results are consistent with these, but

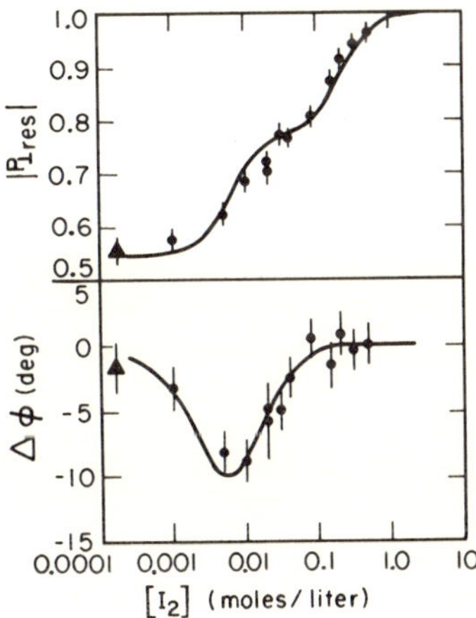

FIG. 16. Residual muon polarization in methanol as a function of the concentration of dissolved iodine. Best fit to the data for the proper muonium mechanism.

TABLE 6

ASYMMETRY NORMALIZATION AND COMPARISON OF FIT QUALITY WITH AND WITHOUT RADICALS

Solvent	Reagent	Field (gauss)	A_0	χ^2/Degree of freedom: Without radicals	χ^2/Degree of freedom: With radicals
CH_3OH	I_2	103	0.25 ± 0.01	0.8	0.4
		1000	0.27 ± 0.01	1.1	same
		4500	0.27 ± 0.01	0.6	same
C_6H_6	Br_2	200	0.271 ± 0.005	27	2.3
	I_2	200	0.272 ± 0.03	11	0.4
H_2O	H_2O_2	100	0.26 ± 0.1	7.4	0.8
	HNO_3	100	0.277 ± 0.01	11	1.0
		4500	0.30 ± 0.01	1.9	same
	$FeCl_3$	4500	0.31 ± 0.01	0.6	same
	$Fe(ClO_4)_3$	100	0.29 ± 0.03	8.3	0.6
		4400	0.30 ± 0.02	0.6	same
	$Fe(NO_3)_3$	11,000	0.35 ± 0.02	7.7	same

are much more conclusive, since the phase dip and plateau are most evident at low fields [see Fig. 11]. The numerical results of these and other fits are listed in Tables 6 and 7.

B. EVIDENCE FOR RADICAL FORMATION IN BENZENE. The muon asymmetry in benzene (C_6H_6) has long been known (Swanson, 1958) to be exceptionally low, implying a hot fraction $h \approx \frac{1}{8}$, as compared to $h \approx \frac{1}{2}$ for methanol or water. This property makes benzene an attractive solvent for studies of muonium chemistry, since the range $(1 - h)$ through which $|P_{\perp \text{res}}|$ can be varied by chemical means is near maximum, and the amplitude of the phase dip is increased accordingly. Bromine was chosen as a muonium scavenger because of its virtually unlimited solubility in benzene and because of the analogy with iodine; the expected reaction in this case is

$$\text{Mu} + \text{Br}_2 \xrightarrow{(k_{\text{mxd}})} \text{MuBr} + \text{Br} \tag{111}$$

Data were taken in a 200-G magnetic field so that the "plateau" would be visible.

However, as can be seen in Fig. 17, the results were in strong disagreement with the predictions of the proper muonium mechanism, the best fit for which is indicated by the dashed lines. There is no discernible plateau,

TABLE 7

RESULTS OF BEST FITS[a]

Solvent	Reagent	Field (gauss)	Radical	ω_r/ω_0	h	Rate constants ($\times 10^{10}$ liter/mole sec)			
						k_{mxd}	(Z)	k_{mzr}	k_{rxd}
CH_3OH	I_2	103	none	—	0.54 ± 0.02	13.4 ± 2	—	—	—
		1000	none	—	0.51 ± 0.02	13 ± 3	—	—	—
		4500	none	—	0.53 ± 0.02	13.4 ± 2	—	—	—
C_6H_6	Br_2	200	$C_6H_6Mu\cdot$	0.095	0.134 ± 0.01	9.4 ± 0.3	S	0.125 ± 0.05	0.36 ± 0.1
	I_2	200	$C_6H_6Mu\cdot$	0.095	0.133 ± 0.07	5.7 ± 1	S	0.054 ± 0.03	0.2 ± 0.1
H_2O	H_2O_2	100	$MuO\cdot$	0.083	0.59 ± 0.01	0.24 ± 0.05	X	0.85 ± 0.1	0.14 ± 0.02
	HNO_3	100	?	0.125 ± 0.05	0.545 ± 0.01	3 ± 1	X	10 ± 5	0.1 ± 0.01
	$Fe(ClO_4)_3$	100	?	$0.010^{+0.040}_{-0.005}$	0.52 ± 0.03	$0.57^{+0.8}_{-0.4}$	X	3.8 ± 0.8	0.02 ± 0.002
	$FeCl_3$	4500	none (?)	—	0.51 ± 0.02	2.1 ± 0.2	—	—	—

[a] Errors are approximate.

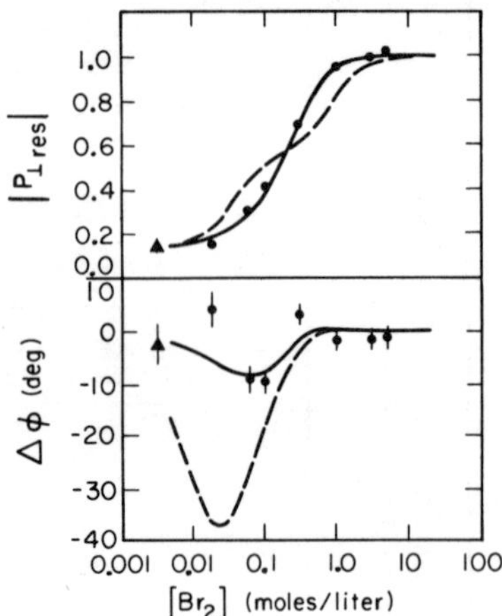

FIG. 17. Residual muon polarization in benzene as a function of the concentration of dissolved bromine. Uncertainties of $|P_{\perp \text{res}}|$ data are less than the dimensions of the points. Dashed curves: best fit without radicals; solid curve: best fit with radicals.

and the phase variation is much less sharp than predicted by the simple theory. This behavior resembles the predictions of the proper muonium mechanism in a stronger magnetic field. Since the criterion for a "strong" field is that it be comparable with the effective hyperfine field, it was this observation that led to consideration of environments similar to muonium but with lower values of the effective hyperfine field (i.e., radicals).

On the basis of other chemical studies (Michael and Hart, 1970; Sauer and Ward, 1967), the following assumptions were made about the chemical processes involved: first, that the reaction

$$\mathrm{Mu} + \mathrm{C_6H_6} \xrightarrow{(k_{\mathrm{msr}})} \mathrm{C_6H_6Mu^{\cdot}} \tag{112}$$

forming the muonium analog of the radical cyclohexadienyl ($C_6H_7^{\cdot}$), is in competition with reaction (111) for muonium. Second, that the radical reacts subsequently with bromine to place the muon in a diamagnetic compound, according to

$$\mathrm{C_6H_6Mu^{\cdot}} + \mathrm{Br_2} \xrightarrow{(k_{\mathrm{rxd}})} \mathrm{D}\ \text{(unidentified)} \tag{113}$$

The isotropic average effective hyperfine field at the unpaired electron due to the extra proton in cyclohexadienyl is 47.71 G (Landolt and Bornstein, 1965), as compared with 1593 G in muonium; thus the ratio of the hyperfine frequency ω_r in $C_6H_6Mu^{\cdot}$ to the hyperfine frequency ω_0 in muonium was taken to be

$$\frac{\omega_r}{\omega_0} \doteq \frac{(47.71 \times \mu_\mu/\mu_p)}{1593} = 0.095 \tag{114}$$

This value was used to obtain the best fit to the data [solid lines in Fig. 17] corresponding to the best values for the fitted parameters, as listed in Tables 6 and 7; a trial and error search for the best empirical value for ω_r/ω_0 gave a minimum χ^2 for $\omega_r/\omega_0 \approx 0.03^{+0.04}_{-0.02}$. A study of the field dependence of the residual polarization in benzene in longitudinal field (Goldanskii and Firsov, 1971) yielded an estimate of $\omega_r/\omega_0 = 0.054 \pm 0.004$. A new fit to the data using this value gives estimates for the rate constants $k(111) \approx 5.5$, $k(112) \approx 0.10$, and $k(113) \approx 0.57 \times 10^{10}$ liter/mole-sec.

Although it was not possible to dissolve enough iodine in benzene to achieve full "repolarization," it was possible to study the dependence of $P_{\perp\,\mathrm{res}}$ upon $[I_2]$ in C_6H_6 over a large enough region to determine that the results were consistent with those observed for Br_2 in C_6H_6. These results are also listed in Tables 6 and 7.

C. MU CHEMISTRY IN AQUEOUS SOLUTIONS. In spite of its rather large hot fraction ($h \approx \frac{1}{2}$), water has proved to be a nearly optimal solvent for muonium chemistry. Most importantly, these results show that H_2O is more or less inert with respect to thermal chemical reactions with Mu—that is, any reaction of Mu + H_2O has a rate constant $<10^7$ liter/mole-sec. Thus all significant thermal reactions of muonium are with the reagent. This situation would be expected to favor many examples of the proper muonium mechanism, but instead we have found a number of more complicated mechanisms, all involving radicals.

1. *Hydrogen peroxide.* Perhaps the most elegant system studied was Mu with H_2O_2 in H_2O. The experimental dependence $P_{\perp\,\mathrm{res}}([H_2O_2])$ at a field of 100 G is shown in Fig. 18 along with the best fits to the data. Again, the dashed curve is the best fit with the proper muonium mechanism, and the solid curve is the best fit with the general mechanism, including radicals. Clearly, radicals are present. In this case the muonium is presumed to react with hydrogen peroxide to form a diamagnetic compound containing the muon, presumably according to

$$\mathrm{Mu} + H_2O_2 \xrightarrow{(k_{\mathrm{mxd}})} \mathrm{HO}^{\cdot} + \mathrm{MuHO} \tag{115}$$

and (competitively) to form a muonic radical, presumably according to

$$\mathrm{Mu} + H_2O_2 \xrightarrow{(k_{\mathrm{mxr}})} \mathrm{MuO}^{\cdot} + H_2O \tag{116}$$

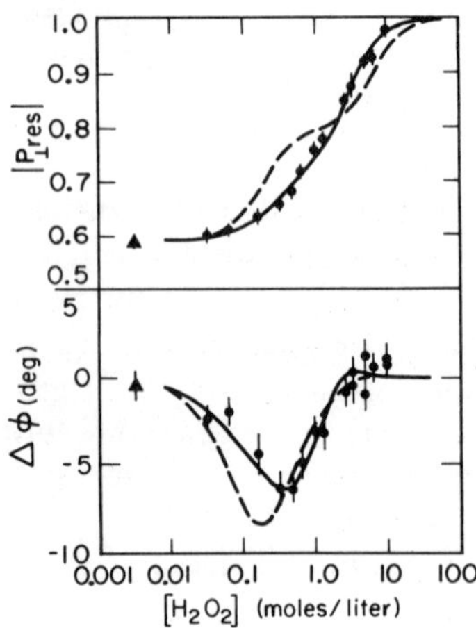

FIG. 18. Residual muon polarization in water as a function of the concentration of dissolved hydrogen peroxide. Dashed curve: best fit without radicals; solid curve: best fit with radicals.

The radical $MuO^{\cdot}$ subsequently reacts with H_2O_2 to leave the muon in a final diamagnetic environment:

$$MuO^{\cdot} + H_2O_2 \xrightarrow{(k_{rxd})} D' \text{ (unidentified)} \tag{117}$$

These assumptions are consistent with most interpretations of *H* atom reactions with H_2O_2, as described later. Nevertheless, it is possible that the radical species has been misidentified. If, for instance, the predominant radical species were $MuO_2^{\cdot}$ rather than $MuO^{\cdot}$, the value assumed for ω_r would be incorrect, possibly introducing errors of as much as an order of magnitude in the rate constants, as discussed previously. However, regardless of possible ambiguities in the identification of chemical species, the conclusion that the presence of radicals is essential to the overall depolarization mechanism is inescapable.

The effective hyperfine field at the unpaired electron due to the proton in the hydroxyl radical $HO^{\cdot}$ is known (Landolt and Bornstein, 1965) to be 41.3 G (isotropic average), which would imply $\omega_r/\omega_0 = 0.0825$ for $MuO^{\cdot}$ [recall Eq. (114)]. This value was used to obtain the results listed in Tables 6 and 7. The empirical value giving a minimum χ^2 was $\omega_r/\omega_0 = 0.175 \pm 0.1$, consistent with the predicted value.

2. *Strong acids.* Preliminary results show a great deal of variety in the reactions of muonium with various acids. In HCl, as noted earlier by

Swanson (1958), there seems to be no "repolarizing" effect at any concentration. The muon precession in 10*M* HCl is virtually indistinguishable from that in pure water. Therefore, no combination of reactions between Mu, H^+, and Cl^- leads to a diamagnetic compound containing the muon in times shorter than about 10 nsec. Similar results in concentrated $MnCl_2$ solutions indicate that these conclusions are relatively independent of pH.

However, addition of nitric acid to water causes marked "repolarization," with a maximal asymmetry reached at about 10*M*. Experimental results for $P_{\perp\,\mathrm{res}}([\mathrm{HNO_3}])$ at 100 G are shown in Fig. 19. It was assumed that HNO_3 dissociates sufficiently that the Mu reacts predominantly with the anion, NO_3^-. Again, the proper muonium mechanism (dashed curve) is a poor fit, but an excellent fit (solid curve) can be obtained if one assumes the following reactions to be significant. First, the usual direct reaction leading to a diamagnetic compound:

$$\mathrm{Mu} + \mathrm{NO_3^-} \xrightarrow{(k_{\mathrm{mxd}})} \mathrm{D\ (unidentified)} \tag{118}$$

In addition, the competitive reaction leading to a muonic radical:

$$\mathrm{Mu} + \mathrm{NO_3^-} \xrightarrow{(k_{\mathrm{mxr}})} \mathrm{R\ (unidentified)} \tag{119}$$

followed by the final reaction of the radical to place the muon in a diamag-

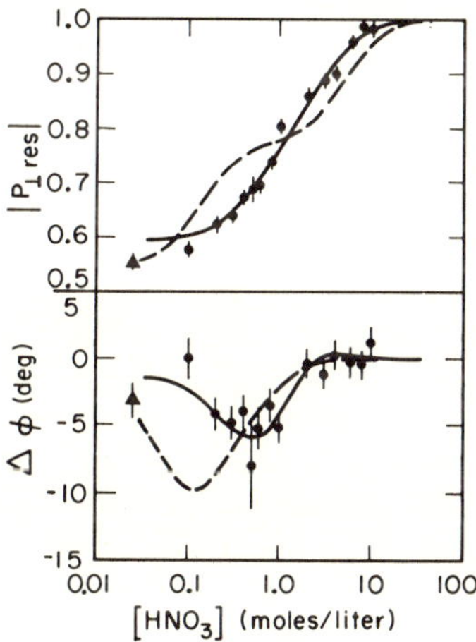

Fig. 19. Residual muon polarization in water as a function of the concentration of dissolved nitric acid. Dashed curve: best fit without radicals; solid curve: best fit with radicals.

netic environment:

$$R + NO_3^- \xrightarrow{(k_{rxd})} D' \text{ (unidentified)} \tag{120}$$

Here there has been no attempt to identify any of the product species, but only the types of processes taking place; all the fitted results listed in Table 7, including ω_r/ω_0, were obtained by minimizing χ^2. Results for $P_{\perp\,res}([HNO_3])$ at a field of 4500 G are consistent with these, but are much less sensitive to the presence of radicals.

Similar results were seen for solutions of $HClO_4$ in water at 4400 G. However, no one has yet undertaken a study of $HClO_4$ at low field, where the results are sensitive to radical formation, and so the existing data are interpreted only in terms of the proper muonium mechanism. Such interpretation predicts a rate constant $k(Mu + HClO_4) \approx 10^9$ liter/mole-sec.

3. *Ferric salts.* The quenching effects of ferric ions on μ^+ depolarization in $Fe(NO_3)_3$ solutions at 11 kG were first interpreted strictly in terms of a strong relaxation of the muonium electron by Fe^{3+} ions, assuming the rate ν_m of that relaxation to be proportional to the square of Fe^{3+} concentration (Schenck, 1970). Although this mechanism may be present, the model considered in Nosov and Yakovleva (1965) included the additional assumption that the mean chemical lifetime of muonium was independent of reagent concentration, which is now known to be incorrect. Results for $FeCl_3$ and $Fe(ClO_4)_3$ at 4500 G were later treated as evidence for the proper muonium mechanism (Brewer *et al.*, 1971), with the assumption that the only important reaction was

$$Mu + Fe^{3+} \xrightarrow{(k_{mxd})} \mu^+ + Fe^{2+} \tag{121}$$

where either the free muon itself or the product of its subsequent reaction with anions in the solution constitutes a diamagnetic environment for the muon. In light of the lack of reaction of muonium with HCl, one might expect the system $Mu + FeCl_3$ in H_2O to provide a good example of the proper muonium mechanism. Results at 4500 G are consistent with this assumption, but low-field measurements must be made to test for the presence of radicals in the depolarizing mechanism.

Results for $Fe(NO_3)_3$ and $Fe(ClO_4)_3$ at high field should not be interpreted strictly in terms of the proper muonium mechanism. The evidence for radical formation in nitric acid suggests that muonium might form radicals in $Fe(NO_3)_3$ solutions as well; again, low-field data may resolve this question. For $Fe(ClO_4)_3$ there is no doubt that radical formation is involved. Figure 20 shows the experimental dependence $P_{\perp\,res}([Fe(ClO_4)_3]$

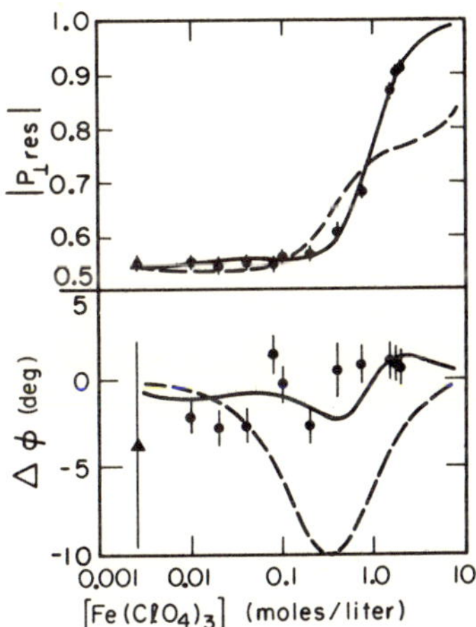

FIG. 20. Residual muon polarization in water as a function of the concentration of dissolved ferric perchlorate. Dashed curve: best fit without radicals; solid curve: best fit with radicals.

at 100 G. The best fit without radicals (dashed curve) is very poor; only by assuming that muonium reacts with dissolved $Fe(ClO_4)_3$ to form a muonic radical can one obtain an acceptable fit (solid curve). The situation here is formally the same as in reactions (118)–(120) for NO_3^-, with the additional process (121) for Fe^{3+}. Again, no attempt was made to identify chemical species. The results listed in Tables 6 and 7 are obtained by minimizing χ^2. The existence of muonic radicals in $Fe(ClO_4)_3$ solutions leads one to expect that radical formation will be found to play an important role in $HClO_4$ as well; low-field measurements should confirm this.

It should be mentioned here again that the muonium "spin-flip" frequency ν_m may not be negligible in solutions of paramagnetic ions. Although the absence of any significant "repolarization" in concentrated Mn^{2+} solutions (Brewer *et al.*, 1971) demonstrates that the repolarizing effect of other paramagnetic reagents is due mainly to the types of chemical mechanisms described above, some concentration-dependent muonium relaxation [as postulated in Schenck (1970)] could serve partially to quench the phase variations and generally mimic the effects of radicals shown in Fig. 20. Consequently, estimates of the contribution of radicals to the depolarization mechanism in paramagnetic solutions are to be regarded as tentative, pending further clarifying experiments.

D. CONCLUSIONS REGARDING THE MODEL. Several of the above results are particularly important in resolving certain controversies about the theory. First, the results for I_2 in CH_3OH at 102 G firmly establish that the residual polarization in pure methanol is due solely to hot-atom chemistry. If, as claimed by Babaev *et al.* (1966), $P_{\perp\,\mathrm{res}}(CH_3OH)$ were nonzero due to thermal chemical reaction of the type

$$\mathrm{Mu} + \mathrm{CH_3OH} \xrightarrow{(k_{\mathrm{msd}})} \mathrm{D}\ (\text{unidentified}) \tag{122}$$

muonium atoms would never remain uncombined long enough to process, and there could be no phase dip. In fact, such reactions must be totally unimportant to the mechanism in order to explain the return of the phase to zero as $[I_2] \rightarrow 0$. Therefore, we can be sure that $k(122) < 10^7$ liter/mole-sec and that the fraction of muonium reacting epithermally with methanol at room temperature is $h(CH_3OH) = 0.53 \pm 0.01$. Similarly, the results for benzene indicate $h(C_6H_6) = 0.13 \pm 0.01$, but are not as conclusive regarding $k_{\mathrm{msd}}(C_6H_6)$, due to the small phase dip. However since the asymmetry in pure benzene is so small, it is still fairly certain that $k_{\mathrm{msd}}(C_6H_6) < 10^8$ liter/mole-sec.

The incomplete depolarization in water is also exclusively due to hot-atom chemistry, as is especially clear from the curves of $P_{\perp\,\mathrm{res}}$ vs hydrogen peroxide concentration in water. The best value for $h(H_2O)$ is 0.55 ± 0.03; the anomalously high value (0.59 ± 0.01) of $h(H_2O)$ obtained in the fit of the H_2O_2 results is probably a reflection of the low value for A_0 in the same instance, which in turn could be due to the low density of concentrated H_2O_2 solutions compared to other concentrated aqueous solutions. A higher-density target gives a slightly increased A_0; such variations of A_0 with density are not allowed for in the fits. This introduces a systematic error of $\sim 5\%$ in the numerical results for A_0 and h, but does not significantly distort the other results.

The second general conclusion to be drawn from these results is that formation of fast-reacting radicals plays a central role in many (if not most) examples of μ^+ depolarization in liquids. If the radicals formed by reactions of Mu were relatively stable, or if radicals were rarely formed at all (proper muonium mechanism), the model formulated by Ivanter and Smilga (1969a) would be completely adequate for analysis of muonium chemistry. It is clear, however, that the more general case derived in Section VI is necessary for most practical applications.

2. *Comparison with Hydrogen Atom Chemistry*

Absolute rates of reaction in solution are difficult to estimate reliably from first principles, due to the complexity of the processes involved. It is possible, however, to make some qualitative predictions of how rates will depend on the mass of one reactant when all other physical parameters are held constant.

As mentioned earlier, rate constants of Mu and H in gases are expected to differ by a factor of 3 due to the kinetic isotope effect in the mean thermal velocity [see Eq. (106)]. Additional factors may arise from "dynamic" isotope effects. Unfortunately, such a treatment is only appropriate for gases, where the mean free path is many molecular radii and the concept of a "collision rate" is well defined. In liquids, each reagent molecule is continually surrounded by a "cage" of solvent molecules, which severely restrict its thermal motion (Benson, 1971). The reactants must diffuse through this crowded environment to find each other, and when they do approach they are apt to stay in each other's presence for some time: the probability of reaction in such a prolonged "encounter" is often close to unity. Such reactions are called "diffusion controlled" (DC), since the rate of reaction depends only on how fast the reactants diffuse through the solvent to meet each other. Since diffusion in liquids proceeds primarily by "squeezing" and "tumbling," such rates are largely determined by the geometrical properties of solvent and reactant molecules, and the mass dependence is generally weaker than in gases.

A rough estimate of the DC rate for reactions of Mu atoms in water or methanol is $k_{DC}(\mathrm{Mu}) \approx 10^{11}$ liter/mole-sec. Most of the measured rate constants for Mu in liquids are near this limiting value. Rate constants less than k_{DC} usually reflect an "activation energy" E_a required to form the activated complex $\mathrm{HX}^{\ddagger}$ in the reaction $\mathrm{H} + \mathrm{X} \rightarrow \mathrm{HX}^{\ddagger} \rightarrow$ products (Glasstone *et al.*, 1941; Wolfsberg, 1972). The rate constant then acquires an exponential temperature dependence via the Boltzmann distribution: $k \propto \exp(-E_a/k_BT)$. The quantity E_a may depend on factors such as the vibrational frequencies of bonds formed in the activated complex, which may in turn depend on the mass of the light atom. Even in the case of DC reactions, the diffusion process itself requires an activation energy (Glasstone *et al.*, 1941; Logan, 1967; Wolfsberg, 1972) that may depend on mass. In addition, quantum-mechanical tunneling, which may be important for many reactions of H (Lewis and Robinson, 1968), can be expected to be quite significant for muonium. Such "dynamic" isotope effects can cause dramatic differences between $k(\mathrm{Mu} + \mathrm{X})$ and $k(\mathrm{H} + \mathrm{X})$.

Table 8 shows a comparison between Mu and H rate constants for some of the more unambiguous reactions studied.

A. RATES NEAR THE DIFFUSION-CONTROLLED LIMIT. The rate constant $k(110) = (1.33 \pm 0.1) \times 10^{11}$ liter/mole-sec for reaction (110) of Mu with I_2 in CH_3OH is near the DC limit for muonium in methanol. The corresponding H atom rate has been measured in aqueous solution (Anbar

TABLE 8

COMPARISON OF OVERALL RATE CONSTANTS OF H AND MU WITH VARIOUS REAGENTS[a]

	Hydrogen		Muonium		
Reagent	Solvent	k_H	Solvent	k_{Mu}	k_{Mu}/k_H (approx.)
CH_3OH	H_2O	$(1.6 \pm 0.1) \times 10^6$	CH_3OH	$<10^7$	—
C_6H_6	H_2O	$(7 \pm 3) \times 10^8$[b]	C_6H_6	$(8^{+5}_{-3}) \times 10^8$	1
H_2O	H_2O	nil	H_2O	$<10^7$	—
I_2	H_2O	4×10^{10}	CH_3OH	$(13.3 \pm 1) \times 10^{10}$	3
			C_6H_6	$(5.7 \pm 1) \times 10^{10}$	1
Br_2	H_2O	$(12 \pm 6) \times 10^{10}$[c]	C_6H_6	$(9.4 \pm 0.3) \times 10^{10}$	1
H_2O_2	H_2O	$(9 \pm 5) \times 10^7$[d]	H_2O	$(1.09 \pm 0.15) \times 10^{10}$	100
NO_3^-	H_2O	$(9 \pm 5) \times 10^6$[e]	H_2O	$(13 \pm 6) \times 10^{10}$	10^4
ClO_4^-	H_2O	nil	H_2O	$\sim 4 \times 10^{10}$	?

[a] Rate constants are in units of liter/mole-sec. Source of values for H is Anbar and Neta (1967), except where otherwise specified.
[b] From Sauer and Ward (1967) and Michael and Hart (1970).
[c] From Farhataziz (1967).
[d] From Sweet and Thomas (1968).
[e] From Navon and Stein (1965).

and Neta, 1967) to be 4×10^{10} liter/mole-sec, in qualitative agreement with this result. The rate constant $(5.7 \pm 1) \times 10^{10}$ liter/mole-sec for Mu + I_2 in C_6H_6 indicates that diffusion of Mu through benzene is about one-half as fast as through methanol, if reaction (110) is truly diffusion controlled. Such an assumption is supported by the fact that the rate constant $k(111) = (9.4 \pm 0.3) \times 10^{10}$ liter/mole-sec for reaction of Mu with Br_2 in C_6H_6 is nearly the same as with I_2. This value agrees well with the measured (Farhataziz, 1967) rate $(12 \pm 6) \times 10^{10}$ liter/mole-sec for H + Br_2 in water.

B. REACTIONS WITH SOLVENTS. The rate constant for H + CH_3OH in aqueous solution is (Anbar and Neta, 1967) $(1.6 \pm 0.1) \times 10^6$ liter/mole-sec. While this result cannot rigorously be compared with the rate for Mu in pure CH_3OH, where diffusion is irrelevant, it does qualitatively corroborate the value $k(122) < 10^7$ liter/mole-sec for Mu + CH_3OH. The reaction rate of thermalized H atoms with benzene to form cyclohexadienyl [analogous to reaction (112)] was measured by pulsed radiolysis in aqueous solution (Sauer and Ward, 1967) to be about $(7 \pm 3) \times 10^8$ liter/mole-sec, whereas for muonium $k(112) = (8^{+5}_{-3}) \times 10^8$ liter/mole-sec in the pure solvent. Again, these two rates in different solvents cannot legitimately be compared in an absolute sense; nevertheless, the fact that they agree constitutes some justification for the assumption that the radical is formed by thermal, rather than "hot-atom," reactions. In water, the results are consistent with $k(\text{Mu} + \text{H}_2\text{O}) < 10^7$ liter/mole-sec. We are not aware of any evidence for fast reactions of H with H_2O.

C. REACTIONS OF MUONIUM IN AQUEOUS SOLUTION.

1. *Hydrogen peroxide.* The basic reaction of H with hydrogen peroxide is thought to be (Sweet and Thomas, 1968)

$$\text{H} + \text{H}_2\text{O}_2 \rightarrow \text{HO}^{\cdot} + \text{H}_2\text{O} \tag{123}$$

The rate constant for this reaction has been measured (Takakura and Ranby, 1968; Sweet and Thomas, 1968) over a range of pH to be $k(123) = (9 \pm 5) \times 10^7$ liter/mole-sec. The reaction is presumed to involve a cleavage of the O–O single bond, but from (123) it is impossible to tell whether the original H atom emerges in the H_2O as in reaction (115) ("OH abstraction") or in the $HO^{\cdot}$ as in reaction (116) ("O abstraction"). For muonium, $k(116)$ is nearly 4 times higher than $k(115)$.

One would expect $k(115) + k(116)$ to be the Mu rate analogous to $k(123)$; instead, $k(115) + k(116) = (1.09 \pm 0.15) \times 10^{10}$ liter/mole-sec, roughly a factor of 100 higher than the corresponding H atom rate. Such a large anomaly is probably attributable to dynamic isotope effects. Even a gross error in the assumed value of ω_r (caused, for instance, by misidentification of the radical) would not explain such a discrepancy between the two rate constants.

2. *Strong acids.* Since HCl, HNO_3, and $HClO_4$ are all highly dissociated in aqueous solutions, their reactions with Mu can be considered primarily in terms of the ionic species H^+, Cl^-, NO_3^-, and ClO_4^-. As mentioned earlier, HCl solutions up to 10*M* do not repolarize the muon; we must conclude that no combination of reactions with H^+ and/or Cl^- leads to a stable diamagnetic environment for the muon in times less than about 10^{-8} sec.

In nitric acid, on the other hand, one finds a net reaction rate $k(\text{Mu} + \text{NO}_3^-) = k(118) + k(119) = (1.3 \pm 0.6) \times 10^{11}$ liter/mole-sec, an essentially diffusion-controlled rate. This result is a factor of 10^4 higher than the measured (Navon and Stein, 1965) H atom rate constant $k(\text{H} + \text{NO}_3^-) = (9 \pm 5) \times 10^6$ liter/mole-sec. Assuming that the rates of the same reactions have been measured, such a dramatic isotopic effect probably reflects a tunneling process (Lewis and Robinson, 1968). As in the case of H_2O_2, such a large discrepancy cannot be explained simply by postulating an error in ω_r.

A solution of concentrated $NaNO_3$ was studied as a test of pH dependence; complete repolarization was observed, as for concentrated HNO_3. While a full curve of $P_{\perp\,\mathrm{res}}([NaNO_3])$ would be necessary to clarify the details of the chemical processes involved, this single measurement was sufficient to indicate that Mu reacts with NO_3^- at approximately the same rate, independent of the presence of H^+. Also, highly reactive species such as NO_2, O_2, and NO_2^- should not have been present in significant concentrations in the freshly prepared $NaNO_3$ solution.

The repolarization of muons by $HClO_4$ in high magnetic field was also studied. A fast reaction is suggested, $k(\mathrm{Mu} + ClO_4 \rightarrow \mathrm{D}) \approx 10^9$ liter/mole-sec, for the complete process leaving the muons in diamagnetic compounds. Until studies are made in low field, the details of the process are uncertain. However, one can predict that radical formation will be important on the basis of results at 100 G with $Fe(ClO_4)_3$, which show a rate constant $k_{\mathrm{mxr}} = (3.8 \pm 0.8) \times 10^{10}$ liter/mole-sec. Here Fe^{3+} is unlikely to react with Mu to produce a radical, so we expect that this rate represents $k(124)$ for the reaction

$$\mathrm{Mu} + ClO_4^- \xrightarrow{(k_{\mathrm{mxr}})} (\text{muonium-containing radical}) \qquad (124)$$

Since reactions of H atoms with ClO_4^- are regarded as virtually nil (Anbar and Neta, 1967), there is again dramatic disagreement between Mu and H rates.

3. *Ferric salts.* The data for $Fe(ClO_4)_3$ at 100 G provide detailed information about the rates and qualitative features of several reactions, but the large number of species involved complicates the extraction of rates of specific reactions of Mu with Fe^{3+} and/or ClO_4^- to produce both diamagnetic and paramagnetic products. In strong fields, even less detail is available from the data (see Fig. 11), and in the case of $Fe(NO_3)_3$ we can only be sure that a fast reaction does take place.

The situation with $FeCl_3$ solutions should be much simpler, since Mu does not appear to react significantly with Cl^-. Interpreting the high-field data on Mu + $FeCl_3$ strictly in terms of reaction (121), one obtains a rate constant $k(121) = (2.1 \pm 0.2) \times 10^{10}$ liter/mole-sec. The H atom rate constant for the direct oxidation–reduction reaction analogous to (121) has been measured to be $(9 \pm 1) \times 10^7$ liter/mole-sec (Anbar and Neta, 1967) in similarly mild acidic solutions. Taken at face value, the muonium rate is 200 times that for hydrogen. However, it is unlikely that the process involved is as simple as reaction (121). Ferric ions are known (Cotton and Wilkinson, 1966) to form complexes in solution, in particular $FeCl^{2+}$ and $FeCl_2^+$, whose rate constants for reaction with H atoms are, respectively, 4.5 and 9.0 $\times$ 10^9 liter/mole-sec (see Anbar and Neta, 1967). It is possible that reactions of Mu with one or both of these species were actually observed.

D. REACTIONS OF RADICALS. The μ^+ depolarization technique also allows measurement of rate constants for reactions of various radicals incorporating muonium.

In comparing these rate constants with the corresponding rates for analogous radicals in which the muon is replaced by a proton, the difference in masses of Mu and H should affect only the "dynamics" of the processes. Even $MuO^\cdot$, the lightest muonic radical envisioned, should diffuse through liquids at the same rate as $HO^\cdot$, its protonic analog. Comparisons of reaction rates of muonic and protonic versions of these radicals should therefore admit of straightforward interpretation in terms of the dynamics of the activated complex.

The most serious difficulty with this interpretation is the uncertainty as to which radical is actually being produced. In the cases of HNO_3 and $Fe(ClO_4)_3$ solutions, for instance, no attempt was made to identify the radical species. The fitted value for ω_r/ω_0, while imprecise, does provide a hint as to likely candidates, suggesting $MuO^\cdot$ in the case of HNO_3 and some species with a weaker hyperfine coupling in the case of $Fe(ClO_4)_3$. However, this cannot be regarded as conclusive evidence, and the products of reactions (119) and (124) must be regarded as unknown. A longitudinal-field technique has given an experimental estimate of ω_r in $C_6H_6Mu^\cdot$ (Goldanskii and Firsov, 1971), and it may prove possible to determine other hyperfine couplings in this way. Such studies would be very helpful.

In some cases it is possible to deduce the identity of the radical, if there is only one species of "reagent" and the products of its reaction with H are well known. In hydrogen peroxide solutions, for instance, it seems most probable that reactions (115) and (116) should dominate (Sweet and Thomas, 1968; Takakura and Ranby, 1968), making $MuO^\cdot$ the most likely radical species. The value for the rate constant for reaction of $MuO^\cdot$ with H_2O_2 is $k(117) = (1.4 \pm 0.2) \times 10^9$ liter/mole-sec. The corresponding rate for $HO^\cdot + H_2O_2$ is (Anbar and Neta, 1967) about $(3 \pm 2) \times 10^7$ liter/mole-sec, a factor of 50 slower. Unless the radical has been misidentified, this difference is almost certainly due to dynamic isotope effects in the $(MuOH_2O_2)^\ddagger$ complex.

The addition of H to benzene to form cyclohexadienyl is also a well established reaction (Michael and Hart, 1970), a fact that lends credence to the assumption that $C_6H_6Mu^\cdot$ is the radical involved in reactions (112) and (113). We are unaware of any measurement of the reaction rates for $C_6H_7^\cdot$ with Br_2 or I_2; these measurements of $k(C_6H_6Mu^\cdot + Br_2) =$

$(3.6 \pm 1.0) \times 10^9$ liter/mole-sec and $k(C_6H_6Mu^{\cdot} + I_2) = (2 \pm 1) \times 10^9$ liter/mole-sec may represent the only information available on these reactions. In view of the large size of the $C_6H_6Mu^{\cdot}$ molecule and the similarity of the rates with Br_2 and I_2, the reaction is probably diffusion controlled in liquids.

E. PROSPECTS FOR MUONIUM CHEMISTRY IN LIQUIDS. In summary, there are evidently a number of startling exceptions to the naive expectation (Brewer *et al.*, 1971; Firsov and Byakov, 1965) that Mu and H should react at similar rates in analogous processes in liquids. The present results, far from settling the issue, call for further investigations, both experimental and theoretical. The accuracy of H atom measurements may also need critical examination. Although experimentally difficult, more attempts should be made to measure radical hyperfine frequencies directly. Such measurements would be quite interesting in their own right, for comparison with hyperfine couplings in protonic versions of the same radicals. It is clear from these results that comparisons of Mu and H atom solution chemistry are feasible, and that one may expect to encounter large differences in rates. The interpretation and final understanding of these differences, presumably in terms of dynamic isotope effects, may be of great significance to the chemical physics community.

3. *Muonium Hot-Atom Chemistry*

The "hot fraction" h has been a source of annoyance in the history of muonium chemistry, largely due to the analysis of many experimental results with the assumption that it was negligible (Goldanskii and Firsov, 1971). Now that epithermal reactions have been shown to be important (and their efficiencies easily measurable), this topic becomes a new area of study.

Unambiguous measurements of h are available in only a few substances:

$$h(H_2O \text{ at } 300°K) = 0.55 \pm 0.03$$

$$h(CH_3OH \text{ at } 300°K) = 0.53 \pm 0.01$$

$$h(C_6H_6 \text{ at } 300°K) = 0.13 \pm 0.01$$

In addition, a preliminary analysis of unpublished $P_{\perp\,\text{res}}([Br_2])$ results in CS_2 at 300°K indicates a hot fraction in CS_2 similar to that in C_6H_6. These numbers tell us only the efficiency for epithermal formation of stable *diamagnetic* compounds containing muonium; paramagnetic products (radicals) may also form, and at this stage we can only put weak upper limits on the efficiency of such reactions. Furthermore, we cannot tell *which* diamagnetic products are formed; however, studies of hot tritium chemistry (Rowland, 1970) can help us make intelligent guesses. These guesses can sometimes be complemented by measurement of "chemical shifts" in the μ^+ precession frequency due to diamagnetic shielding by molecular electrons. Each molecular species will have a characteristic chemical shift that can be estimated from theory and/or proton NMR measurements on analogous species. However, these shifts are very small ($\sim$1 ppm) and their measurement requires extremely precise frequency determinations. Furthermore, when more than one molecular species is present in significant numbers, it is impossible to separate the different frequencies that result. So far the only attempts to sort out such chemical shifts have been in precise measurements of the muon magnetic moment (Crowe *et al.*, 1972a; Hutchinson *et al.*, 1963).

The specification of temperature is not superfluous; although one might naively expect h to be independent of temperature, this is not the case. In water, h is a uniformly increasing function of temperature, showing a spectacular rise as the phase changes from ice at 0°C to water at 0°C (Myasishcheva *et al.*, 1967c) (see Fig. 21). Not surprisingly, these results were first interpreted in terms of thermal reactions. We are now bound to seek an explanation for the phase dependence of hot-atom processes. One possibility is that hydrogen-bonded ice is much more efficient than water at slowing down the epithermal Mu atom, which then has fewer collisions in the energy range where reactions are likely.

Very high asymmetries are seen in halogen-substituted methanes (see Table 3). To test for a "thermal contribution" to the residual polarization

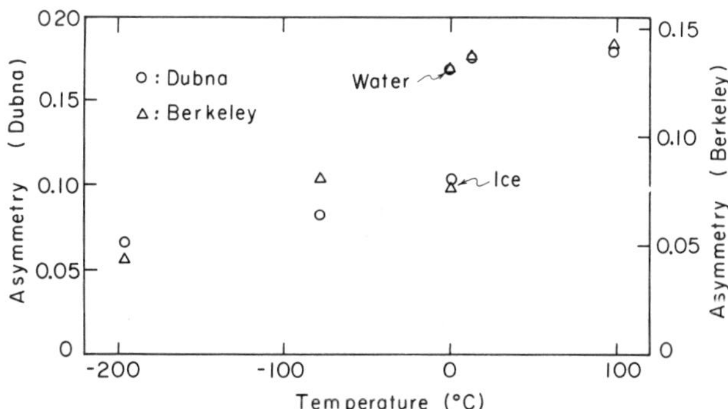

Fig. 21. Temperature dependence of the μ^+ precession signal asymmetry in H_2O at a transverse magnetic field of 100 G. The asymmetry scale for the Berkeley results (right-hand ordinate) is different from that of the Dubna measurements (left-hand ordinate) (Myasishcheva *et al.*, 1967c) because of different experimental setups. The asymmetry is proportional to the muonium hot-atom reaction efficiency in water.

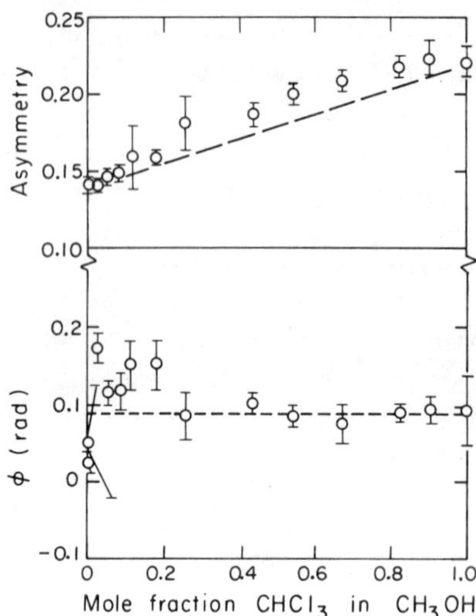

FIG. 22. Asymmetry ($A = P_{\perp \text{res}} A_0$) and overall phase ($\phi = \theta_0 + \Delta\phi$) for mixtures of methanol and chloroform. Mole fraction equals number of $CHCl_3$ molecules per total number of molecules. Dashed line (phase) shows best fit for a constant phase ($\Delta\phi = 0$).

in $CHCl_3$, a series of mixtures of $CHCl_3$ and CH_3OH (known to be thermally inert) were examined; a thermal process would be reflected in a characteristic dip in $\Delta\phi$. The (unpublished) results, shown in Fig. 22, are taken to be good evidence for a purely epithermal reaction with $CHCl_3$. It will also be noticed that the dependence of the asymmetry on the mole fraction of $CHCl_3$ is not linear. This must be a consequence of the dynamics of the hot-atom processes: the cross section for epithermal reaction will have a peak at a different energy for each molecular variety, and the different molecules will have different efficiencies at each energy for slowing down the Mu atoms without reaction. The resultant dependence of the epithermal reaction probability on mixture can have a variety of shapes.

In Table 3 we list the muon asymmetries in a variety of pure substances. On the basis of a single measurement, one cannot determine whether the residual asymmetry is strictly due to hot-atom chemistry or partially due to thermal reactions; however, a good indication of the dominant mechanism can sometimes be obtained by varying the strength of the transverse magnetic field; for a purely epithermal reaction, the asymmetry should be independent of field strength (except for apparatus effects). In the case of purely thermal reactions, more than 50% depolarization in low fields reflects the brief coherent precession of muonium atoms (see Fig. 11); this depolarizing process can be slowed down (increasing the residual asymmetry) by reducing the field strength. Thus a field-independent asymmetry in low fields (20 to 500 G) can be taken as an indication of predominantly epithermal processes (Gurevich *et al.*, 1971b). In this light, Table 3 represents a tremendous amount of information about hot-atom reactions of Mu, none of which has yet been satisfactorily analyzed.

To summarize, we feel that muonium "hot-atom" chemistry is a field of great promise (for example, in comparisons with hot-tritium reactions to help illuminate the dynamics of epithermal processes), which has suffered from profound theoretical neglect.

C. *Reactions of Mu in Solids*

Muons exhibit a much more varied behavior in solids than in gases or liquids. The μ^+ may replace a proton constituent (as in gypsum) (Schenck and Crowe, 1971), or the Mu atom may occupy a large interstitial vacancy (as in quartz) (Myasishcheva *et al.*, 1968), or the μ^+ may acquire a screening cloud of conduction electrons (as in most metals), which has an average charge density similar to that of the e^- in the Mu atom, but in which there is no coherently coupled μ^+–e^- spin system in the sense of muonium. A description appropriate to one case is generally a poor way of treating another, and the concept of chemical reactions takes on a rather broad interpretation, restricted to those media in which quasi-free Mu atoms can be said to exist in the familiar sense. In solids, then, we can define a "chemical reaction" as any process that removes the μ^+ from quasi-free muonium and places it in a diamagnetic environment. Any such reaction is assumed to take place at a constant rate $1/\tau_m$.

Because of this diversity, we will leave most of the "chemical" reactions of Mu in solids to be discussed along with other phenomena under the subject headings of the particular media in the last two sections; however, there are several cases that are appropriately mentioned here.

Epithermal reactions of Mu in frozen liquids (e.g., water) have already been discussed. Thermal reactions of Mu in certain amorphous solids could be studied by "freezing in" varying amounts of a reagent, exactly as in the liquid-phase technique. This sort of investigation involves no new concepts (except perhaps in the interpretation of the results), but has not yet been undertaken.

In sulfur, some very unusual μ^+ behavior has been observed (Gurevich *et al.*, 1968). Since yellow sulfur is an excellent insulator, one might expect to see free Mu precession in crystalline sulfur. On the other hand, Mu might be expected to react rapidly with sulfur, causing Mu precession to be replaced by μ^+ precession with some residual polarization. In fact, one

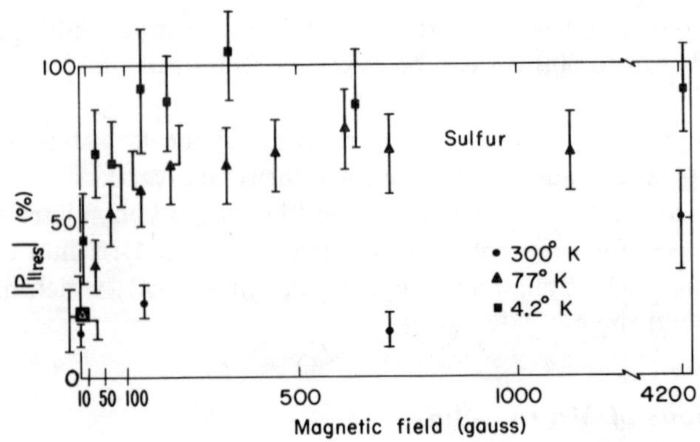

FIG. 23. Quenching of the depolarization in sulfur by longitudinal field at three temperatures (from Eisenstein *et al.*, 1966).

observes μ^+ precession, but there seem to be two components to the signal, one long-lived, with an asymmetry of 0.05, and one rapidly relaxing, with an initial asymmetry of 0.16 and a relaxation time of 30 ± 5 nsec.

Gurevich *et al.* (1968) interpreted this as a case of chemical reaction of Mu with $\nu \gg \omega_0(1 + x^2)^{1/2}$, in which $P_\mu(t) = P_\mu'(t) \exp(-i\omega_\mu t)$, where $P_\mu'(t)$ is given by Eq. (65). They extracted a chemical lifetime of $\tau_m = 130$ nsec and an electron relaxation rate $\nu = 270\omega_0$. This interpretation is entirely consistent with the data, but is subject to many reservations. First, the possible presence of hot-atom processes weakens the necessary assumption that the polarization of the nonrelaxing component is equal to $\tau_1/(\tau_1 + \tau_m)$. Second, one is hard pressed to imagine a mechanism for such rapid relaxation of the Mu electron in so good an insulator.

An alternate model for μ^+ in sulfur can be proposed: muonium reacts epithermally with the S_8 ring structure to form both diamagnetic (constant polarization) and radical products. For the radical (e.g., $MuS_8^\cdot$) one can easily imagine $\nu_r \gg \omega_r$ due to multiple hyperfine couplings. Thus the μ^+ polarization in the radical fraction could well have the time dependence (28). In this model no thermal chemical reactions need to be postulated.

Longitudinal-field studies of $P_{\parallel\,\mathrm{res}}$ in sulfur (Eisenstein *et al.*, 1966) (see Fig. 23) indicate a low-temperature field dependence vaguely like that described by Eq. (13) if an "effective hyperfine field" of ~75 G is assumed in the definition of x. This suggests a depolarizing process including a radical with $\omega_r \sim 0.05\omega_0$. At room temperature, the residual polarization is not completely restored even by strong fields; this is consistent with the hypothesis of a relaxing radical.

Neither of the above models is supported by conclusive evidence and the behavior of μ^+ and/or Mu in sulfur must still be regarded as an unresolved mystery.

A somewhat similar behavior is observed for muonium in a number of ionic crystals; in addition, the process by which positive muons replace protons in $CaSO_4 \cdot 2H_2O$ crystals must be in some sense a hot-atom reaction. However, the mechanisms acting in these cases are so unclear that we cannot pretend to have understood the "chemistry" involved, and so we will reserve discussion for the sections dealing more generally with Mu and μ^+ in solids.

VIII. Muonium in Solids

In many solids, the μ^+ captures a single electron to form a muonium atom in some vacancy or interstitial site. The coupled μ^+ and e^- spins then evolve as described in Section V, subject to a variety of interactions causing depolarization or destruction of the muonium spin system. There are also a number of situations in which the μ^+ does *not* form muonium in this sense; these cases will be discussed in the next section.

We can separate the behavior of Mu in solids into four general categories correlated with the physical characteristics of the solid. In pure nonmagnetic insulators, Mu finds a spacious enough interstitial site that its behavior resembles that of a free muonium atom in vacuum. In other insulators with high densities of paramagnetic impurities, or in ferromagnetic insulators, Mu may find a spacious site, but is subject to rapid "chemical" reactions and depolarizing magnetic perturbations that destroy the Mu precession signal and make direct identification of Mu difficult. In pure semiconductors at low temperature, Mu may be confined to a cramped interstitial site or may have a delocalized electron wave function spanning many lattice sites; in each case it is strongly perturbed by the crystal field and the available conduction electrons. And in warm n-type semiconductors or semimetals with high densities of conduction electrons, muonium suffers such severe perturbations that it exists only by a marginal definition.

Each of these categories will be treated more or less separately in the following discussion.

A. Muonium in Solid Insulators

Muonium precession is directly observable in many pure insulators (Brewer *et al.*, 1973a; Gurevich *et al.*, 1969; Myasishcheva *et al.*, 1968). In quartz, ice, and solid CO_2 the "two-frequency precession" described by Eq. (20) has been observed in moderate fields (see, for example, Fig. 9). In each case, the value of ω_0 extracted from the "beat frequency" is consistent with the hyperfine frequency of muonium in vacuum. Since the hyperfine coupling is given by

$$\hbar\omega_0 = \tfrac{8}{3}\pi g_e \mu_0^e g_\mu \mu_0^\mu \,|\, \psi_s(0) \,|^2 \tag{125}$$

where

$$|\, \psi_s(0) \,|^2 = 1/\pi r_0^3 \tag{126}$$

one can say that the mean radius r_0 of the Mu atom is the same in these insulators as in vacuum. Thus the interstitial sites in such media are both "spacious" (dimensions $>2r_0$) and "empty" (negligible valence electron densities). This is not the case for some semiconductors, as will be seen later.

Whenever Mu precession has been observed, a fairly rapid damping of the oscillations has been noted. This relaxation may be due to "chemical" reaction of Mu ("τ_m effects"), depolarization of the Mu electron ("ν effects"), or random local magnetic fields (RLMF) that "smear out" the Mu precession frequency. Mu precession is ~100 times more sensitive to RLMF than is free μ^+ precession. In quartz (SiO_2), which should be chemically inert, the damping may be due to RLMF from the ^{29}Si nuclei ($\mu = -0.55$ n.m.) comprising 4.7% of the Si atoms (Minaichev *et al.*, 1970a). This hypothesis is supported by the fact that the damping is more rapid at 77 than at 300°K, as shown in Fig. 24 (Myasishcheva *et al.*, 1968). This suggests a "motional narrowing" effect: the enhanced diffusion at higher temperature decreases the time of influence of a given local field. It was also observed that the damping is faster in crystalline quartz than in fused quartz at the same temperature (300°K), as can be seen from Fig. 25.

In corundum (Al_2O_3), no Mu precession can be seen, presumably (Minaichev *et al.*, 1970a) due to the action of RLMF from the ^{27}Al nuclei ($\mu = +3.64$ n.m.). The field dependence of the polarization in longitudinal field, shown in Fig. 26, is consistent with Eq. (13) for muonium with a "normal" (vacuum-like) size, except for very low fields, where the polarization drops suddenly to zero. This is taken as evidence that the strength of the RLMF is about 50 G.

In each of these insulators, a significant free μ^+ precession signal is observed along with the Mu precession. In view of the relatively stable state of muonium, this cannot be due to a fast thermal chemical reaction of Mu. No satisfactory explanation has yet been offered for the general simultaneous presence of stable Mu and μ^+, but there are two attractive conjectures. The μ^+ signal may be due to muons placed in a diamagnetic environment by epithermal processes of an unknown nature, or to a phenomenon analogous to the "Ore gap" for positrons: in some media, a positron must capture an electron while epithermal in order to form positronium; once thermalized, it lacks sufficient ionization potential to steal an electron away from the medium. For muonium, with twice the ionization potential of positronium, such effects could only occur in highly electrophilic media.

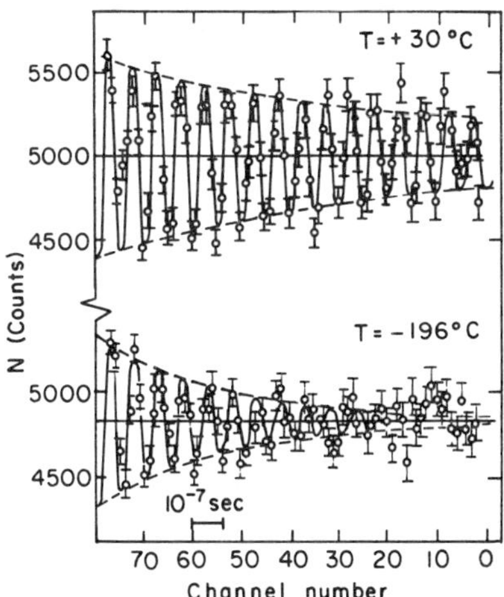

FIG. 24. Muonium precession signal in fused quartz at 30 and −196°C for a transverse magnetic field of 7.2 G. The rate N is corrected for the exponential decay of the muon (from Myasishcheva *et al.*, 1968).

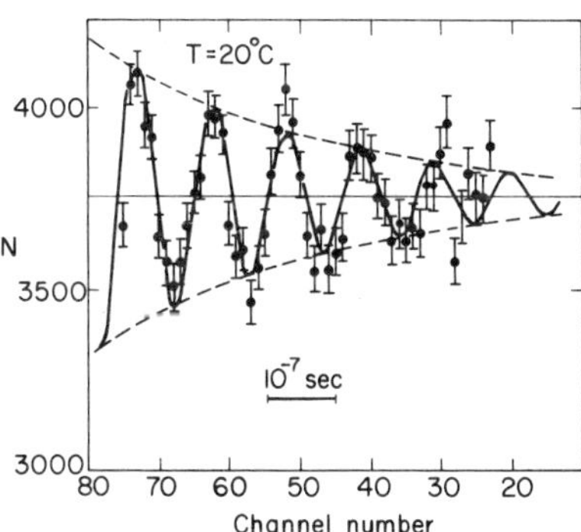

FIG. 25. Muonium precession signal in crystalline quartz at 20°C for a transverse magnetic field of 7.2 G. The rate N is corrected for the exponential decay of the muon (from Myasishcheva *et al.*, 1968).

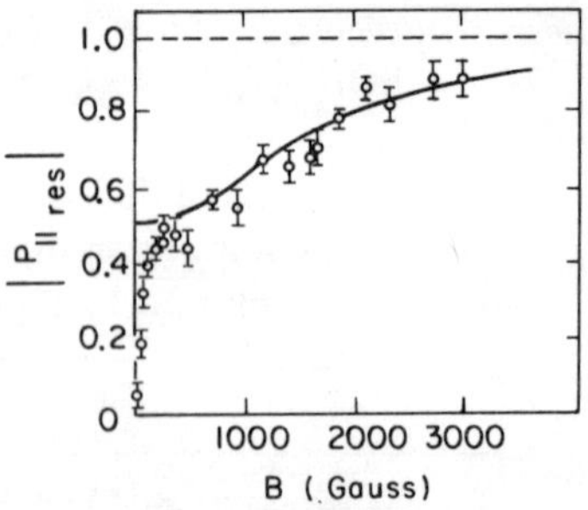

FIG. 26. Observed longitudinal residual muon polarization $P_{\| \text{res}}$ vs field strength in Al_2O_3. The solid curve corresponds essentially to the quenching of the depolarization for vacuum-like muonium [Eq. (13)] (from Minaichev *et al.*, 1970).

As discussed in the previous section, there is no direct evidence for quasi-free Mu in sulfur. Chemical reactions may be expected to play an important role, but the behavior of Mu in sulfur is still an enigma.

1. *Alkali Halides*

The alkali halides have also proved to be a very interesting environment for muonium. No Mu precession signal can be seen in any of these ionic crystals, but longitudinal field measurements show evidence for muonium. Perhaps the best-studied (Ivanter *et al.*, 1972) example is KCl. Here (as in most cases) a small μ^+ precession signal is seen in transverse field. This component of the polarization is presumably due to some epithermal process and has been subtracted out of the residual polarization in longitudinal field, which is shown in Fig. 27 as a function of the field strength. The residual polarization $R_{\|}$ is zero at zero field, but is restored in fields of $\sim$100 G; when a second crystal with higher purity and fewer dislocations was used, this critical field dropped to $\sim$20 G. It is concluded (Ivanter *et al.*, 1972) that the behavior in weak field is due to RLMF from paramagnetic lattice defects. If the high-field behavior is extrapolated back to zero field, a residual polarization $R_{\|}(0) < 0.5$ is obtained. Muonium formation alone can never cause $R_{\|}(0) < 0.5$, so the field dependence cannot be fitted to the form (13) as could the Al_2O_3 data. However, the shape of $R_{\|}(B)$ is clearly that of a hyperfine-coupled spin system, presumably muonium, so it was assumed that muonium itself was being depolarized ($\nu > 0$). This depolarization would always go to completion if it were not cut short by some quasi-chemical reaction of Mu ($1/\tau_m > 0$), and so the data must obey the general form (69). Ivanter *et al.* (1972) found an excellent fit to this model for a wide range of values of ω_0, ν, and

τ_m. Given the additional assumption that conditions (70) hold in KCl, it was possible to extract from the resultant dependence (71) the following values:

$$\omega_0(\text{KCl}) = (0.97 \pm 0.04)\omega_0(\text{vac}) \tag{127}$$

$$\nu\tau_m = 1.81 \pm 0.10 \tag{128}$$

This fit, subject to the conditions (70), is indicated by the solid line in Fig. 27. Unfortunately one cannot obtain separate determinations of ν and τ_m from this fit, since the trade-off between the two is very sensitive to the exact value of ω_0 in the region $\omega_0 \approx \omega_0(\text{vac})$. The fitted value of the hyperfine frequency (and thus the size of the Mu atom) is consistent with the vacuum value, as in other insulators.

The hydrogen analog of muonium in ionic crystals is the U_2 center. In the ESR measurements on U_2 centers in KCl (Spaeth, 1966) one obtains a hyperfine frequency for the hydrogen ground state that is 3% smaller than the vacuum value. If we assume this value for muonium also, we obtain a rough estimate of the separate values of ν and τ_m:

$$\nu \approx 1/\tau_m \approx 10^9 \text{ sec}^{-1}$$

Measurements were also made in low field on a sample of KCl with fewer dislocations (Ivanter *et al.*, 1972). The extrapolated value of the residual polarization at zero field was the same (within errors) as for the earlier

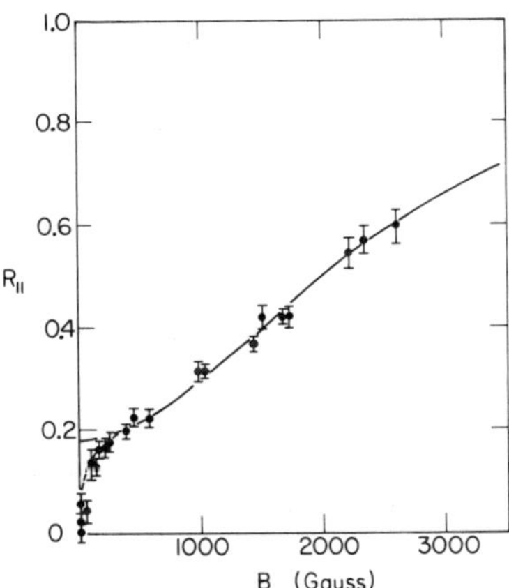

FIG. 27. Longitudinal residual muon polarization $R_{\parallel}$ (excluding contributions from epithermal reaction channels) vs field strength for single crystals of KCl (from Ivanter *et al.*, 1972).

sample. Whenever conditions (70) apply,

$$R_{||}(0) = 1/2(1 + \nu\tau_m)$$

It is concluded that $\nu\tau_m$ is independent of the concentration of lattice defects. There are two plausible explanations for this. In the first picture both ν and $1/\tau_m$ are proportional to the defect density, since Mu interacts with the paramagnetic defects both by spin exchange collisions and by "chemical" reactions, in which Mu becomes bound to a dislocation. In the second picture neither ν nor τ_m is related to defects at all.

If ν is unrelated to defects then it must be simply the inverse of the spin–lattice relaxation time of the muonium electron. As such, it would be surprisingly large. If muonium were in a pure s-state, no coupling of the electron spin to the diamagnetic lattice should occur; that is, there would be no coupling to phonons, and consequently no relaxation. If the electron spin is coupled to some orbital angular momentum (LS coupling), which in turn is sensitive to the electric crystal field, then a spin–lattice interaction is possible, due to the time modulation of the crystal field by phonon modes. Hence, for a spin–lattice relaxation to occur, the ground state of Mu in the crystal must contain some admixtures of excited vacuum states such as the 2p-state. Such mixing is facilitated by the static part of the electric field. The 3% reduction of the hyperfine frequency of hydrogen atoms (U_2 centers) in KCl is an indication of such a mechanism.

In the original work of Ivanter and Smilga (1968) ν is always tacitly assumed to be field independent. This assumption is by no means obvious. Table 9 lists formulas for the relaxation times of paramagnetic impurities in ionic crystals as derived (Manenkov and Orbach, 1966) for different phonon processes: (1) the direct process, a one-phonon process; (2) the Raman process, involving two phonons; and (3) the Orbach process, which involves an excited electronic level. Only the Orbach process is field in-

TABLE 9

FORMULAS FOR RELAXATION TIMES OF PARAMAGNETIC IMPURITY SPINS IN IONIC CRYSTALS DUE TO PHONON PROCESS[a]

Process	Relaxation rate
Direct process: one-phonon exchange	$1/T_1 \approx CB^4T$
Raman process: two-phonon exchange	$1/T_1 \approx D_1T^9 + D_2B^2T^2$
Orbach process	$1/T_1 \approx E\exp(-\Delta_c/kT)$

[a] B is magnetic field, T is temperature.

dependent. Future μ^+ depolarization measurements promise to shed some light on which phenomena actually occur.

Clearly these processes are not an isolated property of the muon but are relevant to many other solid-state phenomena.

Other alkali halides have also been studied with muonium (see Table 3), with similar but not identical results. The residual asymmetry in transverse field, postulated to be due to hot-atom processes, is much larger for LiF than for KCl. This might be expected, since Li and F are so much lighter than K or Cl; however, the "hot fraction" (if such it is) is smaller in NaCl than in KCl. In LiF, one observes a fairly short transverse relaxation time T_2, which decreases with decreasing temperature. A "motional narrowing" effect may be suggested. Uncertainties as to the muon's whereabouts following epithermal processes in these crystals may be alleviated by the fact that similar transverse field results are seen in LiF and LiH, which we may regard as an alkali halide of sorts. The analogy is especially good since the magnetic properties of ^{1}H and ^{19}F nuclei are nearly identical. Since the μ^+ surely replaces protons in LiH, one is tempted to conjecture that the epithermal process in LiF is a substitution of μ^+ for Li^+ or Mu^- for F^- at a lattice site. It is expected that further experiments, including new longitudinal-field studies, will greatly clarify the processes at work in these crystals.

It is evident that a careful analysis of μ^+ depolarization in the alkali halides leads one to ask important questions about the interactions of impurities with the crystal field. Perhaps, as further measurements provide answers to these questions, muonium will help us understand more about the properties of the crystal itself.

2. *Ferromagnetic and Antiferromagnetic Insulators*

So far, we have only considered muonium in a nonmagnetic insulator. Some particularly nice effects can be expected for muonium in a ferromagnetic or antiferromagnetic insulator (Siegmann *et al.*, 1971; Ivanter, 1973). The muonium electron will be coupled by an exchange interaction to ferromagnetically ordered spins of the sample; not so, however, the muon. The relevant Hamiltonian would thus be of the form

$$H = A\mathbf{S}_\mu \cdot \mathbf{S}_e + (g_e \mu_0{}^e \mathbf{B}_{loc} + \sum_i J_i \mathbf{S}_i) \cdot \mathbf{S}_e + g_\mu \mu_0{}^\mu \mathbf{B}_{loc} \cdot \mathbf{S}_\mu \qquad (129)$$

with J_i an exchange integral of the muonium electron with a neighbor electron, labeled i, whose spin is S_i. The electron may then see an effective magnetic field that differs dramatically from the field that acts on the muon. An interesting mechanism for relaxation of the muonium electron is also possible in ferromagnetic insulators—the absorption or emission of magnons, the quanta of ferromagnetic spin waves. The measurement of relaxation times in these substances might lead, for instance, to a determination of magnon scattering cross sections with muonium (Siegmann *et al.*, 1971). Further, the presence of a majority spin on a ferromagnet may lead to unequal initial population of the $^1S_0$ and $^3S_1$ states of muonium. The resultant time dependence of the μ^+ polarization could be very exotic.

So far, very little data exist for muonium in ferro- or antiferromagnetic insulators. It has been noted that both the residual polarization and the transverse relaxation time of the quasi-free μ^+ precession signal in Cr_2O_3 drop dramatically when the powdered sample is cooled below its Néel temperature (307°K). Precession at frequencies other than the free μ^+ Larmor frequency is evidently absent in Cr_2O_3 in a finite external field, but the available data is very limited, and the observation of μ^+ precession in coherent internal fields in Ni and Fe (Foy *et al.*, 1973; Patterson *et al.*, 1974) encourages further investigation.

B. Muonium in Semiconductors

The study of muons and muonium in semiconductors has been a long, interesting, and fruitful undertaking, which still has only just begun. Much elegant experimental and theoretical work has been done in the last 15 years in an attempt to understand the general features of the behavior of Mu in silicon and germanium, to which most of our discussion will be limited. However, recent results have relieved much of the confusion and ambiguity that plagued the early work, and we will present the data in the context of what is now known to occur, at the risk of some injustice to those who opened up this frontier.

1. *Deep-Donor Muonium in Germanium and Silicon*

Perhaps the most illuminating recent advance in this field has been the detection of "two-frequency Mu precession" in germanium (Gurevich *et al.*, 1971a) and p-type silicon (Crowe *et al.*, 1972b; Brewer *et al.*, 1973) at 77°K. These observations not only firmly established the existence of long-lived interstitial muonium atoms in these crystals, but also provided measurements of the hyperfine frequency of the Mu atom in the interstitial site [recall Eq. (20)]. The results were

$$\omega_0(\mathrm{Ge})/\omega_0(\mathrm{vac}) = 0.56 \pm 0.01 \tag{130}$$

and

$$\omega_0(\mathrm{Si})/\omega_0(\mathrm{vac}) = 0.45 \pm 0.02 \tag{131}$$

The latter result is in agreement with that of Andrianov *et al.* (1970), who studied the quenching of the depolarization in longitudinal field for a mildly p-type single crystal of silicon at 300°K. Fitting the field dependence to Eq. (13), they found

$$\omega_0(\mathrm{Si})/\omega_0(\mathrm{vac}) = 0.405 \pm 0.026 \tag{132}$$

This technique is comparable in accuracy with the two-frequency precession method, but is not nearly as unambiguous. In an earlier longitudinal-field study by Eisenstein *et al.* (1966), the field dependence of the polarization in a mildly p-type Si crystal at 300°K was consistent with that observed by Andrianov *et al.*, but at temperatures $\leq$77°K a completely different behavior was seen. The field dependence in n-type Si was also quite different. We will return to this point later.

Recalling Eq. (126), we see that these Mu atoms have a radius about 1.2 times that of muonium in vacuum, $r_0(\mathrm{vac}) = 0.532$ Å; their dimensions are still much smaller than one lattice parameter ($\sim$5 Å). Interstitial muonium is therefore concluded to be a "deep donor" in Si and Ge—that is, the whole atom fits into one interstitial site and has a binding energy of several electron volts. Wang and Kittel (1973) have explained the magnitude of the reduction of ω_0, as well as the small difference between silicon and germanium, in terms of known properties of the crystals. In their model, the potential function for the bound electron is cut off at large radii due to screening by the valence band electrons of the neighboring silicon atoms.

2. *Anomalous Muon Precession in Silicon*

A further clarification has resulted from the observation of a second type of two-frequency precession in silicon, corresponding to a much weaker hyperfine coupling. This phenomenon has been labeled "anomalous muon precession," for lack of a positive identification of its source; like deep-donor Mu precession, it is observed only in cold p-type silicon crystals.

The square of the Fourier transform of an experimental time histogram (corrected for background and μ decay) yields a power spectrum of precession frequencies, in which two-frequency precession is manifested as a pair of lines. Figure 28 shows a comparison between such Fourier spectra for silicon and fused quartz in the same field, demonstrating the absence of anomalous precession in quartz. Whereas the deep-donor muonium frequencies rise approximately linearly with field up to a few hundred gauss, and are independent of the orientation of the crystal in the field, the anomalous frequencies were found to have the field dependence shown in Fig. 29,

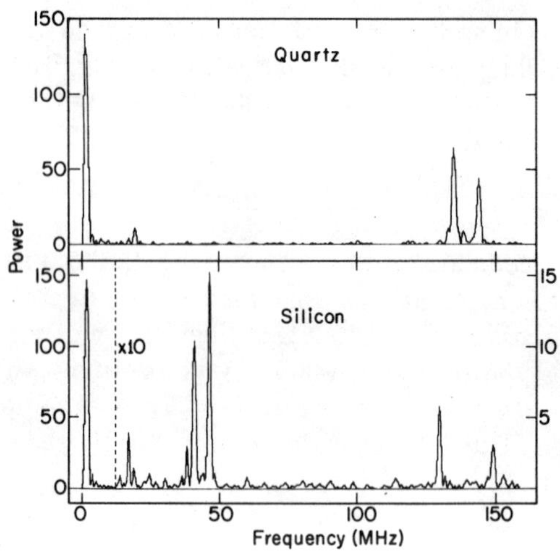

FIG. 28. Frequency spectra (square of the Fourier amplitudes, arbitrary units) of muons in fused quartz at room temperature and in p-type silicon at 77°K. In both cases the applied transverse field is 100 G. The prominent peaks (from left to right) are: the free muon precession signal at 1.36 MHz; a characteristic background signal at 19.2 MHz, due to rf structure in the cyclotron beam; the two anomalous frequencies at 43.6 ± 2.9 MHz (silicon only); and the two 1s muonium peaks centered about 139 MHz. The wider splitting of the two 1s muonium lines in silicon is due to the weaker hyperfine coupling (from Brewer *et al.*, 1973a).

and are slightly anisotropic, as indicated. Deep-donor Mu precession and anomalous precession have lifetimes of about 500 nsec.

The field dependence of the anomalous frequencies is much stronger than that of the free μ^+ precession frequency in weak fields. The muon must therefore be coupled, as in muonium, to a particle or system with a larger magnetic moment than its own. The field dependence of the anomalous frequencies can in fact be fitted to that of transition frequencies ω_{12} and ω_{34} in a modified version of the Breit–Rabi energy levels (9), if the different crystal orientations are treated as separate cases. This can be seen qualitatively from Fig. 7. [As is implicit in the transverse field selection rules (15) and explicit in the equation of motion (19), frequencies ω_{12} and ω_{34} should be dominant when $B \gg B_0$.] However, it is necessary to allow both the hyperfine coupling strength and the g-factor of the electron to vary in order to obtain a fit. For the case of the [111] crystal axis parallel to the field, the best value for $\omega_0/\omega_0(\text{vac})$ is 0.0198 ± 0.0002; for [100] parallel to the field, the best value is $\omega_0/\omega_0(\text{vac}) = 0.0205 \pm 0.0003$. In both cases the best value for g_e is 13 ± 3. Clearly, the spin g-factor of an electron cannot be much different from 2, nor can a pure contact interaction be anisotropic;

this modified Breit–Rabi description is meant only as a phenomenological characterization of the data.

These results can be interpreted in terms of a number of physical models (Brewer *et al.*, 1973). Perhaps the most attractive is shallow-donor muonium. Here the electron wave function is spread over many lattice sites, whereas the entire deep-donor muonium atom fits into one interstitial site. An s-state cannot produce the observed behavior, due to the relatively invariable spin g-factor of the electron. However, in an $l \neq 0$ state the orbital g-factor can be large and anisotropic: the electron wave function for a shallow donor must be a superposition of conduction band states, which may have small anisotropic effective masses. If the spin–orbit coupling for the electron is large, j_e becomes a good quantum number, and $\mathbf{J}_e$ formally replaces $\frac{1}{2}\boldsymbol{\sigma}_e = \mathbf{S}_e$ in the Breit–Rabi Hamiltonian. For $j_e = \frac{1}{2}$ the observed field dependence can easily be reproduced. [It should be noted that other components could be formed with $j_e > \frac{1}{2}$ that might precess at unobserved frequencies or be quickly relaxed (e.g., by transitions to the $j_e = \frac{1}{2}$ level).] We conclude that the postulated shallow-donor muonium state must contain substantial

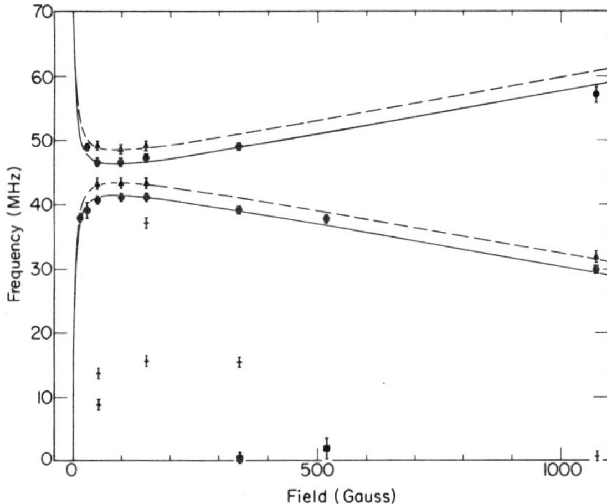

FIG. 29. Dependence of anomalous frequencies in silicon on field strength and crystal orientation. Round points and solid lines are data and best fit for [111] crystal axis along the field; triangular points and dashed lines are data and best fits for [100] axis along the field. Free muon, 1s muonium, and cyclotron background signals are not shown. A number of peaks appear in the spectra in addition to the fitted "proper" anomalous frequencies; these are unexplained. They are indicated by square points (for prominent peaks) and horizontal bars (for weak or questionable peaks). The higher of the "proper" anomalous frequencies is missing at several fields. This is because the spectra showed no statistically significant peaks at those positions.

admixtures of $l \neq 0$ excited vacuum states. The stability of this state (lifetime ~500 nsec) against radiative transitions to the deep-donor 1s state can be explained by the small overlap between shallow- and deep-donor electron wave functions.

A conventional phenomenological description for such behavior is provided by the "effective spin Hamiltonian" formalism often used in ESR work on paramagnetic impurities:

$$H = \mathbf{J}_e \cdot \mathsf{A} \cdot \mathbf{S}_\mu + \mu_0{}^e \mathbf{B} \cdot \mathsf{g}_e \cdot \mathbf{J}_e + \mu_0{}^\mu g_\mu \mathbf{S}_\mu \cdot \mathbf{B} \tag{133}$$

where A and g_e are now tensors and $\mathbf{J}_e$ an effective spin. By adopting this phenomenological Hamiltonian, one can consistently describe quite complicated ESR patterns and their dependence on the orientation of the crystal in the field. A good fit to the data is obtained assuming a scalar g_e and a minimal anisotropy with symmetry about the [111] axis for A, which has then only two independent nonzero elements: $A_{33} = A_{||}$ and $A_{11} = A_{22} = A_\perp$. Postulating $j_e = \frac{1}{2}$, we can express the results as follows:

$$(\mathsf{g}_e)_{ij} = \delta_{ij} \times (13 \pm 3) \tag{134}$$

$$A_{||} = (0.0198 \pm 0.0002) A_0(\text{vac}) \tag{135}$$

and

$$A_\perp = (1.035 \pm 0.02) A_{||} \tag{136}$$

where $A_0(\text{vac})$ is the hyperfine coupling of muonium in vacuum.

Whether the "anomalous precession" in silicon is actually due to shallow-donor muonium or to some other spin system (Brewer *et al.*, 1973a) is not known; however, its behavior is precisely that of a muoniumlike object with $g_e \approx 13$ and $\omega_0 \approx \omega_0(\text{vac})/50$, so we will henceforth refer to "shallow-donor Mu precession" for practical and mnemonic reasons.

3. *Tentative Model*

In all the silicon and germanium samples studied, a quasi-free μ^+ precession signal was observed in transverse field. The asymmetry and relaxation time of this signal vary dramatically with temperature and doping concentration. In the case of cold mildly p-type Si, all three signals (μ^+, deep-donor Mu, and shallow-donor Mu) are present simultaneously. Since all these motions are out of phase with each other within a few nanoseconds, each must represent an independent component of the muon ensemble, starting out simultaneously (within ~10^{-10} sec) as distinct products of the thermalization process. The measured asymmetries in these signals only account for about $\frac{2}{3}$ of the muon polarization; the missing $\frac{1}{3}$ must represent a fraction of the ensemble that relaxes within a few nanoseconds of thermalization or precesses at frequencies too high to be resolved. An unpublished observation of a short-lived ($T_2 \sim 30$ nsec) component in the μ^+ precession signal at 4400 G (p-type Si at 77°K) suggests the following hypothesis: some fourth fraction of the muon ensemble thermalizes as a free μ^+, which soon (but not immediately) captures an electron to form some muonium state. The resultant Mu atoms are out of phase in their subsequent precession and appear depolarized. The long-lived μ^+ signal must represent a fraction of muons that thermalize into fundamentally different circumstances and are immune to the process postulated above.

If we allow ourselves these conjectures, the situation in p-type Si at 77°K can be summarized as follows: a fraction f_+ of the muons thermalizes as free μ^+ and waits in interstitial sites for free electrons to come by, which they promptly capture to form some Mu states. Another fraction f_{dd} thermalizes as deep-donor Mu, and a third fraction f_{sd} thermalizes as shallow-donor Mu. The remaining muons (excluding any channels we may have neglected) undergo some unknown epithermal or otherwise very fast process that places them in a stable diamagnetic environment; we retain the symbol h for this fraction, although a "chemical" interaction is not necessarily indicated. This situation is pictured diagrammatically in Fig. 30.

Such a model is consistent with observations for p-type Si at 77°K, but how does it work for other Si and Ge crystals at different temperatures and with different dopings? Let us for instance consider the longitudinal-field data on p-type Si at 300°K. Andrianov *et al.* (1970) found that the relation-

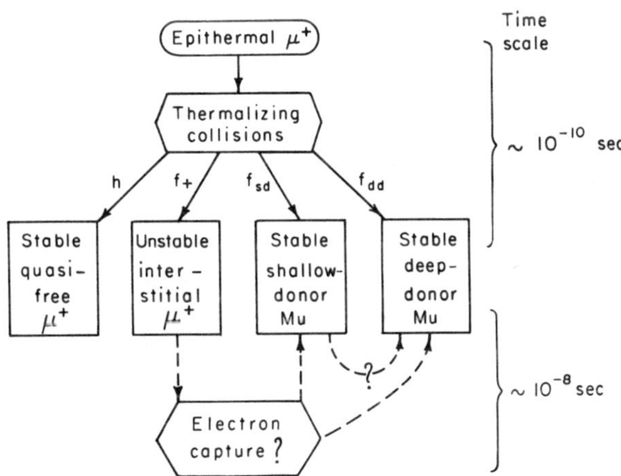

FIG. 30. Hypothetical comprehensive model of the possible fates of muons in mildly p-type silicon crystal at 77°K.

ship between the zero-field polarization $P_{||}(0)$ and the polarization $P_{\perp}$ of the long-lived μ^+-precession component in transverse field was $P_{||}(0) = P_{\perp} + \frac{1}{2}(1 - P_{\perp})$, within about 2%. Assuming that the f_+ fraction is not much more stable at 300°K than at 77°K, we may equate $P_{\perp}$ with h. Thus the fraction $(1 - h)$ that thermalizes in states other than stable quasi-free μ^+ has a polarization of $\frac{1}{2}$ in zero field (averaged over any hyperfine oscillations)—consistent with the assumption that it all forms muonium. The fraction f_+ should be included in this component, since its originally free interstitial muons should quickly capture electrons. (In longitudinal field there are no problems with "dephasing" due to statistically distributed formation times.) All of the polarization is thus accounted for: $P_{\perp} = h$ and $h + f_+ + f_{sd} + f_{dd} \approx 1$.

When a weak longitudinal field is applied, any stable fraction f_{sd}, with its weak hyperfine coupling, should be quickly "repolarized"; no such phenomenon is observed experimentally. In fact, as mentioned earlier, the field dependence is entirely consistent with the assumption that muonium ends up *only* in the deep-donor state at 300°K. One might then expect to see a large deep-donor Mu precession signal in transverse field; in fact, no such signal has been detected, nor is any shallow-donor Mu signal visible at 300°K. It has been fashionable to explain this absence in terms of a rapid relaxation ($\nu \neq 0$) or charge exchange ($\mathrm{Mu} \rightleftarrows \mu^+$) process; however, in this case such a hypothesis is not plausible, due to the behavior in longitudinal field. As discussed for the case of alkali halides, relaxation or charge exchange will lead to complete depolarization even in longitudinal field unless a "chemical" process permanently removes the muon to a diamagnetic environment at a rate $1/\tau_m \sim \nu$. Because no such "extra" depolarization is observed, we are forced to conclude that $1/\tau_m \gg \nu$; the disappearance of any transverse-field Mu precession signal must then be due to "chemical" reactions within a few nanoseconds, and not to relaxation or charge exchange at all. It seems much more plausible in this case to assume that nearly all of the fraction $(1 - h)$ forms initially in the free μ^+ and shallow-donor Mu states, which then undergo transitions to the deep-donor Mu ground state within a few nanoseconds.

Obviously, these arguments cannot be carried much further into the realm of conjecture until more experimental evidence is available. The desirability of an in-depth study of mildly p-type silicon is obvious, for not until one case is thoroughly understood can credible extensions be made to others. However, we are compelled to mention some of these other cases in an attempt to put this field in perspective.

Feher *et al.* (1960) and Eisenstein *et al.* (1966) studied a variety of silicon crystals with various dopings in longitudinal field at 300, 77, and

$\sim$10°K. They also studied the muon polarization in transverse field, but did not watch the precession in the sense of Section II. In both cases a gated scaling technique was used that effectively yielded the *average* polarization over 1 to 4 μsec. Their polarization measurements are thus subject to serious underestimation when even slow relaxation is present. Nevertheless, a few interesting conclusions can be drawn from their data.

In each of their cold ($\leq$77°K) weakly doped silicon samples, Eisenstein *et al.* measured a zero-field polarization $P_{\perp}(0) < 0.5$, considerably smaller than that observed in the same sample at 300°K. The missing polarization was restored by weak longitudinal fields ($\sim$50 G). The low-field "quenching" effect would be expected in p-type Si due to the formation of shallow-donor Mu ($B_0 = 32 \pm 1$ G), but a value of $P_{\parallel}(0) < 0.5$ would not. This is probably due to RLMF from ^{29}Si nuclei (recall the relaxation of Mu precession in quartz): if the measured $\sim$500-nsec lifetime of Mu precession in p-type Si at 77°K is due to local fields of about 1 G, the same relaxation rate would be seen in zero field; such a relaxing signal is averaged over 4 μsec in Eisenstein's technique, yielding a misleading value for $P_{\parallel}(0)$. Thus the low-field quenching effect in cold n-type samples is not necessarily evidence for shallow-donor muonium.

For increasingly n-type Si samples, more and more polarization is lost, even in strong longitudinal fields, up to the point at which electron wave functions from adjacent donors begin to overlap, producing an impurity conduction band. These silicon samples are effectively metallic, and the muons are not depolarized significantly. This behavior is entirely consistent with a model in which muonium electrons spin-exchange with conduction band electrons, producing a relaxation rate ν that increases slowly with the density of conduction band electrons until the silicon goes metallic, at which point $\nu \gg \omega_0$ and the muonium electron is effectively decoupled from the muon. One may ask whether muonium can be said to exist any more at this stage; the distinction is not entirely academic. In some metals, for instance, long-range screening of the μ^+ by conduction electrons will prevent formation of an atomlike charge density about the muon; in others, the interstitial μ^+ may have a very localized screening charge distribution similar to a Mu atom; and in some cases, the muon may acquire a screening charge distribution rather like that of a Mu$^-$ ion (Friedel, 1958). In any case one can think of the μ^+ with the screening charge as a sort of "collective muonium atom," whether deep- or shallow-donor, in which $\nu \gg \omega_0$.

With highly p-type Si, not much polarization is lost even in zero field; most is accounted for in the fraction $(h + f_+)$ observed as μ^+ precession in transverse field. Here one might expect that the depletion of available con-

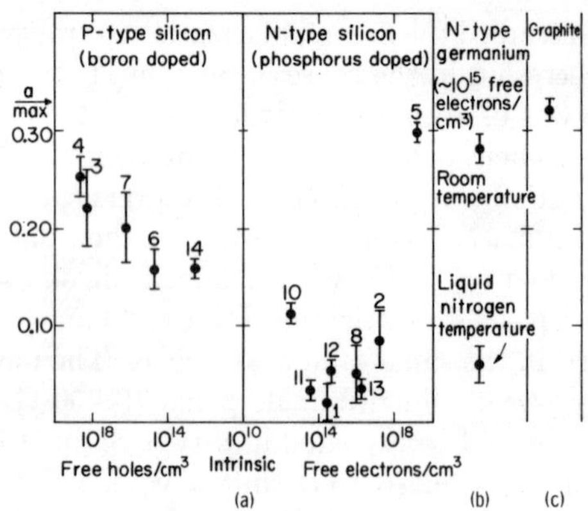

FIG. 31. Experimental values of a, the μ^+ asymmetry parameter: (a) versus free-electron concentration in n-type silicon and free-hole concentration in p-type silicon at room temperature; (b) in one sample of n-type germanium (phosphorus-doped) at room temperature and liquid-nitrogen temperature; and (c) in a graphite sample for which the maximum value of $a = 0.33$ is assumed to correspond to full muon polarization. The abscissas for n-type and p-type silicon have been joined at the value of the intrinsic concentration for room temperature ($\sim 10^{10}$ cm^{-3}). Since the product of the numbers of free holes and electrons in thermal equilibrium with the lattice is constant at a given temperature (i.e., $\sim 10^{20}$ for silicon at room temperature), the entire abscissa represents an increasing free electron concentration to the right (or an increasing hole concentration to the left) (from Feher *et al.*, 1960).

duction band electrons would make the fraction f_+ more stable and perhaps even inhibit formation of the fractions f_{dd} and f_{sd}. If this is the case, then virtually all the variety evident in μ^+ behavior in silicon could be expressed as a function of conduction electron concentration.

The transverse-field results of Feher *et al.* (1960) for μ^+ in silicon samples at room temperature with various dopings are summarized in Fig. 31. Additional data of Eisenstein *et al.* (1966) are given in Table 10.

4. *Germanium*

The behavior of muons in germanium crystals appears to be very similar to that observed in silicon. As mentioned earlier, deep-donor Mu precession has been studied in Ge at 77°K, and has nearly the same properties as that observed in Si. Attempts to detect shallow-donor Mu precession in Ge crystals have so far been unsuccessful.

Andrianov *et al.* (1969) made a very nice study of the μ^+-precession signal in Ge over a range of temperatures from 77 to 360°K, and found that both the initial asymmetry and the relaxation rate were smooth functions

TABLE 10

RESIDUAL POLARIZATION OF POSITIVE MUONS STOPPED IN SAMPLES OF SILICON AT VARIOUS DOPINGS AND TEMPERATURES AND IN OTHER TARGETS[a]

Sample	Resistivity (Ω cm)	Phosphorus donors (per cm³)	Boron acceptors (per cm³)	Residual polarization 300°K	77°K	4.2–10°K
Silicon:						
No. 4	0.05		4×10^{18}	0.90 ± 0.01	0.67 ± 0.08	0.45 ± 0.01
No. 14	3000		3×10^{12}	0.52 ± 0.05	0.24 ± 0.03	0.24 ± 0.08
No. 10	350	4×10^{12}		0.45 ± 0.08	0.15 ± 0.04	0.19 ± 0.05
No. 11	50	4×10^{13}		0.07 ± 0.03	0.10 ± 0.03	0.15 ± 0.03
No. 13	0.3	2×10^{16}		0.09 ± 0.03	0.12 ± 0.03	0.16 ± 0.04
No. 15	0.03	3×10^{17}		0.10 ± 0.05		
No. 16	0.01	1.5×10^{18}		1.00 ± 0.15		
No. 5	0.003	1.5×10^{19}		1.07 ± 0.10	0.92 ± 0.09	0.89 ± 0.10
Germanium		10^{15}		0.92 ± 0.08	0.23 ± 0.07	0.18 ± 0.07
Alumina				0.24 ± 0.16	0.24 ± 0.16	0.12 ± 0.10
Sulfur				0.14 ± 0.01; 0.00 ± 0.02[b]	0.22 ± 0.01	0.20 ± 0.02
LiF				0.52 ± 0.12[b]		
MgO				0.44 ± 0.12[b]		
Red P				0.07 ± 0.14[b]		
Black P				0.72 ± 0.16[b]		

[a] From Eisenstein *et al.* (1966). Measurements were made using a gated scaling technique in transverse field. Except where indicated, data were taken with the samples in a cryostat; corrections have been made only for muons stopping in the walls of the empty cryostat. Residual polarization is expressed relative to aluminum, in which a muon decay asymmetry of $A_0 = 0.27\pm0.01$ was measured.

[b] Indicates measurement outside cryostat, other measurements made with sample in cryostat.

of temperature, as can be seen in Fig. 32. In a metallic n-type Ge crystal, virtually no depolarization was seen, either in the initial polarization or in its time dependence, at any temperature. In a moderately p-type Ge crystal at 77°K, all the polarization can be accounted for in the μ^+ precession asymmetry and the Mu precession asymmetry measured in low field (7 G); the μ^+ precession has no detectable relaxation. As the temperature is raised past ~200°K, the asymmetry and relaxation rate of the μ^+ signal begin to rise simultaneously, until at 360°K it accounts for all of the polarization, and relaxes with a lifetime $T_2 \approx 3.5$ μsec.

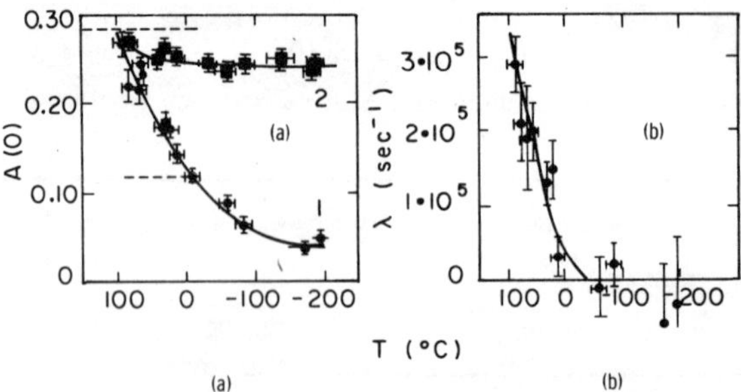

FIG. 32. (a): temperature dependence of the initial experimental asymmetry of the μ^+-precession signal in a transverse field $A(0)$, for undoped Ge single crystals (curve 1) and for As-doped Ge (curve 2). The upper dashed line represents the maximal asymmetry A_0; the lower dashed line corresponds to the residual asymmetry $A_0 \times R_\perp$ measured at low field. (b): temperature dependence of the depolarization rate ($\lambda = 1/T_2$) of the μ^+ in undoped Ge (from Andrianov *et al.*, 1969).

This behavior is consistent with that observed in similarly doped silicon, and suggests a growth of fractions f_+ and h at the expense of fractions f_{dd} and f_{sd}. Andrianov *et al.* suggest that the scattering of electrons by phonons may inhibit muonium formation, or that thermal ionization of Mu may become important at high temperatures. The latter hypothesis is questionable, since even temporary ($>10^{-10}$ sec) muonium formation leads to depolarization of the μ^+-precession signal in transverse field, and if repeated formation and ionization (i.e., charge exchange) takes place, all of the polarization will be quickly lost. It seems unlikely that the μ^+ signal is due to an enhanced fraction f_+, since we have already postulated a rapid relaxation of free interstitial μ^+ precession in cold p-type Si due to capture of conduction electrons; at the same temperature, conduction electrons might be expected to be much more available in Ge, with its smaller band gap. The increase of the relaxation rate with temperature could be due to so many different (and contradictory) mechanisms that we must conclude that it is not understood.

5. *Summary*

We can only reiterate that the study of Mu and μ^+ in semiconductors is by no means finished. Enough evidence has been gathered to suggest that most of the differentiation of the muon ensemble into various components takes place during thermalization, as pictured in Fig. 30, but this model is still subject to great confusion. Some fairly straightforward experiments should suffice to clear up many uncertainties.

The eventual value of a more complete understanding of the behavior of muons in semiconductors cannot be assessed in advance, but the general problem of impurity states is of obvious interest. The most elementary natural impurity is hydrogen, of course, but searches for atomic hydrogen in Si and Ge have yielded negative results, even though H is known to diffuse freely through these crystals. Observations of deep-donor Mu have thus helped to clarify the status of hydrogenlike interstitial impurities in Si and Ge; Wang and Kittel (1973) concluded that more is known about muonium than about H or H_2 in these crystals upon which most of modern solid-state electronics technology depends.

IX. Quasi-Free Muon Precession and Slow Depolarization

In this section we will discuss processes involving muons that end up in diamagnetic environments with their initial polarization unaffected by coherent hyperfine interactions. Examples are the quasi-free μ^+ component h produced in hot-atom reactions, and muons in metals, where the screening of the μ^+ charge is accomplished by the collective motion of conduction electrons rather than by a single muonium electron. In the former case the muon is actually part of a diamagnetic molecule, while in the latter case one can think of the μ^+ as being in a muonium atom with $\nu \gg \omega_0$, so that the hyperfine coupling is "broken." In either case, the μ^+ loses no polarization in reaching its ultimate environment, and any subsequent relaxation can be observed directly.

The behavior of quasi-free muons in condensed matter has been studied principally by means of their precession in a transverse magnetic field. Therefore only this case will be considered here. We will deal first with quasi-free muons in solids and then with muons in paramagnetic solutions.

A. Muons in Solids

1. *Muon Precession in Local Fields*

In a field $\mathbf{B}$, the precession frequency $\boldsymbol{\omega}^\mu$ of a free μ^+ is given by Eq. (5), which can be written

$$\boldsymbol{\omega}^\mu = -\gamma_\mu \mathbf{B}$$

where

$$\gamma_\mu \equiv |g_\mu| \mu_0{}^\mu/\hbar = 0.85 \times 10^5 \text{ rad/sec-G} \tag{137}$$

We consider several cases in which $\mathbf{B}$ is not homogeneous throughout the stopping target:

(a) Assume that the field $\mathbf{B}$ seen by a given muon is constant in time, parallel to the external field ($\mathbf{B} = B\hat{z}$), and has a limited number of discrete magnitudes B_i. One must then replace the cosine in the rate distribution formula (6) by a sum over cosines with different discrete frequencies $\omega_i^\mu = \gamma_\mu B_i$:

$$\cos(\omega^\mu t + \theta_0) \to \sum_i P_i \cos(\omega_i^\mu t + \theta_0) \tag{138}$$

where P_i is the fraction of muons that precess at frequency ω_i^μ. This will obviously result in complicated beat phenomena in the precession pattern.

(b) Next let us assume that the fields seen by individual muons are still constant in time and parallel to the external field, but have a continuous distribution of magnitudes $\mathfrak{D}(B)$, implying a frequency distribution $f(\omega) \equiv (1/\gamma_\mu)\mathfrak{D}(B)$. The cosine in Eq. (6) must then be replaced by an integral:

$$\cos(\omega^\mu t + \theta_0) \to \int_0^\infty d\omega\, f(\omega)\, \cos(\omega t + \theta_0) \tag{139}$$

The frequency spectrum $f(\omega)$ is a distribution function of the probability density for finding muons precessing with frequency ω; in an NMR experiment it would be directly observable as the NMR line shape. Thus the μ^+ precession pattern is simply the Fourier transform of the line shape that would be measured if one could perform a conventional (macroscopic power absorption) NMR experiment with stopped muons (Abragam, 1970). A muon technique—effectively "trigger" detection of magnetic resonance—used by Coffin *et al.* (1958) in measuring the muon's magnetic moment actually yielded the line shape directly.

The integral in Eq. (139) can be expressed in the form

$$\int_0^\infty d\omega\, f(\omega)\, \cos(\omega t + \theta_0) = F(t)\, \cos(\overline{\omega^\mu} t + \theta_0) \tag{140}$$

When the variance ΔB of the local field is small compared with the average local field $B_\mu = \overline{\omega^\mu}/\gamma_\mu$, the envelope function $F(t)$ can often be approximated by a Gaussian or exponential decay with relaxation time T_2. Thus $F(t)$ describes a slow relaxation and can be identified with $A(t)/A(0)$ as introduced in Eq. (6).

The field distribution $\mathfrak{D}(B)$ often represents an average over a spatial field distribution $\mathfrak{D}(B, \mathbf{r})$ weighted according to the spatial muon distribution $\rho_\mu(\mathbf{r})$:

$$\mathfrak{D}(B) = \int_{\text{target volume}} d^3r\, \rho_\mu(\mathbf{r})\mathfrak{D}(B, \mathbf{r})$$

In the case where the field strength is a simple function of position, $B(\mathbf{r})$, one can write $\mathfrak{D}(B, \mathbf{r}) = \delta[B - B(\mathbf{r})]$ (Dirac delta function); if, in addition, the muons are uniformly distributed throughout the target volume, $\rho_\mu(\mathbf{r}) = 1$ and one can rewrite substitution (139) as

$$\cos(\omega^\mu t + \theta_0) \to \int_{\text{target volume}} dr^3 \cos[\gamma_\mu B(\mathbf{r})t + \theta_0]$$

This description is expected to apply in the case of the "fluxoid lattice" in type II superconductors (Ivanter and Smilga 1969b). Since the observed time dependence is just the Fourier transform of the field distribution, a μ^+-precession experiment can be used in much the same way as γ–γ angular correlation measurements (Alonso and Grodzins, 1968) to study the fluxoid lattice.

(c) If the *direction* of the constant local field seen by a given μ^+ is not fixed, the situation can become quite complicated. However, some qualitative features are evident: the *frequency* of precession does not depend on the orientation, but only on the strength of the local field; the apparent *amplitude* and the initial *phase* do, however, depend critically on the relative orientations of the field, the initial muon polarization, and the observation direction (the axis of symmetry of the positron counter telescope). Thus the net effect of a distribution of field directions will be to reduce the amplitudes and change the phases of different frequency components. If we note that

$$\int_0^\infty d\omega\, f(\omega) \cos(\omega t + \theta_0) = \operatorname{Re} \int_0^\infty d\omega\, f(\omega) \exp[i(\omega t + \theta_0)] \quad (141)$$

we can express these effects as follows. Distributions of field orientations contribute an imaginary part to $f(\omega)$ and reduce the net amplitude:

$$\int_0^\infty d\omega \,|\, f(\omega) \,| \leq 1$$

In the simplest case, an isotropic distribution of field directions, the effect is simply to reduce $f(\omega)$ to $\frac{2}{3}[\cos\theta_0 \exp(-i\theta_0)]$ times its corresponding value when $\mathbf{B} = B\hat{z}$.

(d) Finally, we relax the assumption that the field seen by an individual muon is constant in time. This is most often the result of diffusion of the muons in a medium where the local field varies with position. We will not attempt a quantitative derivation of the consequences of such behavior, but the main qualitative features are obvious: if the muon moves from a position with one local field to a position with a different local field in a time much shorter than the difference between its precession periods in the two fields, it will "see" an adiabatic average field

$$B_\mu \to \int \mathfrak{D}(B) B \, dB \Big/ \int \mathfrak{D}(B) \, dB$$

For somewhat slower diffusion, the behavior is more complicated, but can generally be approximated by Eq. (140), where the damping described by the envelope function $F(t)$ is slower for faster rates of diffusion (generally the consequence of higher temperature). This effect, known in NMR work as motional narrowing (Abragam, 1970), has been observed for muons in copper, where the transverse relaxation time T_2 has a marked temperature dependence (Gurevich *et al.*, 1972).

2. *The Magnetic Field Measured via μ^+ Precession*

The magnetic field $B_\mu = \overline{\omega^\mu}/\gamma_\mu$ determined by a measurement of the mean precession frequency is the average local field experienced by the muon at its site. This local field need not be identical with the applied external field; in ferromagnetic materials, for instance, the external field may have very little net effect. In general, the local field at some position in the crystal can be broken down into the following contributions (Shirley *et al.*, 1968; Hellwege, 1970):

$$\mathbf{B}_\mu = \mathbf{B}_{\text{ext}} + \mathbf{B}_{\text{DM}} + \mathbf{B}_{\text{L}} + \mathbf{B}_{\text{dip}} + \mathbf{B}_{\text{hf}} \tag{142}$$

where $\mathbf{B}_{\text{ext}}$ is the external applied field, $\mathbf{B}_{\text{DM}}$ the demagnetization field, $\mathbf{B}_{\text{L}}$ the Lorentz field, $\mathbf{B}_{\text{dip}}$ the field due to nearby magnetic dipoles, and $\mathbf{B}_{\text{hf}}$ the corrected hyperfine field.

The external field $\mathbf{B}_{\text{ext}}$ and the demagnetization field $\mathbf{B}_{\text{DM}}$ (determined by the geometry and bulk permeability of the sample) describe the familiar macroscopic features of the field inside a sample. The microscopic features of the magnetic field are accounted for in the other terms, which describe the contributions from the immediate neighborhood of the field probe. Following standard practice, we consider for magnetic media a spherical "Lorentz cavity" centered about the probe (muon), and calculate the effects of its surface and volume field sources upon that probe. The imagined sphere should have a diameter of at least several lattice spacings, in order to include all of the important dipole sources in the volume contribution, but should fit within a single domain. The surface contribution is the Lorentz field, $\mathbf{B}_{\text{L}} = (4\pi/3)\mathbf{M}$, where $\mathbf{M}$ is the sample magnetization. The volume contribution $\mathbf{B}_{\text{dip}}$ is just the net field due to all the local magnetic dipoles within the Lorentz cavity. Finally, the corrected hyperfine field $\mathbf{B}_{\text{hf}}$ is the effective field due to contact interactions with polarized electrons.

Let us now examine the various contributions to $\mathbf{B}_\mu$ for a few specific cases.

A. INSULATORS. In a diamagnetic crystal $\mathbf{B}_{DM} + \mathbf{B}_L$ is vanishingly small; contributions to $\mathbf{B}_{dip}$ arise only from nuclear moments (see the example below). In a paramagnetic crystal $\mathbf{B}_{DM} + \mathbf{B}_L$ is still very small and may in many cases be neglected; $\mathbf{B}_{dip}$ will consist of contributions from the various paramagnetic ions inside the cavity. In some cases these contributions will cancel due to the symmetry of the site, leaving $\mathbf{B}_{dip} = 0$. For more details see Narath (1967).

B. NONMAGNETIC METALS. In a metal there is a contribution to $\mathbf{B}_{hf}$ from the contact interaction with conduction electrons, which are polarized by an external magnetic field; this field causes the Knight shift (Abragam, 1970; Narath, 1967). In this case $\mathbf{B}_{hf}$ is given by the expression

$$\mathbf{B}_{hf} = \mathbf{B}_{cep} = K\mathbf{B}_{ext} \tag{143}$$

where K is the Knight shift parameter (or tensor)

$$K = \tfrac{8}{3}\pi\langle|\, u(0)\, |^2\rangle\chi \tag{144}$$

Here $\langle|\, u(0)\, |^2\rangle$ is the conduction electron density at the muon site, averaged over all states at the Fermi level, and χ the Pauli paramagnetic susceptibility (per atom).

Conventional measurements of the Knight shift require the performance of NMR with a metal probe. Due to the skin effect, the rf field will only penetrate into a thin surface region, and it may sometimes be questionable whether one measures the bulk Knight shift of the probe material or some surface properties. With the muon one can measure real bulk Knight shifts (Hutchinson *et al.*, 1963a). However, such studies are tedious, requiring accurate measurement of shifts of $\sim$100 ppm in the muon precession frequency. The only known measurements were made in conjunction with the first high-precision determination of the magnetic moment of the muon (Hutchinson *et al.*, 1963a). The Knight shift at the muons was consistently an order of magnitude smaller than that measured or predicted at the lattice nuclei—not surprisingly, since the μ^+ is presumably located at interstitial sites, where enhancement of conduction electron wave functions is weak (Hutchinson *et al.*, 1963a). Singlet annihilation rates of positrons stopped in metals (also proportional to the electron probability density) are consistent with this explanation. An unexpectedly large positive frequency shift observed in graphite is unexplained; μ^- precession measurements in the same sample (Hutchinson *et al.*, 1963b) yielded the expected frequency.

C. FERROMAGNETIC AND ANTIFERROMAGNETIC METALS. Here $\mathbf{B}_{\text{dip}}$ is the usual sum over the dipole fields of the magnetic ions. This term will disappear if the muon occupies a site with cubic symmetry. The hyperfine field seen by the muon is thought to be decomposed as follows (Kossler, 1973):

$$\mathbf{B}_{\text{hf}} = \mathbf{B}_{\text{cep}} + \mathbf{B}_{\text{vep}} \tag{145}$$

where $\mathbf{B}_{\text{cep}}$ is the above-mentioned contribution due to the conduction electron polarization, and $\mathbf{B}_{\text{vep}}$ a positive field produced by the polarized valence electrons shielding the probe charge in its vicinity. The latter term ($\mathbf{B}_{\text{vep}}$) should be small for the muon. As the ferromagnetism of transition elements like Ni, Fe, Co originates mainly from the conduction electrons, $\mathbf{B}_{\text{cep}}$ may be expected to account for most of $\mathbf{B}_{\text{hf}}$ in these substances.

D. OTHER SOLIDS. Muons may be expected to provide a useful probe for a number of hitherto neglected or poorly understood solids. The μ^+ precession signal in semiconductors, for instance, has been studied in detail, as discussed earlier; however, the state of the quasi-free muon in these crystals is not well enough understood to warrant discussion in this section. The behavior of quasi-free muons in ferromagnetic and antiferromagnetic insulators has not yet been systematically investigated, but preliminary studies of Cr_2O_3 show a marked difference in the μ^+ signal above and below the Néel temperature (see the previous section). Exceptionally small relaxation times of the μ^+ precession signal have been found in sulfur (see discussion at the end of Section VII), silicon (see Section VIII), $GdNO_3 \cdot 6H_2O$, and $Fe(NO_3)_3 \cdot 6H_2O$ (Table 3). None of these results are satisfactorily understood and further investigations are needed.

3. *Examples*

We turn now to descriptions of various recent experiments that can be understood or analyzed in terms of the framework outlined above.

A. SLOW μ^+ DEPOLARIZATION IN A SINGLE CRYSTAL OF GYPSUM. This experiment in gypsum ($CaSO_4 \cdot 2H_2O$) (Schenk and Crowe, 1971) helped to clarify the mechanism for slow μ^+ depolarization in solid crystals. It can be completely understood in terms of muon precession in local fields as described in Section IX,A,1, in perfect analogy with NMR experiments on the protons of the water molecules in the hydrated form of $CaSO_4$. The interpretation is based on the assumption that the quasi-free μ^+ precession signal comes from muons that replace protons in waters of hydration via hot-atom reactions of muonium. Each observable muon thus occupies the lattice site of a proton in one of the two water molecules in the unit cell of the crystal. This assumption draws support from the results in aqueous solutions.

The neighboring proton will create a magnetic dipole field at the site of the muon, given by the expression (Schenk and Crowe, 1971)

$$\delta B_\mu = \pm(\mu_p/r^3)(3\cos^2\theta - 1) \qquad (146)$$

where μ_p is the magnetic moment of the proton, θ the angle between the magnetic moment vector of the proton and the muon–proton radius vector, and r the muon–proton distance $= 1.55$ Å.

Depending on whether the proton spin is parallel or antiparallel to an external field, the dipole field will either add to or subtract from the external field. Since there are two H_2O molecules oriented differently with respect to the crystal axes, one expects up to four different muon precession frequencies, as shown in Fig. 33. ω_2 and ω_3 belong to the first pair; ω_1 and ω_4 belong to the second pair. In addition, the muon will feel the field components due to protons (and perhaps to magnetic impurities) farther away, which will lead to an inhomogeneous broadening of the frequency distribution about each ω_i. From NMR measurements it is inferred that this distribution is Gaussian in shape, with full width $\Delta\omega$.

Taking the Fourier transform of this field or frequency distribution, one obtains the following expression for $F(t)$ [defined as in Eq. (140)]:

$$F(t) = \exp(-t^2/T_2^2)\cos(\tfrac{1}{2}\,\delta\omega_1 t)\cos(\tfrac{1}{2}\,\delta\omega_2 t) \qquad (147)$$

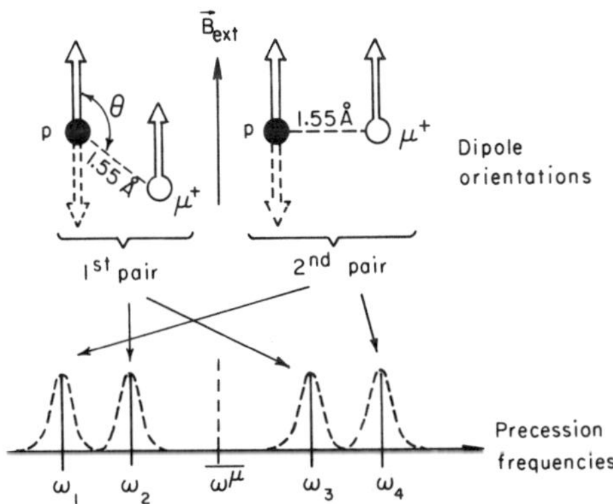

Fig. 33. Muon–proton dipole–dipole interaction in a single crystal of gypsum: schematic representation of the effect of the muon–proton situation relative to the magnetic field direction on the μ^+ spin precession. $\overline{\omega^\mu}$ corresponds to the precession frequency unperturbed by dipole–dipole interactions; it is split by that interaction into two symmetrically shifted frequencies (one per proton spin orientation for each μ^+–p pair. The line broadening produced by the magnetic dipoles farther away is indicated by the dashed curves.

with

$$\delta\omega_1 = \omega_2 - \omega_1 = \omega_4 - \omega_3, \quad \delta\omega_2 = \tfrac{1}{2}(\omega_4 - \omega_1) + \tfrac{1}{2}(\omega_3 - \omega_2), \quad T_2 = 4/\Delta\omega$$

The average frequency $\overline{\omega^\mu}$ [Eq. (140)] is the central frequency

$$\overline{\omega^\mu} = \tfrac{1}{2}(\omega_1 + \omega_4) = \tfrac{1}{2}(\omega_2 + \omega_3)$$

There are actually two beat frequencies, $\delta\omega_1$ and $\delta\omega_2$, and a Gaussian damping function with a relaxation time T_2. The values of the beat frequencies $\delta\omega_1$ and $\delta\omega_2$ depend on the crystal orientation in the external field and can be calculated without difficulty (Pake, 1948).

Figure 34 shows data for $F(t)$ for two different crystal orientations. The solid lines are calculated curves, not fits. The beat behavior as well as the damping are clearly visible. In Fig. 34b the agreement between the data and the calculated curve is rather poor; however, in this case the crystal orientation in the external field was not accurately known.

The points in Fig. 34 were obtained by dividing the experimental histogram into 500-nsec sections and performing a Fourier analysis on each section. This leads to a determination of the amplitude and phase of the precession signal at the central frequency $\overline{\omega^\mu}$. For all crystal orientations the experiment showed that $\tau_\mu < 2/(\delta\omega_1 + \delta\omega_2)$—that is, that the beating effect was not important for the most statistically significant part of the histograms. Thus the analysis could be simplified. Equation (147) can be written:

$$F(t) = \exp(-t^2/T_2^{*2}) \tag{148}$$

with formally

$$\frac{1}{T^{*2}} = \frac{1}{T_2^{\,2}} - \frac{1}{t^2}\ln\left[\cos\left(\frac{\delta\omega_1}{2}t\right)\cos\left(\frac{\delta\omega_2}{2}t\right)\right] \tag{149}$$

For early times ($t \ll 2/\delta\omega_1$ or $2/\delta\omega_2$),

$$\frac{1}{t^2}\ln\left[\cos\left(\frac{\delta\omega_1}{2}t\right)\cos\left(\frac{\delta\omega_2}{2}t\right)\right]$$

is approximately time independent. A value of T_2^* is therefore obtained by fitting Eq. (148) to the measured histogram. Table 11 shows the calculated muon NMR spectra, the calculated T_2^* values, and the corresponding values obtained from the fits for the crystal orientations studied. The line width $\Delta\omega = 4/T_2$ is extracted from the data for the third crystal orientation, where the spectrum effectively consists of only one line, implying $T_2 \approx T_2^*$. The other calculated T_2^* values are in good agreement with the measured ones.

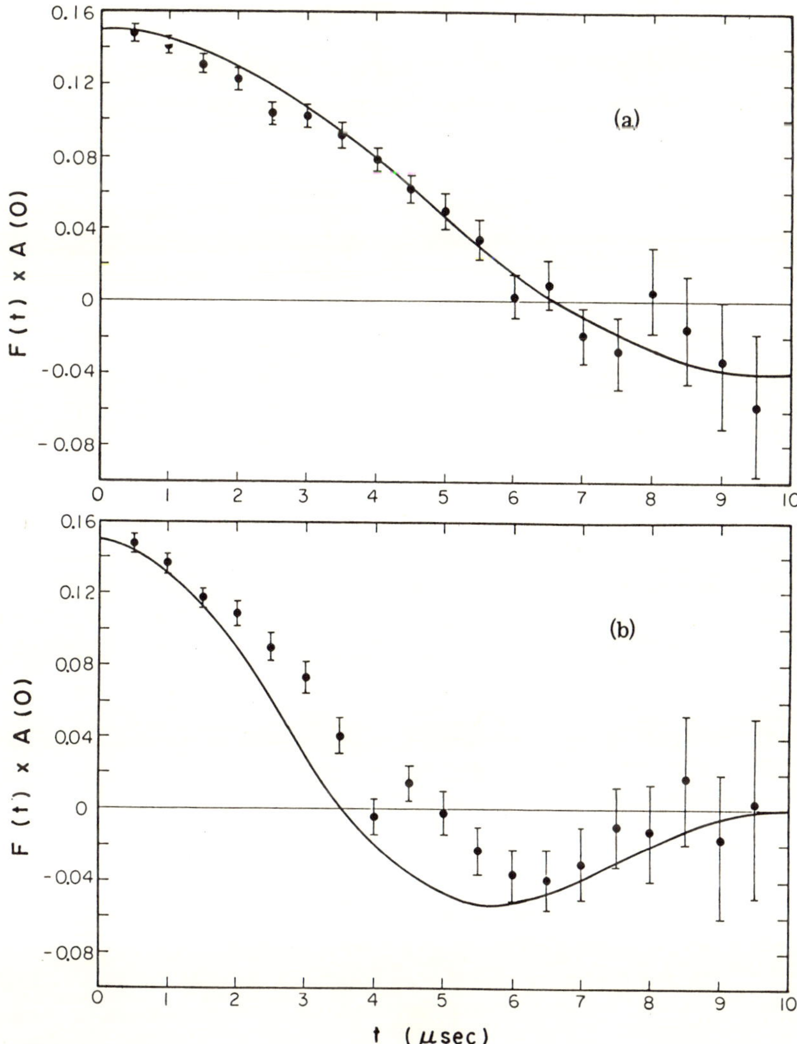

FIG. 34. (a) Observed μ^+ precession signal asymmetry $A(0)F(t)$ for a gypsum crystal orientation showing two NMR lines. The asymmetry values are determined for 0.5-μsec intervals. The solid curve represents the theoretical $A(0)S(t)$ dependence. (b) For a second crystal orientation showing four NMR lines, the agreement between the observed asymmetry values and the calculated curve is marginal, suggesting that the actual crystal orientation is slightly off the assumed one. The determination of these angles is uncertain to ~10° (from Schenck and Crowe, 1971).

TABLE 11

COMPARISON BETWEEN RELAXATION MEASUREMENTS AND CALCULATIONS FOR DIFFERENT ORIENTATIONS OF THE GYPSUM CRYSTAL[a]

Position	Crystal orientation (degrees)	Muon—NMR spectrum (scale = Gauss)	Calculated T_2^* from NMR spectrum	Measured T_2^* (μsec)
1	$\mathbf{B}, \hat{z} = 90$ $\mathbf{B}, \hat{x} = 94$ $\mathbf{B}, \hat{y} = 57.5$	5	5.6	5.50±0.50
2	$\mathbf{B}, \hat{z} = 90$ $\mathbf{B}, \hat{x} = 146$ $\mathbf{B}, \hat{y} = 62.5$	1.8	9.0	10.45±1.00
3	$\mathbf{B}, \hat{z} = 90$ $\mathbf{B}, \hat{x} = 176$ $\mathbf{B}, \hat{y} = 32.5$	0.3	10.50	10.50±1.00
4	$\mathbf{B}, \hat{z} = 0$ $\mathbf{B}, \hat{x} = 90$ $\mathbf{B}, \hat{y} = 90$	5.3	5.3	4.80±0.30
5	$\mathbf{B}, \hat{z} = 40$ $\mathbf{B}, \hat{x} = 142$ $\mathbf{B}, \hat{y} = 111$	ω_1 ω_2 ω_3 ω_4 6.5 10.3 14.2	2.9	3.09±0.20
6	$\mathbf{B}, \hat{z} = 24$ $\mathbf{B}, \hat{x} = 90$ $\mathbf{B}, \hat{y} = 66$	1.2 7.2	5.60	5.50±0.40
Polycrystal powder			5.3	5.30±0.20
anhydrous			—	60.0±20.0

[a] The crystal axes are given in the Onorato convention: (100) = $\hat{x}$, (010) = $\hat{y}$, and $\hat{z} \perp (\hat{x}, \hat{y})$. The measured T_2^* are independent of magnetic field strength.

The consistency of the results of this experiment shows that slow μ^+ depolarization in solids is often a consequence of the local field distribution (dipolar line broadening). In this special case it is further demonstrated that quasi-free muons are indeed most likely to be found in proton sites.

B. SLOW DEPOLARIZATION AND MUON DIFFUSION IN COPPER. In an experiment by Gurevich *et al.* (1972), slow μ^+ depolarization in copper was shown to be due to inhomogeneous local fields from the nuclear magnetic moments. Natural copper consists of the two isotopes ^{63}Cu and ^{65}Cu with abundances of 69.1 and 30.9%, respectively. Each has spin $I = \frac{3}{2}$; the magnetic moments are +2.23 and 2.38 n.m., respectively. Thus the two isotopes have almost identical effects upon the muon.

When the external field is much stronger than the dipole field at the muon site and when the muon's gyromagnetic ratio is much greater than that of the nuclei (in this case, $\gamma_\mu/\gamma_I = 11.7$), $F(t)$ is given by the following expression (Gurevich *et al.*, 1972):

$$F(t) = \exp(-\Omega^2 t^2) \tag{150}$$

where

$$\Omega^2 = \tfrac{1}{6} I(I+1) \sum_{j=1}^{m} \omega_j^2 \tag{151}$$

and

$$\omega_j^2 = (\hbar/r_j^3)\gamma_\mu \gamma_I (1 - 3\cos^2\theta_j) \tag{152}$$

Here $r_j = |\mathbf{r}_j|$ and θ_j are, respectively, the distance from the muon to the jth nucleus and the angle that the vector $\mathbf{r}_j$ makes with respect to the direction of the external field. For a polycrystalline sample this formula must be averaged over all crystal orientations.

In the derivation of Eq. (150) it is assumed that the muon occupies a fixed position. If the muon diffuses through the crystal at a significant rate, $F(t)$ will be modified due to motional narrowing. If $F(t)$ has a Gaussian shape in the absence of diffusion [as in Eq. (150)], the inclusion of motional effects leads (Abragam, 1970) to the form

$$F(t, \tau_c) = \exp\{-2\sigma^2\tau_c^2[\exp(-t/\tau_c) - 1 + t/\tau_c]\} \tag{153}$$

where σ is the relaxation rate in the absence of diffusion, and can be identified with Ω [Eq. (151)] properly averaged over all crystal orientations. The correlation time τ_c represents the average time the muon takes to cross a unit cell by diffusion.

Gurevich *et al.* observed that at 77°K the damping of the muon precession signal had an essentially Gaussian shape; their fit to the data yielded $\sigma = 0.219 \pm 0.008\ \mu\text{sec}^{-1}$. This implies that diffusion of muons at liquid-nitrogen temperature is slow compared with the muon decay rate

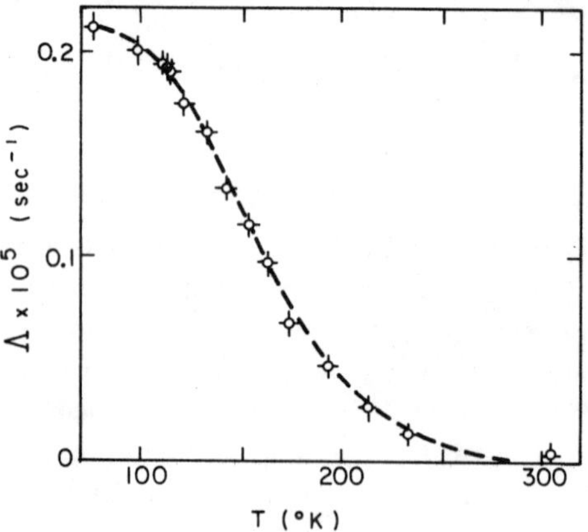

FIG. 35. Temperature dependence of the μ^+ depolarization rate Λ in copper. $\Lambda = 1/\tau_r$, where τ_r is the time at which the measured $F(t)$ had decreased by a factor of e (from Gurevich *et al.*, 1972).

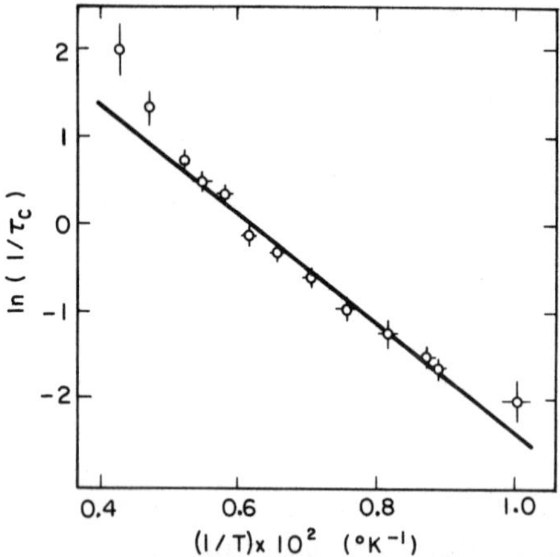

FIG. 36. Dependence $\ln(1/\tau_c) = f(1/T)$ obtained from $F(t)$ measurements (Gurevich *et al.*, 1972) of the μ^+ spin relaxation in copper. In the graph the "hopping time" values τ_c are entered in microseconds. The straight line corresponds to the theoretical relation (154) with $E/k_B = 540°K$.

and is of no importance in the slow depolarization. The measured σ thus gives a good account of the static field distribution from the magnetic moments of the Cu nuclei. However, at higher temperatures the relaxation rate is considerably slower. Figure 35 shows the relaxation rate $\Lambda = 1/\tau_r$ as a function of temperature (τ_r is the time at which the muon polarization had decreased by a factor of e). Only diffusion can account for these results, as the relaxation rate for the spins of the copper nuclei is practically constant over the temperature region investigated. By comparing the experimental damping curves for different temperatures with Eq. (153), using $\sigma = 0.219\ \mu\text{sec}^{-1}$, the correlation time τ_c as a function of temperature was obtained. Figure 36 shows $\ln(1/\tau_c)$ versus $1/T$. The plotted data may be interpolated by a straight line. This dependence is to be expected if the diffusion mechanism is governed by an activation energy E representing the potential barrier between adjacent sites for the muon: the muon is hopping from one crystal site to another. The correlation time should thus be described by

$$\frac{1}{\tau_c} \approx \frac{D_0}{a^2} \exp\left(-\frac{E}{k_B T}\right) \tag{154}$$

where $D_0 \exp(-E/k_B T)$ is the diffusion coefficient and a the lattice parameter of copper. One can interpret D_0 as the vibration frequency of the muon in its temporary interstitial site.

From the slope of the straight line in Fig. 36 one can deduce an activation energy corresponding to $E/k_B = 540°\text{K}$. The deviation from the straight line in Fig. 36 may be due to a temperature dependence of the preexponential factor.

It would be very interesting to compare the diffusion coefficients of protons, deuterons, and tritons with that of muons for the purpose of studying isotope effects in diffusion processes. For instance, it is known that $D_T < D_H < D_D$ for palladium (Sicking, 1972) but no study has yet been made of muons in Pd. The study of positive muons in metals should in fact be regarded as a vital new part of the expanding field of hydrogen in metals (e.g., Sicking, 1972), with all the concomitant implications for such technological problems as corrosion and embrittlement.

C. MUON PRECESSION IN NICKEL AND IRON. Muon precession studies in ferromagnetic materials are particularly interesting inasmuch as they promise to yield new information about the various contributions to the local field. Ferromagnetism is not yet perfectly understood on a microscopic level, and it is likely that the use of the positive muon as a probe may help to clarify some hitherto unanswered questions. Kossler and collaborators (Foy *et al.*, 1973) have observed muon precession in polycrystalline nickel and iron over a temperature region encompassing both ferromagnetic and paramagnetic states. The temperature dependence of the precession frequency and the relaxation rate were studied. Similar results were obtained by Crowe *et al.* (Patterson *et al.*, 1974) for a single crystal of nickel and over a larger range of external fields.

The observed field in *paramagnetic Ni* is essentially equal to the applied external field. No fast depolarization processes seem to affect $P_{\perp\,\text{res}}$, but

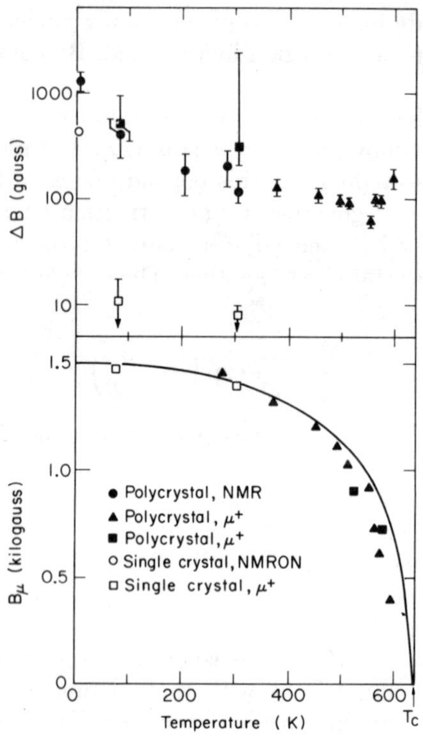

FIG. 37. Local field B_μ at the muon site and line widths ΔB observed in polycrystalline and single-crystal nickel by various techniques. The solid curve is from normalized magnetization data in Ni of Weiss and Forrer (1926).

slow depolarization with a relaxation time of $T_2 \approx 4$ μsec is evident. $P_{\perp \text{res}}$ and B_μ are essentially independent of temperature in the paramagnetic region studied ($T_c = 631°\text{K} < T < 705°\text{K}$) (Foy *et al.*, 1973); however, preliminary studies of T_2 just above T_c show evidence for critical phenomena. No measurements were made in paramagnetic iron ($T > T_c = 1043°\text{K}$).

All the existing measured values of B_μ for an unsaturated sample of *ferromagnetic Ni* are shown as a function of temperature in the lower part of Fig. 37. For each sample no more than a single B_μ value is observed at each temperature. The solid line in the figure is the magnetization curve for Ni (Weiss and Forrer, 1926), normalized to a saturation field of 1500 G at $T = 0°\text{K}$. Thus B_μ has approximately the same temperature dependence as the magnetization of the material. The measured local field is in the same direction as the external field. In order to interpret these results, one must first determine the most probable position of the muon in the unit cell of the crystal. Octahedral and tetrahedral interstitial sites are considered the only likely possibilities, with some indication that the former type would be favored (Foy *et al.*, 1973).

For B_{ext} too weak to saturate the sample, B_μ is found to be essentially independent of B_{ext}. This is due to the high permeability of nickel: magnetic domain walls move in such a way as to screen out the external field, implying that in Eq. (142) $\mathbf{B}_{ext} + \mathbf{B}_{DM} = 0$. Since the dipole fields from Ni cores cancel by symmetry at both octahedral and tetrahedral sites ($B_{dip} = 0$), the only remaining contributions are $\mathbf{B}_\mu = \mathbf{B}_L + \mathbf{B}_{hf}$. The larger contribution in this case is $\mathbf{B}_L = (4\pi/3)\mathbf{M}_s$, where $M_s = |\mathbf{M}_s|$ is the saturation magnetization, which has the same value inside every domain. One must, of course, choose the size of the Lorentz cavity smaller than the dimensions of a magnetic domain to preserve the above definition. Below saturation the orientation of $\mathbf{M}_s$ is obviously different from one domain to another, but $\mathbf{B}_L$ and $\mathbf{B}_{hf}$ are always essentially collinear (Shirley *et al.*, 1968). Let us now consider the particular case of the single crystal at 77°K, where $B_\mu = +1.48$ kG is measured and $B_L = +2.14$ kG is calculated. The difference $B_{hf} = -0.66$ kG must be due to a contact interaction with 4s and 3d conduction electrons. If the muon charge were closely screened by fully polarized 4s electrons, we would expect a hyperfine field of +160 kG as in the Mu atom. The reduced value of the measured field arises partly because the screening electrons are only partially polarized.

Neutron diffraction studies yield an unperturbed interstitial magnetization of $M_i = -0.85 \times 10^{22} \mu_B/\text{cm}^3$ (Mook, 1966), i.e., $M_i = -0.079$ kG. For a naive picture in which screening of the muon charge is completely ignored (effectively treating the μ^+ as an uncharged probe), one can compute a contact field $8\pi M_i/3$ of exactly -0.66 kG. It would be unwise to consider this agreement to be other than fortuitous until a detailed calculation of the screening of the interstitial muon charge by the band electrons is available.

For B_{ext} strong enough to saturate the sample, no further domain wall motion can occur and B_{DM} stays at its maximum value, DM_s, where D is the sample demagnetizing factor. In this case B_μ rises linearly with B_{ext} with unit slope. This is demonstrated in Fig. 38. The "knee" in the experimental curve occurs at the calculated saturation field of the sample.

The temperature dependence of the residual polarization is of some interest. Figure 39 shows results of Foy *et al.* (1973) for nickel. Since the Ni specimen actually consists of a large number of magnetic domains, the value of $P_\perp$ depends on the alignment of domains transverse to the initial muon polarization. Whenever the directions of magnetization of the domains differ from the initial muon polarization direction, the muon

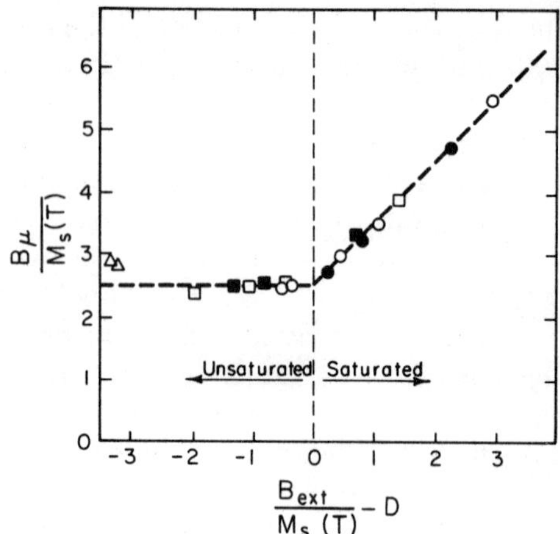

FIG. 38. Local field B_μ at the muon site in nickel versus B_{ext}, the external field measured with target out. D is the sample demagnetizing factor. The points denoted by triangles are from data on an approximately spherical single crystal with the [111] axis parallel to B_{ext} at 300 and 77°K. Other points are from data on a polycrystalline ellipsoid (4.5 × 2 × 0.5 in.). Solid and open circles refer to the 4.5-in. axis parallel to B_{ext} (D = 0.69) at 523 and 573°K, respectively. Solid and open squares refer to the 2-in. axis parallel to B_{ext} (D = 2.24) at 523 and 573°K. B_μ and B_{ext} have been normalized to bring measurements at various temperatures to the same vertical level and to exhibit saturation for all samples at zero on the horizontal scale.

spins will precess on "cones" with different axes and aperture angles. As discussed in Section IX,A,1,c, the overall amplitude of the precession is reduced by $\frac{2}{3}$ when the magnetization directions are isotropically distributed.

The solid line in Fig. 39 is a measure of this domain alignment; it is extracted by dividing the permeability by the macroscopic magnetization and normalizing to the residual polarization. As can be seen, this line accurately represents most of the data. The drop in $P_\perp$ near the Curie temperature is somewhat artificially introduced by assuming a sudden randomization of orientation near this temperature. This assumption, however, needs experimental confirmation. Such measurements may make possible a study of the formation of domains and clusters around the Curie temperature in the absence of an external field (Kossler, 1973).

Slow depolarization rates in Ni were also measured by the LBL and SREL groups. Essentially no slow relaxation was observed in the single crystal, even at 77°K (Patterson *et al.*, 1974). By contrast, a very fast relaxation is observed in polycrystalline samples, especially at lower temperatures (Foy *et al.*, 1973). The absence of this relaxation in single-crystal nickel rules out the possibility that slow depolarization is caused primarily by such dynamical mechanisms as creation or destruction of single magnons or multiple magnon processes, such as suggested by Kossler (1973). Other relaxation mechanisms, such as electron scattering from the μ^+, lead to very long relaxation times, invisible on the scale of the muon lifetime.

The disappearance of precession in a polycrystalline sample at low temperature must arise from an extremely short depolarization time caused by a large inhomogeneous line width ΔB. T_2 and ΔB are related by $\Delta B = 2/\gamma_\mu T_2$. (A depolarization time of 0.1 μsec implies a ΔB of about 200 G.) Figure 37 shows values of ΔB calculated from muon T_2 measurements, compared with ^{61}Ni NMR line widths in polycrystalline nickel from Streever and Bennett (1963). The line widths inferred from μ^+ precession signals in the single-crystal sample at 300 and 77°K ($\Delta B_{77} < 18$ G, $\Delta B_{300} < 10$ G) are an order of magnitude smaller than the narrowest absolute line widths previously observed in nickel (Patterson *et al.*, 1974).

The approximate equality of ΔB as measured with muons and NMR is significant considering that the field at the nickel nuclei is larger than that at the μ^+ by a factor of 50. If the broadening were due to strains in the sample, one would expect the line width to scale approximately with the field measured. It is well known that considerable line broadening occurs in polycrystalline ferromagnets because of the inhomogeneous fields of randomly oriented crystallites. The general trend of ΔB in Fig. 37 toward larger values at lower temperatures may be qualitatively understood in terms of an increase in the anisotropy energy with decreasing temperature (Aubert, 1968), causing an increase in magnetization inhomogeneity.

Kossler's ΔB data show a sharp rise about 550°K. Since $B_\mu(T)$ becomes

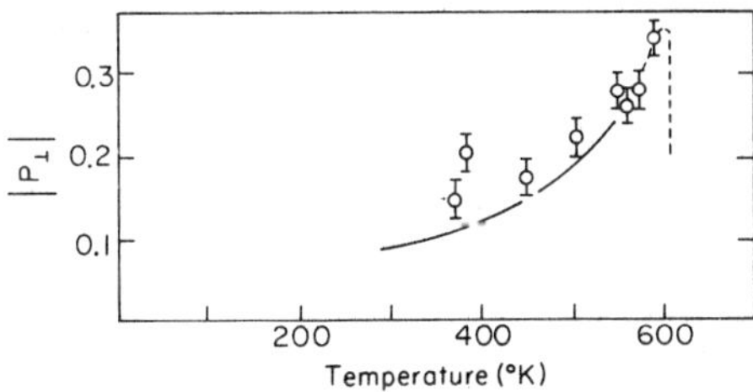

FIG. 39. Temperature dependence of the μ^+ residual polarization $P_\perp$ in polycrystalline Ni. The solid line corresponds to the measured sample permeability divided by B_μ and normalized to the data. The dashed line indicates the assumed drop in $P_\perp$ near the Curie temperature (from Foy *et al.*, 1973).

very steep near the Curie temperature $T_c = 631°K$, imperfect temperature regulation and/or homogeneity would produce an effective internal field inhomogeneity that gets worse closer to T_c. Kossler's ΔB data near T_c can be explained if one assumes an experimental temperature spread $\sim 10°K$.

In *ferromagnetic iron*, Foy *et al.* (1973) have measured B_μ at a number of temperatures between 300 and 675°K. Iron has a bcc crystal structure, so the μ^+ may be found at the interstitial face-centered sites. There, unlike in the case of Ni, B_{dip} would have one of the values $+18.8$ or -9.4 kG, depending on whether the nearest iron cores are situated on a line perpendicular or parallel to the magnetization directions in the domain (Stearns, 1973). It is known that hydrogen in Fe has a diffusion (hopping) time of about 10^{-9} sec at 300°K (Choi, 1970). The μ^+ will probably diffuse even faster from one face-centered site to another. Since there are twice as many of the second type of site as of the first, the muon will feel an average dipolar field that exactly cancels, provided that the "hopping time" is very small compared with the period of precession ($\sim 10^{-8}$ sec) in the local fields. For the temperature range studied, a unique B_μ is always observed, as in nickel. No detailed information on slow depolarization in Fe is presently available. Low-temperature studies of single-crystal Fe may thus be expected to clarify the situation greatly and perhaps to reveal exciting new phenomena.

The interstitial magnetization in iron is known to be a highly structured function of position (Shull and Mook, 1966). Thus the value $B_{hf} = -3.0$ kG extracted from the Fe data using $B_{dip} = 0$ and $B_L = +7.1$ kG cannot be interpreted very meaningfully until a great deal more is understood about the muon's screening charge and position distribution in iron. However, the result is still qualitatively consistent with the naive picture of contact interactions with unperturbed electron bands.

It seems very likely that a great variety of μ^+ precession phenomena await discovery and study in iron; one can imagine a similar wealth of information available from many other such metals. The future seems rich indeed for this latest application of the positive muon as a probe of matter.

B. Slow Depolarization in Paramagnetic Liquids

The NMR line in liquids generally has a Lorentzian shape, the Fourier transform of which is an exponential decay function describing the relaxation of spin components perpendicular to the external magnetic field. Slow muon depolarization in transverse fields is thus expected to obey an exponential decay law:

$$F_\perp(t) = \exp(-t/T_2) \tag{155}$$

where T_2 is the transverse or spin–spin relaxation time. Relaxation in this case means the loss of phase coherence among the spins of the precessing muons. Slow depolarization in a longitudinal field arrangement is also described by an exponential decay law:

$$F_{\parallel}(t) = \exp(-t/T_1) \tag{156}$$

with T_1 the longitudinal or spin–lattice relaxation time. Relaxation in this case consists of transitions between the Zeeman states of the muons, caused by interactions with the lattice, which provides or absorbs the energy quanta involved. No longitudinal slow depolarization has yet been measured in paramagnetic solutions.

As emphasized before, the muons that precess at the free-muon Larmor frequency must be in a diamagnetic environment—placed there either by hot-atom reactions or by thermal reactions of muonium. From results on hot-tritium chemistry in aqueous solutions it is known that the molecule THO is formed preferentially (Wolfgang, 1965). It can be assumed that the same mechanism dominates in hot-atom reactions of muonium, so that most of the "hot fraction" in water [$h(H_2O) = 0.55 \pm 0.03$] represents formation of MuHO molecules—that is, about half of the muons replace protons in water molecules during epithermal collisions. In solutions where thermal Mu does not react rapidly, these are the only muons observed in a precession experiment.

Since the muon takes the place of a proton, it can be assumed that the muon spin will be relaxed in the same way as the proton spin. Proton spin relaxation phenomena in aqueous solutions have been investigated quite extensively, using NMR (Abragam, 1970). Particularly susceptible to study with muons are slow relaxation phenomena in paramagnetic aqueous solutions, in which T_2 is usually short enough to be detected within the muon lifetime (Poulis and Hass, 1962). From the theory of relaxation (Abragam, 1970) one expects that

$$1/T_2 \approx (\mu)^2 \tag{157}$$

where μ is the magnetic moment of the particle involved. When comparing proton and muon relaxation times in the same solution under the same conditions, one thus expects the ratio

$$T_2(\mathrm{p})/T_2(\mu) = \mu_\mu^2/\mu_\mathrm{p}^2 \approx 10 \tag{158}$$

Figure 40 shows T_2 data for muons in paramagnetic Fe^{3+} solutions (Schenck, 1970); T_2 is plotted versus paramagnetic ion concentration. The lower data points, connected by a solid straight line, were obtained from solutions of $Fe(NO_3)_3$. The upper solid line represents proton NMR results in the same

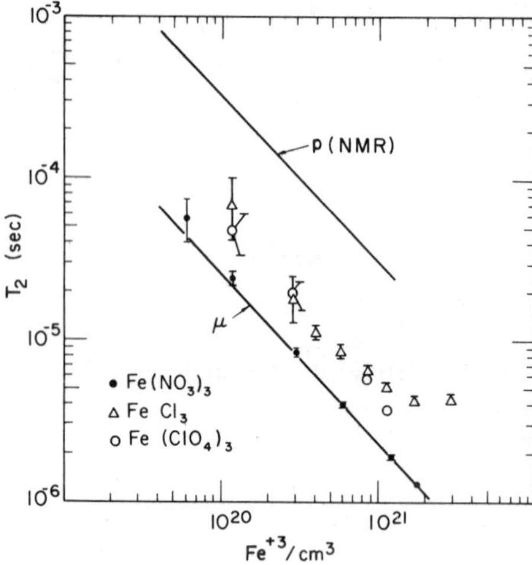

FIG. 40. Concentration dependence of the transverse relaxation time for μ^+ depolarization in $FeCl_3$, $Fe(ClO_4)_3$, and $Fe(NO_3)_3$ solutions. For the $Fe(NO_3)_3$ solution, the T_2 dependence for proton NMR is indicated by the upper solid line (from Schenck, 1970).

solution. The experimental ratio of proton to muon relaxation times is indeed about 10. No NMR data are available for comparison with the results obtained for muons in $FeCl_3$ and $Fe(ClO_4)_3$ solutions. Differences between the data from $Fe(NO_3)_3$ solutions and data from $Fe(ClO_4)_3$ and $FeCl_3$ solutions are not fully understood and need further investigation.

The T_2 results for muons in $Gd(NO_3)_3$ solutions of varying concentration were similarly consistent with proton measurements in the same solutions.

As is well known, Fe^{3+} ions usually form $Fe(H_2O)_6{}^{3+}$ complexes in water, and protons (muons) incorporated into the sphere of hydration are relaxed by the paramagnetic ion. However, the exchange rate for water molecules in the sphere of hydration is slow ($\sim 3 \times 10^3$ sec^{-1}) (Hunt, 1963) compared with the relaxation rate for muons ($\sim 10^5$ sec^{-1}), yet all of the muons appear to relax—not just those "stuck" in the complexes. Furthermore, in concentrated Fe^{3+} solutions a significant fraction of polarized muons reach diamagnetic environments through thermal reactions [recall reaction (121)]. The subsequent state of these muons is unknown, and so part of the depolarization mechanism in concentrated Fe^{3+} solutions is not understood. In an attempt to clarify some of these uncertainties, muon precession has been studied in aqueous solutions of $K_3Fe(CN)_6$ (Fleming and Brewer, 1973). In this case the tightly bound complex $Fe(CN)_6{}^{3-}$ prevents water molecules from closely approaching the Fe^{3+} ion. At the same time, the strong ligand fields cause Hund's rule to be broken, reducing the spin state of Fe^{3+} from $d(\frac{5}{2})$ to $d(\frac{1}{2})$; thus, in $Fe(CN)_6{}^{3-}$ the Fe^{3+} ion has only $\sim\frac{1}{5}$ as large a magnetic moment as in $Fe(H_2O)_6{}^{3+}$. In keeping with expectations, the resultant muon relaxation rate in up to $0.8M$ $K_3Fe(CN)_6$ solutions is undetectable in experimental histograms with modest statistics ($T_2 \gtrsim 20$ μsec). A fast thermal reaction of Mu apparently occurs in these solutions, and so we can conclude that the ligand fields prevent depolarization of muons both in the epithermally formed MuHO molecules and in the diamagnetic products of the thermal reaction. Further investigations may help to clarify the basic mechanisms involved.

It is important to note that NMR measurements were generally done in solutions with concentrations of $<10^{21}$ paramagnetic ions/cm^3, because of problems regarding line width, signal strength, rf power, and other experimental considerations. Proton NMR measurements are thus effectively limited to cases with relaxation times $T_2 \gtrsim 10^{-5}$ sec. However, it is easy to measure much shorter relaxation times with muons, and so to study much more concentrated solutions. In principle, transverse or longitudinal relaxation times can be measured down to $\sim10^{-8}$ sec.

In very concentrated solutions new effects may show up. This is demonstrated by measurements of T_2 for muons in $MnCl_2$ solutions with Mn^{2+} concentration of up to $5M$ ($\approx 3 \times 10^{21}$ ions/cm^3) (Schenck *et al.*, 1972). These studies also provide a good example of the sorts of effects involved and the variety of information that may be extracted from experimental results. At lower concentrations, $MnCl_2$ solutions have been extensively studied with proton NMR by many authors, particularly Bloembergen and Morgan (1961).

Paramagnetic Mn^{2+} ions, like Fe^{3+} ions, are surrounded by six water molecules forming a hydration sphere. In this case the water exchange rate is fast ($\sim 3 \times 10^7$ sec^{-1}) (Hunt, 1963), allowing a straightforward treatment of the relaxation process for the entire quasi-free μ^+ ensemble. Protons (or muons) in this hydration sphere are subject to two time-dependent magnetic interactions: the dipole–dipole interaction between the magnetic moments of the paramagnetic ion and the proton (or muon), and a scalar coupling or spin–exchange interaction caused by the nonvanishing electronic wavefunction of the ion at the site of the proton (or muon) in the hydration sphere. These interactions lead to the following expression for the transverse relaxation time T_2 (Bernheim *et al.*, 1959; Bloembergen and Morgan, 1961):

$$\frac{1}{T_2} = \frac{4}{60}\frac{1}{r^6} S(S+1)\gamma_{p(\mu)}^2\gamma_{\text{ion}}^2 \hbar^2[7\tau_c + 13\tau_c(1+\omega_s{}^2\tau_c{}^2)^{-1}]P$$
$$+ \tfrac{1}{3}S(S+1)A_{p(\mu)}^2\hbar^{-2}[\tau_e + \tau_e(1+\omega_s{}^2\tau_e{}^2)^{-1}]P \qquad (159)$$

where S is the ion spin ($\frac{5}{2}$ for Mn^{2+}), r the internuclear distance between ion and proton (muon), $\gamma_p(\gamma_\mu)$ and γ_{ion} the gyromagnetic ratios, $A_\mu = 3.18A_p$ the coupling constant for the exchange interaction, ω_s the Larmor precession frequency of the ion, τ_c and τ_e the respective correlation times for dipole–dipole and exchange interactions, and P the probability of finding a proton (muon) in a hydration sphere, given by

$$P = \frac{\text{(number of water molecules in hydrogen spheres)}}{\text{(total number of water molecules)}}$$

or

$$P = 0.108 \times (\text{molal concentration of } Mn^{2+})$$

The first term on the right-hand side of Eq. (159) is the contribution due to the dipole–dipole interaction; the second term accounts for the spin-exchange interaction.

The correlation times τ_c and τ_e are a measure of the time dependence of the respective interactions; each may be thought of as the duration of the coherent effects of the interaction involved. The time dependence of the dipole–dipole interaction may be caused by rotational diffusion of the Mn^{2+} complex, by chemical exchange of the H_2O (MuHO) molecules, and by spin relaxation of the paramagnetic ion, parametrized by the respective correlation times τ_r, τ_h, and τ_s:

$$\frac{1}{\tau_c} = \frac{1}{\tau_r} + \frac{1}{\tau_h} + \frac{1}{\tau_s} \tag{160}$$

and

$$\frac{1}{\tau_e} = \frac{1}{\tau_h} + \frac{1}{\tau_s} \tag{161}$$

(The scalar coupling is not influenced by rotational diffusion.)

The temperature dependence of the correlation times τ_r and τ_h is described by a type of Arrhenius law:

$$\tau = \tau_0 \exp(V/RT) \tag{162}$$

where V is the activation energy for rotational diffusion or chemical exchange. The temperature dependence of τ_s is more complicated and involves the mechanisms leading to electronic relaxation (Bloembergen and Morgan, 1961).

The information contained in Eqs. (159)–(162) is typical of what can be obtained from relaxation studies in paramagnetic solutions.

Figure 41a shows the experimental data for $T_2(\mu^+)$ as a function of Mn^{2+} concentration along with the overall T_2 (solid curve), the dipole–dipole term (top dashed curve), and the spin–exchange term (lower dashed curve), obtained from Eq. (159) by using reasonable values for the respective correlation times and correcting for the larger magnetic moment of the muon. For concentrations below 10^{20} Mn^{2+}/cm^3 there is good agreement between the data and predictions scaled for muons. At higher concentrations the data break sharply away from the predicted curve, and there seems to be a quenching of the relaxation mechanism. These deviations from Eq. (159) can be understood if we assume that some of the correlation times become concentration dependent at higher concentrations, due to intermolecular interactions of Mn^{2+} complexes. In particular, spin–spin interactions among Mn^{2+} ions might lead to concentration-dependent correlation times. Indeed, ESR measurements by Garstens and Liebson, and Hinckley and Morgan (1966), show a concentration-dependent line width in concentrated Mn^{2+} solutions. Their data (Fig. 41b) can be approximated by (solid line in the figure):

$$\tau_s^* = \frac{1.24 \times 10^{-9}}{N^2} + 1.27 \times 10^{-11} \text{ sec}$$

where τ_s^* is now used as an additional effective correlation time in the proton (muon)–ion interactions. Here N is the ion concentration in moles/liter. The temperature dependence of τ_s^* can also be obtained from Hinckley

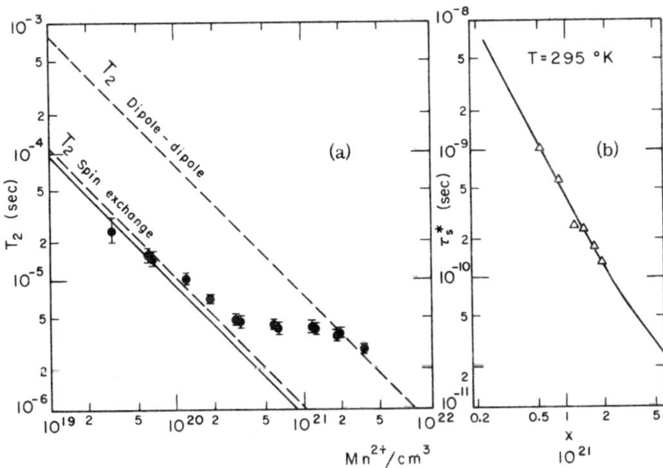

FIG. 41. (a) Transverse muon relaxation times in $MnCl_2$ solutions. The dashed lines represent the spin–exchange and dipole–dipole terms. The solid line is the combined result. (b) Plot of τ_s^* vs Mn^{2+} concentration at 295°K. The solid line was obtained from Hinckley and Morgan (1966).

and Morgan (1966). For a $3M$ solution, one finds

$$1/\tau_s^* = 1.76 \times 10^7[710 - 2.8 \times 10^3 \exp(-1.26 \times 10^3/RT)] \quad (163)$$

The total correlation time τ_e for the spin–exchange interaction is thus given by

$$\frac{1}{\tau_e} = \frac{1}{\tau_s} + \frac{1}{\tau_s^*} + \frac{1}{\tau_h} \quad (164)$$

where τ_s is the usual electron spin relaxation time and τ_h is the mean time for the muon to remain in the hydration sphere. Similarly, the total correlation time τ_c for the dipole–dipole interaction is given by

$$\frac{1}{\tau_c} = \frac{1}{\tau_r} + \frac{1}{\tau_h} + \frac{1}{\tau_s} + \frac{1}{\tau_s^*} \quad (165)$$

where τ_r is the rotational correlation time. At room temperature (Bloembergen and Morgan, 1961)

$$\tau_s \approx 3 \times 10^{-9}, \qquad \tau_h \approx 2 \times 10^{-8}, \qquad \tau_r \approx 3 \times 10^{-11}\ \text{sec}$$

In Fig. 42a the data from Fig. 41a are shown with the linear concentration dependence (P) divided out. If the correlation times were concentration independent, $1/T_2P$ would be constant. By inserting the total correlation times τ_e and τ_c [Eqs. (164) and (165)] into the general expression [Eq. (159)] with the other parameters taken from Bernheim *et al.* (1959), one obtains the solid line in Fig. 42a, which is an excellent fit to the data. The dashed lines in Fig. 42a represent spin–exchange and dipole–dipole contributions separately.

However, using Eq. (159) together with Eqs. (162) and (163) and the temperature dependence of τ_r and τ_h from Bloembergen and Morgan (1961), one obtains the dotted curve in Fig. 42b for 11 kG, which—as is clearly evident—does not adequately describe the measured temperature dependence of T_2 in a $3M$ solution.

If Eq. (163) correctly describes the temperature dependence of τ_s^*, and if τ_h behaves normally, we are forced to adopt parameters different from the ones in Bloembergen and Morgan (1961) in the expression

$$\tau_r = \tau_r^0 \exp(V_r/RT) \quad (166)$$

where V_r is the activation energy of the rotational motion of the Mn^{2+} complex. Using $V_r = 8.5$ kcal/mole and $\tau_r^0 = 1.73 \times 10^{-17}$ sec, we obtain for $1/T_2P$ versus temperature the lower solid curve at 11 kG and the upper one at 4.5 kG external field strength (Garrett and Morgan, 1966).

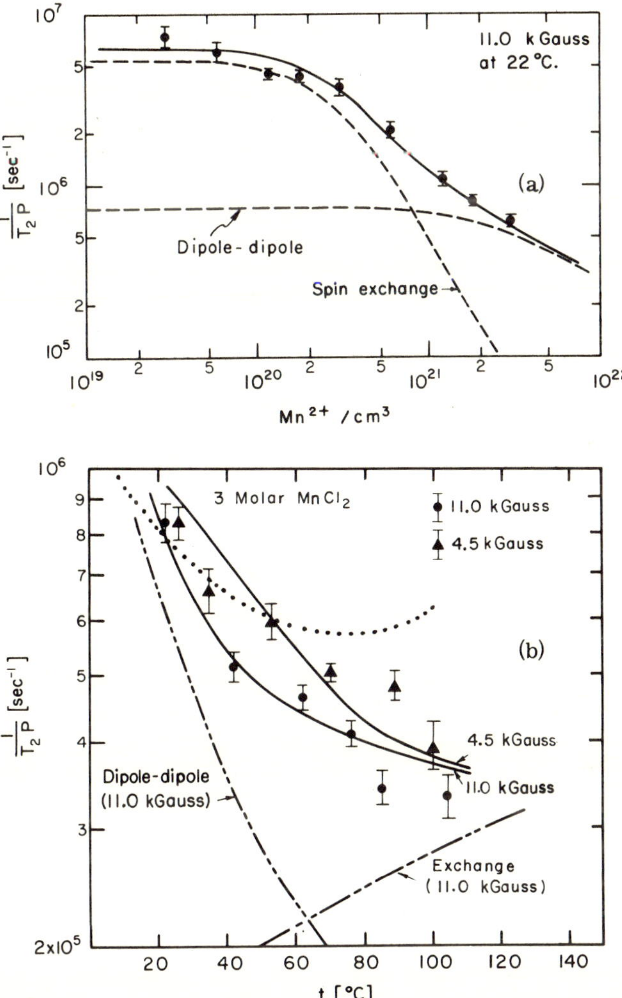

FIG. 42. (a) Plot of $1/T_2P$ vs Mn^{2+} concentration. The solid curve is obtained by combining NMR and ESR results. The dashed curves show separately the contributions from spin–exchange and dipole–dipole interactions. (b) Plot of $1/(T_2P)$ vs temperature. The dotted curve is the fit obtained for 11 kG without modifications. The solid curves result from further analysis at 4.5 and 11 kG (considering a new V_r value). The dashed curves represent the spin–exchange and dipole–dipole contributions separately for the analysis at 11 kG with the modified V_r value.

The large value for the activation energy at $3M$ concentrations (as compared with $V_r = 4.5$ kcal/mole at low concentrations) seems reasonable in view of the greatly increased viscosity of a $3M$ $MnCl_2$ solution [$\eta(3M) \approx$ 3.2 centipoises]. It would be of great interest to establish more detailed experimental relationships between these observations and the dynamics of this liquid (Garrett and Morgan, 1966).

There are unfortunately several questionable assumptions in the foregoing analysis, which we will now discuss.

(1) The results of Hinckley and Morgan (1966) for ESR line widths were obtained in an external field of 3 kG. We have neglected any possible field dependence of the ESR line widths and assumed the same values in fields of 4.5 and 11 kG. This is justified only if the relevant correlation time τ satisfies $\tau\omega_s$ (11 kG) < 1—that is, $\tau < 5 \times 10^{-12}$ sec.

(2) The results of Hinckley and Morgan (1966) were obtained in $Mn(ClO_4)_2$ solutions, whereas the muon relaxation studies were made in $MnCl_2$ solutions.

(3) Although it was necessary to change V_r and τ_r^0 in order to fit the temperature dependence of a $3M$ solution, we had to assume that τ_r remains relatively independent of concentration at 295°K in order to obtain the fit in Fig. 42a.

(4) In view of the quality of the fit shown in Fig. 42a, τ_h has been assumed to be concentration independent. This assumption needs further justification. In particular, a concentration-dependent activation energy for chemical exchange might reduce the value of V_r to less than 8.5 kcal/mole.

(5) The whole analysis depends on the basic assumption that $Mn(H_2O)_6^{2+}$ formation continues almost unchanged up to the strongest concentrations.

The testing of these assumptions in further muon-depolarization studies may contribute to our knowledge of the structure and dynamics of fluids. The experimental program in Mn^{2+} solutions should include measurements of relaxation times in longitudinal as well as transverse fields, as a function of varying field strength and temperature, and finally in solution with different anions.

X. Concluding Remarks

A. μSR—A "Trigger" Detection Technique for Magnetic Interactions

Studies of the precession and depolarization of positive muons and muonium in matter share many characteristics with the associated techniques of NMR and ESR. The important difference is that the more standard resonance techniques involve absorption of macroscopic amounts of power by macroscopic spin ensembles ($\gtrsim 10^{15}$ nuclei or $\gtrsim 10^{12}$ electrons), while as few as 10^6 muons imbedded in the medium can be used passively to detect coherent local fields (through the precession frequency) and random local fields (through the relaxation time). This is possible because the asymmetric decay so efficiently converts information about the muon polarization into an easily detectable external phenomenon (the positron direction). These special features are also shared by a variety of "trigger" detection techniques (Abragam, 1970) such as Mössbauer effect; perturbed angular correlations (PAC) (Shirley and Haas, 1972); oriented nuclei; and others. The "coming of age" of such techniques is often marked by the adoption of a suggestive mnemonic acronym; in this context, the applications of muons outlined in this chapter might be referred to as "μSR" studies, where the acronym stands for Muon (or Muonium) Spin Relaxation, Rotation, Resonance, etc., and is intended to suggest the analogy with NMR and ESR.

While other "trigger" detection techniques have advantages similar to those of μSR (e.g., sensitivity), each has its own set of difficulties. Many, for instance, can only be used to study certain substances into which the probe nuclei can be incorporated; ion implantation can sometimes help in this regard, but generates new problems such as radiation damage. Furthermore, in NMR studies involving nuclei other than hydrogen isotopes, the electron core of the probe causes disturbing effects, such as core polarization, which mask or even change the local fields to be measured. Many nuclei also have electric quadrupole moments, whose interactions complicate measurements of local magnetic fields and contact interactions. The muon, on the other hand, is a bare Dirac particle interacting only through its electric charge and magnetic moment. Thus its "feedback" effects upon local field properties are often small and calculable. The general problems of implanted ion techniques are shared to some extent by the μSR method: the μ^+ or Mu atom may not spontaneously occupy the position one wants to investigate; and radiation damage on a local scale (at the end of a given muon's range) may cause observable effects—although the small number of muons used ($\sim 10^6$) and their distribution throughout an extended bulk sample make cumulative radiation damage a negligible problem.

The μSR technique can be used to study relaxation phenomena over a rather broad range of time scales. In μ^+ precession studies it is possible directly to observe relaxation rates from $\sim 10^4$ to $\sim 10^8$ sec^{-1}, while the relaxation rates ν of the muonium electron can be observed or inferred (with varying degrees of difficulty) over a range from $\sim 10^6$ ($\nu \ll \omega_0$) to

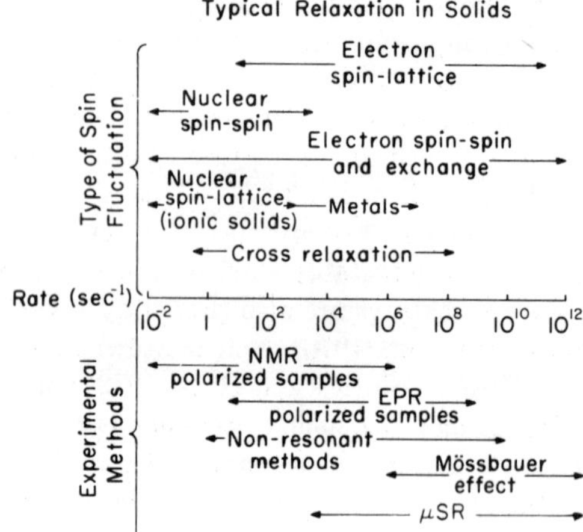

FIG. 43. Ranges of various relaxation rates and sensitive regions for different detection techniques.

$\sim 10^{14}$ sec^{-1} ($\nu \gg \omega_0$), and in principle even higher [see Eq. (28)]. In Fig. 43, a comparison (from Wickman, 1968) of time scales accessible to different techniques is supplemented by an indication of the range available with μSR methods. Obviously, the main shortcoming of the μSR technique is caused by the 2.2-μsec muon lifetime, which makes measurement of relaxation times longer than a few lifetimes progressively difficult. (It should be emphasized that this limitation is merely a problem of counting statistics, since the exponential decay in the experimental histogram [see Eq. (6)] can be exactly divided out.) In this aspect, NMR and ESR have a tremendous advantage. With NMR one can measure relaxation times on a scale of minutes.

B. The State of the Art

The use of μSR techniques in the study of "ordinary" matter has progressed to that tantalizing stage where its potential in many fields is becoming clear, but where each field of application still suffers from serious gaps of understanding or measurement. We have attempted to survey the current situation in many of these fields; we conclude with brief summaries of a few of them.

1. *Muonium Chemistry*

Recent advances in theory (Ivanter and Smilga, 1968, 1969a,b, 1971, 1972; Brewer *et al.*, 1973a; Fischer, 1973) and experimental technique

(Brewer *et al.*, 1974) have opened up the field of muonium chemistry to extensive and productive study. In the liquid phase, spectacular and unpredicted differences between reaction rates of Mu and H have already been observed (Brewer *et al.*, 1974); it is hoped that further study will generate a deeper understanding in terms of dynamic isotope effects. These studies have also revealed the important roles played by hot-atom reactions of muonium, as well as formation and subsequent reaction of radicals containing muonium, in the "fast" depolarization of positive muons in liquids. Two more fields of chemical research have thus become accessible to μSR techniques.

New techniques (Kendall, 1972) also make the study of Mu chemistry in gases a practical reality. In the gas phase, absolute rates of chemical reactions can be treated theoretically (Glasstone, 1941; Wolfgang, 1972) with much more confidence than in liquids. This nearly untouched field may be expected to bloom with definitive experiments in the next few years.

2. *Muons in Solids*

The applications of μSR to solid-state physics are growing at a nearly explosive rate. Muons have been fruitfully implanted in insulators, semiconductors, metals, and superconductors; in glasses, powders, and crystals; in diamagnetic, paramagnetic, and ferromagnetic media. In each case somewhat different phenomena are accessible to study, and in each case a slightly different analysis technique must be applied. We have allowed ourselves a good deal of conjecture throughout the text regarding new directions in which solid-state μSR studies are likely to expand, but it is most likely that the next turn in this field will be just as unexpected as those that preceded it, and just as exciting.

Acknowledgments

No review of such a wide-ranging field as μSR applications could be compiled without the generous assistance of experts in associated fields. We are deeply indebted to many solid-state physicists, in particular, Charles Kittel, Alan Portis, and Carson Jeffries of the University of California at Berkeley, for stimulating discussions and helpful suggestions regarding μSR in solids. We are also grateful to David Walker of the University of British Columbia, Dr. M. Anbar of the Stanford Research Institute, Prof. F. S. Rowland of the University of California at Irvine, and several members of the Berkeley Chemistry Department for help with the chemical interpretation of μSR results in liquids. Our colleagues Donald Fleming (UBC), Bruce Patterson (LBL), Richard Johnson (LBL) and Walter Fischer (SIN) deserve special appreciation for their help with this chapter as well as their judicious critical comments. We are all grateful to Don Fleming for tolerating J. H. Brewer's preoccupation with this writing during the critical stages of preparation for a μSR program at TRIUMF, as well as to Dr. J. P. Blaser from SIN for making it possible for F. N. Gygax to extend his stay at LBL on a SIN grant. One of us (JHB) would like to give special thanks to his wife Suzanne for enduring several months of neglect during the writing of this review.

References

Abragam, A. (1970). "The Principles of Nuclear Magnetism." Oxford Univ. Press (Clarendon), London and New York.

Allison, S. K. (1958). *Rev. Mod. Phys.* **30,** 1137.

Allison, S. K., and Garcia-Munoz, M. (1962). *In* "Atomic and Molecular Processes" (D. R. Bates, ed.), Chapter 19. Academic Press, New York.

Alonso, J., and Grodzins, L. (1968). *In* "Hyperfine Structure and Nuclear Radiations" (E. Matthias and D. A. Shirley, eds.), p. 549. North-Holland Publ., Amsterdam.

Anbar, M., and Neta, P. (1967). *Int. J. Appl. Radiat. Isotopes* **18,** 493.

Andrianov, D. G., Myasishcheva, G. G., Obukhov, Yu. V., Roganov, V. S., Firsov, V. G., and Fistul, V. I. (1969). *Sov. Phys.-JETP* **29,** 643.

Andrianov, D. G., *et al.* (1970). *Sov. Phys.-JETP* **31,** 1019.

Aubert, G. (1968). *J. Appl. Phys.* **39,** 504.

Babaev, A. I., Balats, M. Ya., Myasishcheva, G. G., Obukhov, Yu. V., Roganov, V. S., and Firsov, V. G. (1966). *Sov. Phys.-JETP* **23,** 583.

Bartal, L. J., Nicholas, J. B., and Ache, H. J. (1972a). *J. Phys. Chem.* **76,** 1124.

Bartal, L. J., Nicholas, J. B., and Ache, H. J. (1972b). *Radiochim. Acta* **17,** 205.

Benson, S. W. (1960). "The Foundations of Chemical Kinetics." McGraw-Hill, New York.

Berg, H. C. (1965). *Phys. Rev.* **137,** A1621.

Bernheim, R. A., Brown, T. H., Gutowsky, H. S., and Woessner, D. W. (1959). *J. Chem. Phys.* **30,** 950.

Bloembergen, N., and Morgan, L. O. (1961). *J. Chem. Phys.* **34,** 842.

Bowen, T., *et al.* (1972). Univ. of Arizona, private communication.

Bradt, H. V., and Clark, G. W. (1963). *Phys. Rev.* **132,** 1306.

Brewer, J. H., Crowe, K. M., Johnson, R. F., Schenck, A., and Williams, R. W. (1971). *Phys. Rev. Lett.* **27,** 297.

Brewer, J. H., *et al.* (1973a). *Phys. Rev. Lett.* **31,** 143.

Brewer, J. H., Gygax, F. N., and Fleming, D. G. (1973b). *Phys. Rev.* **A8,** 77.

Brewer, J. H., Crowe, K. M., Gygax, F. N., Johnson, R. F., Fleming, D. G., and Schenck, A. (1974). *Phys. Rev.* **A9,** 495.

Choi, J. Y. (1970). *Metal Trans.* **1,** 911.

Coffin, T., Garwin, R. C., Penman, S., Lederman, L. M., and Sachs, A. M. (1958). *Phys. Rev.* **109,** 973.

Cotton, F. A., and Wilkinson, G. (1966). "Advanced Inorganic Chemistry." Wiley (Interscience), New York.

Crowe, K. M., *et al.* (1972a). *Phys. Rev.* **D5,** 2145.

Crowe, K. M., Johnson, R. F., Brewer, J. H., Gygax, F. N., Fleming, D. G., and Schenck, A. (1972b). *Bull. Amer. Phys. Soc.* **17,** 594.

Culligan, G., Lundy, R. A., Telegdi, V. L., Winston, R., and Yovanovitch, D. D. (1964). *In Rep. Conf. High Energy Cyclotron Improvement, College of William and Mary, Williamsburg, Virginia.*

Eisenstein, B., Prepost, R., and Sachs, A. M. (1966). *Phys. Rev.* **142,** 217.

Fano, U. (1957). *Rev. Mod. Phys.* **29,** 74.

Farhataziz (1967). *J. Phys. Chem.* **71,** 598.

Feher, G., Prepost, R., and Sachs, A. M. (1960). *Phys. Rev. Lett.* **5,** 515.

Fermi, E., and Teller, E. (1947). *Phys. Rev.* **72,** 399.

Firsov, V. G., and Byakov, V. M. (1965). *Sov. Phys.-JETP* **20,** 719.

Fischer, W. E. (1973). SIN-preprint.

Fleming, D. G., and Brewer, J. H. (1973). *Bull. Amer. Phys. Soc.* **18,** 1571.

Ford, G. W., and Mullin, C. J. (1957). *Phys. Rev.* **108,** 477.

Foy, M. L. G., Heiman, N., Kossler, W. J., and Stronach, C. E. (1973). *Phys. Rev. Lett.* **30,** 1064.

Friedel, J. (1958). *Nuovo Cimento Suppl.* **7,** 287.

Garrett, B. B., and Morgan, L. (1966). *J. Chem. Phys.* **44,** 890.

Glasstone, S., Laider, K. J., and Eyring, H. (1941). "The Theory of Rate Processes." McGraw-Hill, New York.

Goldanskii, V. I., and Firsov, V. G. (1971). *Ann. Rev. Phys. Chem.* **22,** 209.

Gurevich, I. I., *et al.* (1968). *Sov. Phys.-JETP* **27,** 235.

Gurevich, I. I., *et al.* (1969). *Phys. Lett.* **29B,** 387.

Gurevich, I. I., *et al.* (1971a). *Sov. Phys.-JETP* **33,** 253.

Gurevich, I. I., *et al.* (1971b). *JETP Lett.* **14,** 297.

Gurevich, I. I., *et al.* (1972). *Phys. Lett.* **40A,** 143.

Hellwege, K. H. (1970). "Einführung in die Festkorperphysik," Vol. II, Chapter 22.1. Springer-Verlag, Berlin and New York.

Hinckley, C. C., and Morgan, L. O. (1966). *J. Chem. Phys.* **44,** 898.

Hughes, V. W., McColm, D. W., Ziock, K., and Prepost, R. (1960). *Phys. Rev. Lett.* **5,** 63.

Hunt, J. P. (1963). "Metal Ions in Aqueous Solutions," p. 86. Benjamin, New York.

Hutchinson, D. P., Menes, J., Shaprio, G., and Patlach, A. M. (1963a). *Phys. Rev.* **131,** 1351.

Hutchinson, D. P., Menes, J., and Shapiro, G. (1963b). *Phys. Rev.* **131,** 1362.

Ivanter, I. G. (1973). *Sov. Phys.-JETP* **36,** 990.

Ivanter, I. G., and Smilga, V. P. (1968). *Sov. Phys.-JETP* **27,** 301.

Ivanter, I. G., and Smilga, V. P. (1969a). *Sov. Phys.-JETP* **28,** 796.

Ivanter, I. G., and Smilga, V. P. (1969b). *Sov. Phys.-JETP* **28,** 286.

Ivanter, I. G., and Smilga, V. P. (1971). *Sov. Phys.-JETP* **33,** 1070.

Ivanter, I. G., and Smilga, V. P. (1972). *Sov. Phys.-JETP* **34,** 1167.

Ivanter, I. G., *et al.* (1972). *Sov. Phys.-JETP* **35,** 9.

Kendall, K. R. (1972). Thesis, Univ. of Arizona.

Kossler, W. J. (1973). Private communication.

Landolt-Bornstein (1965). *In* "Numerical Data and Functional Relationships in Science and Technology" (K. H. Hellwege, ed.), Group II, Atomic and Molecular Physics, Vol. I, Magnetic Properties of Free Radicals. Springer, New York, 1965.

Lewis, E. S., and Robinson, J. K. (1968). *J. Amer. Chem. Soc.* **90,** 4337.

Logan, S. R. (1967). *Trans. Faraday Soc.* **63,** 1713.

Manenkov, A. A., and Orbach, R., (eds.) (1966). "Spin-Lattice Relaxation in Ionic Solids." Harper, New York.

Massey, H. S. W., and Burhop, E. H. S. (1952). "Electronic and Ionic Impact Phenomena," Chapter VIII. Oxford Univ. Press (Clarendon), London and New York.

Michael, B. D., and Hart, J. (1970). *J. Phys. Chem.* **74,** 2878.

Minaichev, E. V., Myasishcheva, G. G., Obukhov, Yu. V., Roganov, V. S., Savel'ev, G. I., and Firsov, V. G. (1970a). *Sov. Phys.-JETP* **31,** 849.
Minaichev, E. V., Myasishcheva, G. G., Obukhov, Yu. V., Roganov, V. S., Savel'ev, G. I., and Firsov, V. G. (1970b). *Sov. Phys.-JETP* **30,** 230.
Mobley, R. M. (1967). Thesis, Yale Univ.
Mobley, R. M., Bailey, J. M., Cleland, W. E., Hughes, V. W., and Rothberg, J. E. (1966). *J. Chem. Phys.* **44,** 4354.
Mobley, R. M., Amato, J. J., Hughes, V. W., Rothberg, J. E., and Thompson, P. A. (1967). *J. Chem. Phys.* **47,** 3074.
Moelwyn Hughes, E. A. (1971). "The Chemical Statics and Kinetics of Solutions." Academic Press, New York.
Mook, H. A. (1966). *Phys. Rev.* **148,** 495.
Myasishcheva, G. G., Obukhov, Yu. V., Roganov, V. S., and Firsov, V. G. (1967a). *High Energy Chem.* **1,** 340.
Myasishcheva, G. G., Obukhov, Yu. V., Roganov, V. S., Suvorov, L. Ya., and Firsov, V. G. (1967b). *High Energy Chem.* **1,** 343.
Myasishcheva, G. G., Obukhov, Yu. V., Roganov, V. S., and Firsov, V. G. (1967c). *High Energy Chem.* **1,** 337.
Myasishcheva, G. G., Obukhov, Yu. V., Roganov, V. S., and Firsov, V. G. (1968). *Sov. Phys.-JETP* **26,** 298.
Myasishcheva, G. G., Obukhov, Yu. V., Roganov, V. S., and Firsov, V. G. (1969a). *High Energy Chem.* **3,** 463.
Myasishcheva, G. G., Obukhov, Yu. V., Roganov, V. S., Suvorov, L. Ya., and Firsov, V. G. (1969b). *High Energy Chem.* **3,** 460.
Myasishcheva, G. G., Obukhov, Yu. V., Roganov, V. S., and Firsov, V. G. (1970). *High Energy Chem.* **4,** 398.
Narath, A. (1967). *In* "Hyperfine Interactions." (A. J. Freeman and K. B. Frankel, eds.), p. 287. Academic Press, New York, 1967.
Navon, G., and Stein, G. (1965). *J. Phys. Chem.* **69,** 1384.
Neta, P., Fessenden, R. W., and Schuler, R. H. (1971). *J. Phys. Chem.* **75,** 1654.
Nosov, V. G., and Yakovleva, I. V. (1963). *Sov. Phys.-JETP* **16,** 1236.
Nosov, V. G., and Yakovleva, I. V. (1965). *Nucl. Phys.* **68,** 609.
Pake, G. E. (1948). *J. Chem. Phys.* **16,** 327.
Patterson, B. D., Crowe, K. M., Gygax, F. N., Johnson, R. F. Portis, A. M., and Brewer, J. H. (1974). *Phys. Lett.* **46A,** 453.
Phillips, L. F., and Schiff, H. I. (1962). *J. Chem. Phys.* **37,** 1233.
Poulis, N. J., and Hass, W. P. A. (1962). "Landolt-Bornstein, Zahlenwerte und Funktionen," Vol. II, Part 9, Magnetic Properties I. Springer-Verlag, Berlin and New York.
Rowland, F. S. (1970). *In* "Molecular Beams and Reaction Kinetics" (C. Schlier, ed.), p. 108 (*Proc. Int. School Phys. Enrico Fermi* Course XLIV). Academic Press, New York.
Sauer, M. C., and Ward, B. (1967). *J. Phys. Chem.* **71,** 3971.
Schenck, A. (1970). *Phys. Lett.* **32A,** 19; unpublished results.
Schenck, A., and Crowe, K. M. (1971). *Phys. Rev. Lett.* **26,** 57.
Schenck, A., Williams, D. L., Brewer, J. H., Crowe, K. M., and Johnson, R. F. (1972). *Chem. Phys. Lett.* **12,** 544.
Shavitt, I. (1959). *J. Chem. Phys.* **31,** 1359.
Shirley, D. A., and Haas, H. (1972). *Ann. Rev. Phys. Chem.* **23,** 385.
Shirley, D. A., Rosenblum, S. S., and Matthias, E. (1968). *Phys. Rev.* **170,** 363.

Shull, C. G., and Mook, H. A. (1966). *Phys. Rev.* **16,** 184.
Sicking, G. (1972). *Int. Meeting Hydrogen Metal, Jülich* **II,** 408.
Siegmann, H. C., Strassler, S., and Wachter, P. (1971). *Proc. Meeting Muons Solid State Phys., Burgenstock, Switzerland* 2. Teil, p. 91.
Spaeth, J. M. (1966). *Z. Phys.* **192,** 107.
Stearns, M. B. (1973). Private communication.
Streever, R. L., and Bennett, L. H. (1963). *Phys. Rev.* **131,** 2000.
Swanson, R. A. (1958). *Phys. Rev.* **112,** 580.
Sweet, J. P., and Thomas, J. K. (1964). *J. Phys. Chem.* **68,** 1363.
Takakura, K., and Ranby, B. (1968). *J. Phys. Chem.* **72,** 164.
Thrush, B. A. (1965). *Progr. Reaction Kinet.* **3,** 65.
Wang, J. S.-Y., and Kittel, C. (1973). *Phys. Rev.* **B7,** 713.
Wangsness, R. K., and Bloch, F. (1953). *Phys. Rev.* **89,** 728.
Weiss, P., and Forrer, R. (1926). *Ann. Phys.* **5,** 153.
Wentzel, G. (1949). *Phys. Rev.* **75,** 1810.
Wickman, H. H. (1968). *In* "Hyperfine Structure and Nuclear Radiations" (E. Matthias and D. A. Shirley, eds.). North-Holland Publ., Amsterdam.
Wolfgang, R. (1965). *Progr. Reaction Kinet.* **3,** 99.
Wolfsberg, M. (1972). *Accounts Chem. Res.* **5,** 225.

Section 2

MESOMOLECULAR PROCESSES INDUCED BY μ^- AND π^- MESONS

S. S. GERSHTEIN

Theoretical Department
Institute of High Energy Physics
Serpukhov, USSR

L. I. PONOMAREV

Laboratory of Theoretical Physics
Joint Institute for Nuclear Research
Dubna, USSR

I. Introduction

Recently the various physical phenomena associated with processes occurring when negative charged mesons are stopped and absorbed in matter have been extensively investigated. The study of these processes is of inherent interest. In addition, mesoatomic and mesomolecular processes affect essentially the course of nuclear reactions in matter, such as, e.g., absorption of mesons by nuclei, catalysis of nuclear reactions by μ^- mesons in hydrogen, excitation of nuclear levels of heavy mesonic atoms, etc. (Zeldovich and Gershtein, 1960; Burhop, 1969; Wu and Wilets, 1969; Devons and Duerdoch, 1969).

The present review consists of essentially two parts: The first part is a systematic description of μ mesoatomic phenomena occurring in mixtures of the hydrogen isotopes and small contamination of heavy atoms. The subject of this part is commonly known because the basic mechanisms of mesoatomic processes were established in the early 1960s. The second part is devoted to a discussion of the results of recent experimental and theoretical studies in which the molecular structure dependence of the processes of atomic and nuclear capture of mesons in matter has been discovered (see review papers by Gershtein *et al.*, 1969; Ponomarev, 1973). So far the mechanisms of these phenomena are not quite clear. However, the phenomena themselves are so characteristic that they can already be used for the qualitative and quantitative analyses of substances and for the solution of various problems of structure chemistry (Petrukhin *et al.*, 1967b; Zinov *et al.*, 1972).

A. Successive Stages of Stopping and Absorption of Negative Mesons in Matter

In the process of slowing down and capture of negative mesons in matter, we can distinguish several stages that differ from one another by their duration and the nature of the physical phenomena. For a detailed discussion of these processes see the recent book by Massey *et al.* (1974).

1. *Slowing Down*

These processes were studied experimentally and theoretically in a number of earlier papers (Conversi *et al.*, 1947; Cosyns *et al.*, 1949; Fermi *et al.*, 1947; Fermi and Teller, 1947; Wheeler, 1947; Fröhlich *et al.*, 1948; Hubby, 1949; Rosenberg, 1949). The time needed for the negative mesons to be decelerated down to velocities of the order of the atomic electron velocities $v_0 \sim \alpha c$ is defined by the rate of their ionization loss and, in condensed substances, amounts to 10^{-9} to 10^{-10} sec.

2. *Capture of Mesons into Mesoatomic Levels*

Very little is known about the deceleration of mesons at velocities smaller than the velocities of atomic electrons, and also about the adiabatic way in which mesons go from the continuous into the discrete spectrum. Energy loss and the transition of the mesons to the bound state of the mesonic atom or mesonic molecule are related to the transfer of small amounts of energy. Therefore this process depends on the properties of the target material. The course of this process is sufficiently obvious only in metals where the meson can give up its energy to the conduction electrons (Fermi *et al.*, 1947; Fermi and Teller, 1947). The time of slowing down of mesons in conductors from $v \sim \alpha c$ to the capture into the discrete spectrum is $\sim 10^{-13}$ to 10^{-14} sec.

In dielectrics and gases, where other mechanisms of slowing down and capture are predominant, the total time of deceleration from an energy $\mathcal{E} \sim 10$–100 keV to the meson capture into the mesoatomic orbital is $\sim 10^{-12}$ sec.

All the theoretical studies confirm the assertion that the capture of negative mesons occurs at the highly excited mesoatomic levels with transfer of energy to the Auger electron of the atom, the dimensions of the mesonic atoms produced being comparable to the atomic dimensions. At the same time, the problem of the initial population of the mesonic-atom levels with different quantum numbers n and l is so far unclear (Burbridge and de Borde, 1953; de Borde, 1954; Baker, 1960; Gershtein, 1960; Mann and Rose, 1961; Martin, 1963; Au-Yang and Cohen, 1968; Russel, 1965, 1970; Haff and Tombrello, 1974; Condo, 1974).

Owing to the complexity of the problem, all the estimates of the probability for meson capture from the continuous to the discrete spectrum are based on the Fermi–Teller model (1947), in which the process of energy loss by the meson is regarded as the motion of the meson in a degenerate gas of the electron shells of the atom, the distribution of the electron-gas density being described by the Tomas–Fermi model. From these assumptions the so-called Fermi–Teller Z-law (or more accurately the $Z^{2/3}$-law) follows:

In a mixture of substances or in a chemical compound the probabilities of meson capture into the levels of different mesonic atoms are proportional to the values Z of the charges of the nuclei.

3. *Deexcitation Process in Mesonic Atoms*

In an isolated mesonic atom, the transition of the meson from highly excited states ($n \gg 1$) to states with lower energy occurs with the emission of γ quanta (radiative transitions) or the ejection of an electron of the atomic shell (Auger transitions). The relative contribution of these processes is a function of the transition energy ΔE: the probability for radiative transitions is $\omega_R \sim (\Delta E)^3$, while the probability for Auger transitions is $\omega_A \sim (\Delta E)^{-1/2}$. The Auger transitions are thus the main transitions in the case of transitions between highly excited mesonic atom levels, whereas for large ΔE the radiative transitions are predominant.†

On the whole, the radiative transitions are the most important ones in heavy atoms ($\omega_R \sim Z^4$) for $n \lesssim 7$, while the Auger transitions prevail in the light atoms. A detailed bibliography of earlier experimental and theoretical studies devoted to the X-ray spectra of mesonic atoms is given in the review papers of Comac *et al.* (1955), de Benedetti (1956), West (1958), and Stearns (1957), as well as in the papers of Fafarman and Shamos (1955), Demuer (1956), Stearns and Stearns (1957), Lathrop *et al.*

† The $Z\mu$ mesonic atoms that have lost some of their electrons in the Auger transitions of a μ meson are equivalent to atomic ions with atomic number $Z - 1$. In collisions with neutral Z atoms these ions can reconstruct their electron shells if the ionization potential for Z atoms is less than that for the atoms formed with atomic number $Z - 1$.

(1961b), Pevsner *et al.* (1961), Quitman *et al.* (1964), Bjorkland *et al.* (1965), Stearns *et al.* (1969), Beresin *et al.* (1970a,b), and Fetkovich *et al.* (1971).

The selection rules with respect to the orbital angular momentum l and its projection m ($\Delta l = \pm 1$, $\Delta m = \pm 1$ and 0) limit the value of ΔE and cause the deexcitation of the mesonic atom to proceed via cascade transitions. The particular selection rules that govern the most probable deexciting transitions tend to drive the mesonic atoms toward the so-called "circular orbits," i.e., states for which $l = n - 1$. Detailed calculations of the cascades in various mesonic atoms have been performed (see, for example, Eisenberg and Kessler, 1961a,b, 1963a,b; Kim, 1962; Suzuki, 1967; Srivasan and Sundarasan, 1968; Rook, 1963a,b, 1969; Atarashi and Narumi, 1971) under different assumptions concerning the initial distribution of mesons over the states with different l. The results of these calculations are in satisfactory agreement with experiment, but they fail to explain certain particularities of μ-mesonic X-ray spectra in chemical compounds, which are discussed in more detail in Section V.

The total time of the cascade transitions in different atoms is 10^{-12} to 10^{-10} sec.

It is worth noting that although the electromagnetic properties of μ^-, π^-, and K^- mesons are identical, the structure of the X-ray series of the corresponding mesonic atoms differs strongly, since in pπ and pK atoms the nuclear capture of mesons from the excited states of mesonic atoms is significant.

4. *Molecular Structure Effects*

All the foregoing considerations are based on the assumption that the processes of slowing down and absorption of mesons are independent of the molecular structure of matter, but experiments performed over the past years disprove this assumption. It was shown that the structure of the mesonic X-ray series depends on the type of chemical bond of the molecules (Zinov *et al.*, 1965a,b, 1967; Kessler *et al.*, 1967; Tausher *et al.*, 1968; Grin and Kunselman, 1970; Konin *et al.*, 1974).

The effect of the chemical bond is especially clearly demonstrated in the study of π^- meson capture in hydrogeneous substances (Panofsky *et al.*, 1951; Ammiraju and Lederman, 1956; Dunaitsev *et al.*, 1962, 1964; Charbe *et al.*, 1963; Bartlett *et al.*, 1964; Krumshtein *et al.*, 1968a,b). These effects have been satisfactorily explained within the framework of the model of large mesonic molecules (Ponomarev, 1965, 1967a) and are discussed in more detail in Section V and in the review papers by Gershtein *et al.* (1969) and Ponomarev (1973).

B. Some Features of Meson Capture in Hydrogen

The slowing down and absorption of μ^- mesons in hydrogen are characterized by a number of particular features. In particular, the energy loss of slow mesons in hydrogen is defined by the mechanism of adiabatic ionization (Fermi and Teller, 1947), according to which the electron leaves the atom, and the meson is captured into the highly excited orbits of the pμ atom.

On the basis of this idea, Wightman (1950) has calculated the time of slowing down of μ^- mesons from $v \sim c$ to $v \sim 0$, which in liquid hydrogen was found to be $\sim 10^{-9}$ sec.

The time of the radiative transitions of the μ^- meson in an isolated pμ atom from the highly excited state to the ground state is very large. However, in matter other ways of deexciting the pμ mesonic atom are possible. In particular, the pμ mesonic atom can lose its excitation energy inducing the dissociation of the hydrogen molecule, according to the reaction $(p\mu)^{**} + H_2 \rightarrow (p\mu)^* + H + H$, acquiring a kinetic energy of 1 eV (Panofsky *et al.*, 1950).

The singularities of the deexcitation of the pμ mesonic atom are associated with its electric neutrality, due to which it penetrates freely inside the electron shell of atoms. In such "deep" collisions there occurs an "external" Auger effect on the electrons of "another's" atoms. Leon and Bethe (1962) and Vermeuler (1969) have shown that, due to the H_2 dissociation and the external Auger effect, the μ^- (π^-) meson of the mesonic hydrogen atom reaches the $n = 4$ level during a time $\sim 10^{-12}$ sec.

When the mechanism of nuclear absorption of π^- mesons from $n \leq 5$ orbitals (Day *et al.*, 1959) is taken into account, this value is in agreement with the measured lifetime of π^- mesons in liquid hydrogen, $\approx 2.10^{-12}$ sec (Fields *et al.*, 1960; Bierman *et al.*, 1963; Block *et al.*, 1963; Doede *et al.*, 1963). Such an agreement points to the decisive role of collisions for the processes of deexcitation of mesonic hydrogen atoms. Similar conclusions are drawn from experiments with K^- mesons (Knopp *et al.*, 1965; Cresti *et al.*, 1965) and from theoretical estimates (Adair, 1959; Russel and Shaw, 1960).†

† An additional confirmation can be obtained from experiments on the measurement of the lifetime of pions in helium, $\tau_{He} \approx 10^{-10}$ sec, which was found to be two orders of magnitude longer than that of pions in hydrogen (Fetkovich and Pewitt, 1963; Block *et al.*, 1965; Berezin *et al.*, 1969, 1970a; Condo, 1964). This fact is precisely explained by the Coulomb repulsion, which does not permit the $(He\pi)^+$ system to penetrate inside the electron shell of the atoms. The lifetime of K^- mesons in helium is of the same order of magnitude as for π^- mesons (Fetkovich *et al.*, 1970).

Thus, we have every reason to assume that the time of deexcitation of the pμ mesonic atom up to the $n = 4$ level is about 10^{-12} sec in a condensed substance. Next, the μ meson is transferred to the ground level of the pμ mesonic atom due to the radiative transitions, contrary to the pπ mesonic atoms, where the nuclear capture of pions from $n = 4$, 5 levels is predominant. The X-ray series of the pμ mesonic atoms was observed by Placci *et al.* (1970a) and by Budick *et al.* (1971). The results of these experiments indicate that the pμ mesonic-atom cascades in liquid and gaseous hydrogen differ from those calculated for the free pμ mesonic atoms.

The above-mentioned mechanisms of deexcitation lead to the conclusion that the μ meson is able to go over to the K-orbit of the pμ mesonic atom during a time much shorter than its lifetime and then to participate in this state in many physically interesting reactions. A detailed description of these reactions is presented in Sections II–IV.

II. Experimental Study of Mesoatomic and Mesomolecular Processes in Hydrogen

A. General Characteristics of the Processes

The electroneutrality and small dimensions of pμ mesonic atoms permit them, like neutrons, to penetrate freely through the electron shells of the atoms and come to the nuclei within a distance on the order of the mesoatomic unit of length, $a_\mu = 2.56 \times 10^{-11}$ cm. This leads to a number of phenomena that are, in many respects, analogous to the phenomena of atomic and molecular physics.

Figure 1 presents schematically processes occurring when negatively charged mesons are stopped in a mixture of hydrogen isotopes with a small admixture of heavier elements with atomic number Z. In such a mixture, in addition to the decay of the free muon

$$\mu^- \rightarrow e + \nu_e + \nu_\mu \tag{1}$$

which takes place with a rate $\lambda_0 = 0.45 \times 10^6$ sec^{-1}, there occur the processes described below.

1. *Elastic Collisions of Mesonic Hydrogen Atoms*

These processes determine the diffusion of mesonic atoms from the point of stopping of the μ meson, which is directly observable in hydrogen

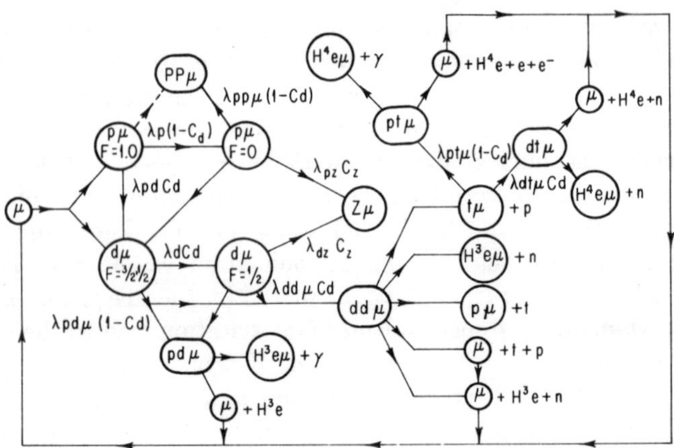

FIG. 1. Scheme of processes occurring when negative μ mesons are stopped and captured in a mixture of hydrogen and deuterium in the presence of impurities of elements with atomic number Z. For notation see Section II, A, 6.

chambers. The natural scale of the values of the cross sections for these processes is determined by the sizes of mesonic atoms

$$\sigma \sim 4\pi a_\mu^2 \approx 10^{-20}\ \text{cm}^2$$

However, in a number of cases, due to specific quantum effects, the cross sections turn out to be either essentially smaller (dμ + p scattering at low energy, pμ + p scattering in the lower hyperfine splitting (hfs) state of the pμ atom) or essentially larger (pμ + Z, dμ + Z scattering) than the value indicated.

2. *Transitions between hfs Levels in Collisions of Mesonic Atoms*

There exists a special mechanism indicated by Zel'dovich that provides a fast transition of μ mesonic hydrogen atoms from the upper to the lower levels (transitions $F = 1 \to F = 0$ and $F = \frac{3}{2} \to F = \frac{1}{2}$ for pμ and dμ mesonic atoms, respectively) (Gershtein, 1958a,b, 1961). This mechanism consists essentially in the transfer of the μ meson from one proton (or deuteron) to another with opposite spin orientation (Fig. 2). It should be noted that this transfer is due not to weak spin–orbit or spin–spin interactions, but to a very strong (in this resonance situation) exchange interaction, which is precisely the cause of high transfer rates. The corresponding transfer rates in liquid hydrogen are on the order of magnitude of $\lambda_p \sim 10^9$, $\lambda_d \sim 10^7$, and $\lambda_t \sim 10^9\ \text{sec}^{-1}$ for pμ, dμ and tμ mesonic atoms, respectively

(Section III,D). Transitions to lower hfs states result in an almost complete depolarization of the μ meson in hydrogen and change by several times the probability of capture of the μ^- mesons by protons and deuterons (Zeldovich and Gershtein, 1958a,b; Gershtein, 1961). They can also noticeably affect the catalysis of nuclear reactions by the μ mesons in the pdμ mesonic molecule (Section IV,D).

3. *Formation of Mesonic Molecules and Catalysis of Nuclear Fusion*

Mesonic molecules are formed in collisions of mesonic hydrogen atoms with hydrogen molecules. The rates of formation of ppμ, pdμ, and ddμ mesonic molecules in liquid hydrogen are $\sim 10^6$ sec^{-1} (Section III,G).

The formation of pdμ, ddμ, ptμ, and dtμ mesonic molecules may lead to subbarrier nuclear fusion of the hydrogen isotopes composing the mesonic molecule (Frank, 1947; Sakharov, 1948; Zel'dovich, 1954). The fact that in the majority of cases the reactions of nuclear synthesis proceed from the mesomolecular states was experimentally proved as early as 1957 by Alvarez *et al.* (1957). In the reaction $\mathrm{p} + \mathrm{d} \rightarrow {}^3\mathrm{He}$ they observed that the energy released was transferred to the muon of the pdμ mesonic molecule (see Section IV).

When the μ^- meson becomes free after the nuclear reaction, it can once more participate in the processes as a catalyst and thus can induce a new reaction of synthesis (Section IV). The formation of ppμ mesonic molecules does not lead to the reaction $\mathrm{p} + \mathrm{p} \rightarrow \mathrm{d} + e^+ + \nu_e$ since its probability is defined by the weak interaction. However, the process of the ppμ mesonic molecule formation essentially affects the capture of the meson by the proton according to the reaction $\mathrm{p} + \mu \rightarrow \mathrm{n} + \bar{\nu}_\mu$ (see Zavattini, Chapter V, Section 5, in Volume II).

4. *Transfer of μ^- Mesons from Light to Heavy Hydrogen Isotopes*

Because of the difference of the reduced masses of the μ mesonic atoms of different hydrogen isotopes, the process of transfer of the μ meson from a

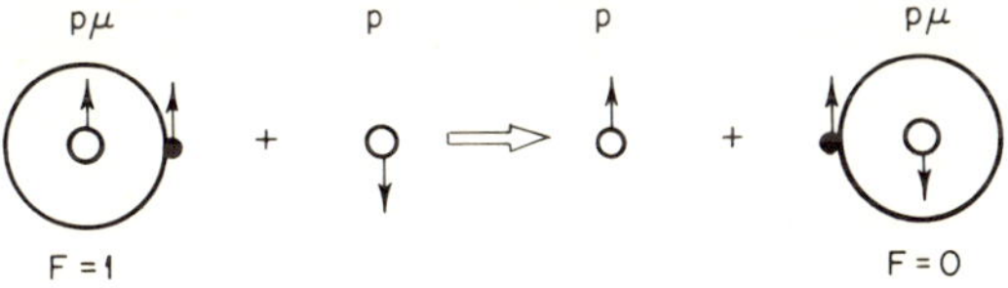

FIG. 2. Scheme of the mechanism of transitions between the hfs levels of mesonic hydrogen atoms in collisions.

FIG. 3. The "gap" between the point of stopping and the point of emission of a μ meson after the sequence of reactions (a) $p\mu + d \rightarrow d\mu + p$, (b) $d\mu + p \rightarrow d\mu + p$, (c) $d\mu + p \rightarrow pd\mu$, and (d) $pd\mu \rightarrow He^3 + \mu^-$. The "gap" is due to the anomalously small cross section for elastic scattering (b). The figure is taken from the paper of Dzhelepov *et al.* (1966).

light isotope to a heavy isotope is irreversible under the ordinary conditions due to energy reasons (Alvarez *et al.*, 1957). The transfer rates λ_{pd}, λ_{pt}, and λ_{dt} for the reaction

$$p\mu + d \rightarrow d\mu + p \tag{2}$$

and analogous to it in the order of magnitude are, respectively, $\lambda_{pd} \sim \lambda_{pt} \sim 10^{10}\ \text{sec}^{-1}$, $\lambda_{dt} \sim 10^{8}\ \text{sec}^{-1}$ (Section III,C). With such large transfer rates, an $\sim$0.1% deuterium concentration in liquid hydrogen leads to an almost complete transfer of mesons, according to reaction (2) (see Fig. 3). The $d\mu$ mesonic atom produced in this reaction leaves with an energy $\mathcal{E} \approx 45$ eV and, due to the small cross section for scattering of $d\mu$ by protons ($\sim 10^{-20}\ \text{cm}^2$), in a liquid-hydrogen chamber covers a noticeable distance, 1 mm. Thus, there appears a "gap," which separates the point of stopping of the μ meson from the point of its decay by reaction (1) or from the place of the catalysis process in the $pd\mu$ mesonic molecule. Since $\lambda_{pd} \gg \lambda_{pd\mu}$, then in collisions $p\mu + d$ reaction (2) proceeds first and only afterwards are the $pd\mu$ mesonic molecules formed according to the reaction $d\mu + p \rightarrow pd\mu$.

5. *Transfer of μ^- Mesons from Hydrogen to Nuclei of Heavy Elements*

When hydrogen is enriched in elements with atomic number $Z > 1$ there can occur the irreversible transition

$$\mathrm{p}\mu + Z \rightarrow Z\mu + \mathrm{p} \tag{3}$$

The cross section for such processes is, as a rule, rather large, $\lambda_{\mathrm{p}Z} \sim 10^{10}Z$ sec^{-1}, with the exception of the transitions to helium ($\lambda_{\mathrm{pHe}} \sim 10^6$ sec^{-1}). Therefore, even a small admixture of Z elements ($\sim$0.01% in liquid hydrogen) causes an almost total transfer of the μ mesons to nuclei Z.

The transitions (3) proceed mainly to highly excited levels of the mesonic atom $Z\mu$ and are therefore accompanied by emission of X-ray quanta and Auger electrons in the deexcitation of the $Z\mu$ mesonic atom, which may serve as an indicator of reaction (3).

In considering all the indicated mesoatomic processes, one can neglect the nuclear capture of μ mesons by protons (or deuterons), since its probability is about 10^{-3} of the decay probability of the free μ meson according to reaction (1).

6. *List of Notation*

The main characteristics of the mesonic hydrogen atoms are given in Table 1. The notation used is given in the following list.

$N_\mathrm{p}, N_\mathrm{d}, N_Z$: number of the nuclei of protium, deuterium, and the element with atomic number Z in cubic centimeters

$N_\mathrm{p}^0 = 4.2 \times 10^{22}$ sm^{-3}: number of the protium nuclei in liquid hydrogen in cubic centimeters

$$C_\mathrm{p} = \frac{N_\mathrm{p}}{N_\mathrm{p} + N_\mathrm{d} + N_Z},$$

$$C_\mathrm{d} = \frac{N_\mathrm{d}}{N_\mathrm{p} + N_\mathrm{d} + N_Z},$$

$$C_Z = \frac{N_Z}{N_\mathrm{p} + N_\mathrm{d} + N_Z}:$$ atomic concentrations of protium, deuterium, and the element Z

$$\frac{\rho}{\rho_0} = \frac{N_\mathrm{p} + N_\mathrm{d} + N_Z}{N_\mathrm{p}^0}:$$ ratio of the total density ρ to the density ρ_0 of liquid hydrogen

ISOTOPIC EXCHANGE PROCESSES AND TRANSITIONS OF μ MESONS FROM THE HYDROGEN ISOTOPES TO THE Z ELEMENT

$$\left.\begin{array}{l} \mathrm{p}\mu + \mathrm{d} \to \mathrm{d}\mu + \mathrm{p} \\ \mathrm{p}\mu + Z \to Z\mu + \mathrm{p} \end{array}\right. \left\{\begin{array}{l} \sigma_{\mathrm{pd}} \\ \sigma_{\mathrm{p}Z} \end{array}\right\} \text{ cross sections } \left\{\begin{array}{l} \omega_{\mathrm{pd}} \\ \omega_{\mathrm{p}Z} \end{array}\right\} \text{ rates}$$

$$\omega_{\mathrm{pd}} = \sigma_{\mathrm{pd}} v N_{\mathrm{d}} = \sigma_{\mathrm{pd}} v N_{\mathrm{p}}^{0} \frac{N_{\mathrm{p}} + N_{\mathrm{d}} + N_Z}{N_{\mathrm{p}}^{0}} \frac{N_{\mathrm{d}}}{N_{\mathrm{p}} + N_{\mathrm{d}} + N_Z}$$

$$= \lambda_{\mathrm{pd}} \frac{\rho}{\rho_0} C_{\mathrm{d}} = \lambda_{\mathrm{pd}}' C_{\mathrm{d}}$$

$\lambda_{\mathrm{pd}} = \sigma_{\mathrm{pd}} v N_{\mathrm{p}}^{0}$: isotopic exchange constant normalized to the number of atoms N_{p}^{0} in liquid hydrogen

$\lambda_{\mathrm{pd}}' = \lambda_{\mathrm{pd}} \dfrac{\rho}{\rho_0}$: the constant fitted to the conditions of the experiment

v: relative velocity of colliding nuclei.

In a similar way,

$$\omega_{\mathrm{p}Z} = \lambda_{\mathrm{p}Z}' C_Z, \qquad \lambda_{\mathrm{p}Z}' = \lambda_{\mathrm{p}Z} \frac{\rho}{\rho_0}, \qquad \lambda_{\mathrm{p}Z} = \sigma_{\mathrm{p}Z} v N_{\mathrm{p}}^{0}$$

ELASTIC SCATTERING OF MESONIC ATOMS

$$\left.\begin{array}{l} \mathrm{p}\mu + \mathrm{p} \to \mathrm{p}\mu + \mathrm{p} \\ \mathrm{d}\mu + \mathrm{d} \to \mathrm{d}\mu + \mathrm{d} \end{array}\right. \left\{\begin{array}{l} \sigma_{1,0} \\ \sigma_{3/2,1/2} \end{array}\right\}$$ cross sections for colliding energies exceeding the value of hyperfine splitting of mesonic atoms $\left\{\begin{array}{l} \sigma_{0} \\ \sigma_{1/2} \end{array}\right\}$ cross sections in the lower hfs state

$$\left.\begin{array}{l} \mathrm{d}\mu + \mathrm{p} \to \mathrm{d}\mu + \mathrm{p} \\ \mathrm{p}\mu + Z \to \mathrm{p}\mu + Z \end{array}\right. \left\{\begin{array}{l} \sigma_{\mathrm{dp}}^{\mathrm{el}} \\ \sigma_{\mathrm{p}Z}^{\mathrm{el}} \end{array}\right\}$$ cross sections for elastic scattering of mesonic hydrogen atoms by the nuclei of the elements

TRANSITIONS BETWEEN THE HFS LEVELS OF MESONIC ATOMS

$$\left.\begin{array}{l} \mathrm{p}\mu(F = 1) + \mathrm{p} \to \mathrm{p}\mu(F = 0) + \mathrm{p} \\ \mathrm{d}\mu(F = \tfrac{3}{2}) + \mathrm{d} \to \mathrm{d}\mu(F = \tfrac{1}{2}) + \mathrm{d} \end{array}\right. \left\{\begin{array}{l} \sigma_{1\to 0} \\ \sigma_{3/2\to 1/2} \end{array}\right\}$$

TABLE 1

MAIN CHARACTERISTICS OF THE MESONIC HYDROGEN ATOMS[a]

	Particle mass (a.u.)	Energy of mesonic atoms, E_μ (eV)	Hyperfine splitting, ΔE_{hfs} (eV)	Isotopic splitting of levels (eV)
μ	206.769			
p	1836.109	2528.52	0.183	$\Delta E_{pd} = 134.71$
d	3670.398	2663.23	0.049	$\Delta E_{dt} = 48.04$
t	5496.753	2711.27	0.241	$\Delta E_{pt} = 182.75$

[a] The values indicated are calculated on the basis of the data by Taylor *et al.* (1969b) and Selinov (1970). In mesoatomic units, $\mathcal{E}_\mu = 5626.53$ eV, $a_\mu = 2.55927 \times 10^{-11}$ cm.

Transition rates:

$$\omega_{1\to 0} = \lambda_p \frac{\rho}{\rho_0} C_p, \qquad \lambda_p = \sigma_{1\to 0} v N_p{}^0$$

$$\omega_{3/2\to 1/2} = \lambda_d \frac{\rho}{\rho_0} C_d, \qquad \lambda_d = \sigma_{3/2\to 1/2} v N_d{}^0$$

FORMATION OF MESONIC MOLECULES

$$\left.\begin{array}{l} p\mu + p \to pp\mu \\ d\mu + p \to pd\mu \\ d\mu + d \to dd\mu \end{array}\right. \left\{\begin{array}{c} \sigma_{pp\mu} \\ \sigma_{pd\mu} \\ \sigma_{dd\mu} \end{array}\right\} \text{formation cross section} \quad \left\{\begin{array}{l} \omega_{pp\mu} = \lambda_{pp\mu} \dfrac{\rho}{\rho_0} C_p \\ \omega_{pd\mu} = \lambda_{pd\mu} \dfrac{\rho}{\rho_0} C_p \\ \omega_{dd\mu} = \lambda_{dd\mu} \dfrac{\rho}{\rho_0} C_d \end{array}\right\} \text{formation rates}$$

LIFETIMES OF THE $p\mu$ AND $d\mu$ MESONIC ATOMS IN THE MIXTURE OF THE HYDROGEN ISOTOPES WITH ADMIXTURE OF Z NUCLEI

$$\frac{1}{\tau_{p\mu}} = \lambda_{p\mu} = \lambda_0 + \lambda'_{pp\mu} C_p + \lambda_{pd}' C_d + \lambda_{pZ}' C_Z$$

$$\frac{1}{\tau_{d\mu}} = \lambda_{d\mu} = \lambda_0 + \lambda'_{pd\mu} C_p + \lambda'_{dd\mu} C_d + \lambda_{dZ}' C_Z$$

NUCLEAR REACTIONS IN THE pdμ MESONIC MOLECULE FROM THE STATE WITH TOTAL MOMENTUM J

$$\left.\begin{array}{l} \mathrm{pd}\mu \to \mathrm{He}^3 + \gamma \\ \mathrm{pd}\mu \to \mathrm{He}^3 + \mu \end{array}\right. \left\{\begin{array}{c} \lambda_\gamma^J \\ \\ \lambda_\mu^J \end{array}\right\} \text{reaction rates} \quad \left\{\begin{array}{c} Y_\gamma \\ \\ Y_\mu \end{array}\right\} \text{reaction yield per one stopped meson}$$

Probabilities for different channels of the molecular reaction of synthesis in the mesonic molecule pdμ:

$$\Lambda_\gamma = \sum_J \omega_J \frac{\lambda_\gamma^J}{\lambda_0 + \lambda_f^J}, \qquad \Lambda_\mu = \sum_J \omega_J \frac{\lambda_\mu^J}{\lambda_0 + \lambda_f^J}, \qquad \lambda_f^J = \lambda_\gamma^J + \lambda_\mu^J$$

where ω_J is the population of the levels of the pdμ mesonic molecule in states with total momentum J.

SOME RELATIVE RATES OF MESOMOLECULAR PROCESSES

$$X_{\mathrm{pd}} = \frac{\lambda_{\mathrm{pd}}}{\lambda_0 + \lambda_{\mathrm{pp}\mu}}, \qquad X_{\mathrm{p}Z} = \frac{\lambda_{\mathrm{p}Z}}{\lambda_0 + \lambda_{\mathrm{pp}\mu}}, \qquad X_{\mathrm{d}Z} = \frac{\lambda_{\mathrm{d}Z}}{\lambda_0 + \lambda_{\mathrm{pd}\mu}}$$

B. Methods of Experimental Determination of Main Characteristics of Mesomolecular Processes in Hydrogen

To determine experimentally the rates of mesoatomic and mesomolecular processes, use is made of a number of directly observable effects, such as nuclear reactions in mesonic molecules, transitions of μ^- mesons from the hydrogen nuclei to the nuclei of the elements with atomic number $Z > 1$, diffusion of mesonic atoms, etc. The study of these phenomena under various conditions makes it possible to determine the characteristics of other mesomolecular processes that compete with the former and the direct measurement of which is rather difficult (for example, formation of ppμ mesonic molecules).

1. *Determination of the Ratio* $X_{\mathrm{pd}} = \lambda_{\mathrm{pd}}/(\lambda_0 + \lambda_{\mathrm{pp}\mu})$

The technique of determination of the mesoatomic constants can conveniently be seen by the example of the catalysis of nuclear fusion in the pdμ mesonic molecule in liquid hydrogen at a low deuterium concentration, $C_{\mathrm{d}} \ll 1$, when the direct "landing" of μ mesons on the K orbit of the dμ mesonic atom can be disregarded. The ratio X_{pd} can be found by measuring the yields Y_γ and Y_μ of the reaction:

$$\mathrm{pd}\mu \to \mathrm{He}^3\mu + \gamma \tag{4a}$$

or

$$\mathrm{pd}\mu \to \mathrm{He}^3 + \mu \tag{4b}$$

The indicated yields are connected with the probabilities n_γ and n_μ per act of reaction (4) by the relations

$$Y_\gamma = \frac{n_\gamma}{1 - n_\mu}, \qquad Y_\mu = \frac{n_\mu}{1 - n_\mu} \tag{5}$$

taking into account the μ-meson regeneration in reaction (4b). In liquid hydrogen the quantities n_γ and n_μ are, respectively,

$$n_\gamma = \frac{\lambda_{\mathrm{pd}} C_{\mathrm{d}}}{\lambda_0 + \lambda_{\mathrm{pp}\mu} + \lambda_{\mathrm{pd}} C_{\mathrm{d}}} \frac{\lambda_{\mathrm{pd}\mu}}{\lambda_0 + \lambda_{\mathrm{pd}\mu}} \Lambda_\gamma \tag{6a}$$

$$n_\mu = \frac{\lambda_{\mathrm{pd}} C_{\mathrm{d}}}{\lambda_0 + \lambda_{\mathrm{pp}\mu} + \lambda_{\mathrm{pd}} C_{\mathrm{d}}} \frac{\lambda_{\mathrm{pd}\mu}}{\lambda_0 + \lambda_{\mathrm{pd}\mu}} \Lambda_\mu \tag{6b}$$

In formulas (6a) and (6b) it is assumed that $C_{\mathrm{d}} \ll 1$ and $\lambda_{\mathrm{pd}\mu} \gg \lambda_{\mathrm{dd}\mu} C_{\mathrm{d}}$. The first factor on the right-hand side equals the $\mathrm{p}\mu \to \mathrm{d}\mu$ transfer probability, and the second one the probability for formation of the mesonic molecule in the reaction $\mathrm{d}\mu + \mathrm{p} \to \mathrm{pd}\mu$.

Since the probability for the μ-meson regeneration is small ($n_\mu < 2.8\%$) (see Section IV,C), the reaction yields $Y_\gamma \approx n_\gamma$ and $Y_\mu \approx n_\mu$.

The C_{d} dependence of Y_γ and Y_μ is defined by the $\mathrm{p}\mu \to \mathrm{d}\mu$ transfer probability. For $\lambda_{\mathrm{pd}} C_{\mathrm{d}} \gg \lambda_{\mathrm{pp}\mu}, \lambda_0$, which is true even at rather small $C_{\mathrm{d}} \sim 10^{-3}$, the $\mathrm{p}\mu \to \mathrm{d}\mu$ transfer probability is close to unity and a "saturation" of the catalysis process can be observed, as indicated by Alvarez *et al.* (1957): the values of Y_γ and Y_μ remain unaffected by further increasing C_{d}.†

According to formulas (5) and (6) the quantities $1/Y_\gamma$ and $1/Y_\mu$ depend linearly on $1/C_{\mathrm{d}}$:

$$1/Y_{\gamma(\mu)} = \mathrm{const}\left(1 + \frac{\lambda_0 + \lambda_{\mathrm{pp}\mu}}{\lambda_{\mathrm{pd}}} \frac{1}{C_{\mathrm{d}}}\right) \tag{7}$$

This dependence is confirmed experimentally and makes it possible to determine the ratio $X_{\mathrm{pd}} = \lambda_{\mathrm{pd}}/(\lambda_0 + \lambda_{\mathrm{pp}\mu})$ knowing the slope of the

† At high deuterium concentration ($C_{\mathrm{d}} \sim 1$), the quantities Λ_μ and Λ_γ change since, in this case, the population ω_J of the hfs levels of the $\mathrm{pd}\mu$ mesonic molecule also changes (Section IV, D).

corresponding straight lines.† The quantity X_{pd} was measured by Alvarez *et al.* (1957) and Schiff (1961) from the yield Y_μ and by Bleser *et al.* (1962, 1963) from the yield Y_γ:

$$\frac{1}{X_{pd}}\begin{cases} \approx 0.91 \times 10^{-4} & \text{(Alvarez } et\ al.\text{, 1957)} \\ = \left(1.12 \begin{matrix} +0.77 \\ -0.46 \end{matrix}\right) \times 10^{-4} & \text{(Schiff, 1961)} \\ = (1.59 \pm 0.05) \times 10^{-4} & \text{(Bleser } et\ al.\text{, 1963)} \end{cases} \tag{8}$$

2. *Direct Measurement of the Rate of Process* $p\mu + d \to d\mu + p$

The rate λ_{pd} was measured by Dzhelepov *et al.* (1962a) in a hydrogen diffusion chamber at a pressure of 25 atm and a deuterium concentration $C_d = 0.44\%$. Under conditions of the diffusion chamber, the "gap" between the point of stopping of a μ meson and the point of its decay after diffusion of the $d\mu$ mesonic atom produced is a few millimeters. It was possible to neglect the probability for formation of $pp\mu$ mesonic molecules; however, it was necessary to take into account the transfer of μ mesons from protons to carbon and oxygen nuclei entering into the compound of alcohol that is present in the chamber. The probability for such transfers was determined by different methods, which had led to the same results. By measuring the rate λ_{pd} and the ratio (8), it is possible to determine the quantity $\lambda_{pp\mu}$.

3. *Measurements of* $\lambda_{pp\mu}$, $\lambda_{pd\mu}$, *and* λ_f *by the Time Distribution of* γ *Quanta from the Reaction* $p + d \to {}^3He + \gamma$ *in the* $pd\mu$ *Mesonic Molecule*

It is convenient to demonstrate the indicated method by the example of catalysis of nuclear synthesis in liquid hydrogen with a deuterium admixture $C_d \approx 1 \div 2\%$, i.e., under conditions where all the μ mesons are fast transferred from the protons to the deuterons. Starting from methodical considerations, we assume the nuclear reaction probability p + d to be characterized by definite averaged constants $\bar{\lambda}_\gamma$ and $\bar{\lambda}_\mu$ and we neglect the formation of mesonic molecules $dd\mu$ and the possibility of the reappearance in the cycle of a meson that has given rise to the nuclear reaction. Under these assumptions, the kinetics of the process is described by the

† The quantity $1/X_{pd} = \bar{C}_d$ is equal to the deuterium concentration at which half of the μ mesons in liquid hydrogen are transferred from the protons to the deuterons, according to reaction (2). This is easily established from the relation

$$W(p\mu \to d\mu) = \frac{\lambda_{pd}C_d}{\lambda_0 + \lambda_{pp\mu} + \lambda_{pd}C_d} = \frac{X_{pd}C_d}{1 + X_{pd}C_d}$$

following system of equations:

$$\frac{dn_{d\mu}}{dt} = -(\lambda_0 + \lambda_{pd\mu})n_{d\mu}$$

$$\frac{dn_{pd\mu}}{dt} = \lambda_{pd\mu}n_{d\mu} - (\lambda_0 + \bar{\lambda}_\mu + \bar{\lambda}_\gamma)n_{pd\mu} \tag{9}$$

$$\frac{dn_\gamma}{dt} = \bar{\lambda}_\gamma n_{pd\mu}$$

It follows from this system that the yield of γ quanta per unit time per μ meson is†

$$\frac{dn_\gamma}{dt} = \frac{\bar{\lambda}_\gamma \lambda_{pd\mu}}{\lambda_{pd\mu} - \bar{\lambda}_f} \exp(-\lambda_0 t)[\exp(-\bar{\lambda}_f t) - \exp(-\lambda_{pd\mu} t)] \tag{10}$$

where

$$\bar{\lambda}_f = \bar{\lambda}_\gamma + \bar{\lambda}_\mu$$

The time distribution of γ quanta from the reaction $p + d \to {}^3He + \gamma$ under conditions of saturation ($C_d = 1.8\%$) was first measured by Ashmore *et al.* (1958), who established the presence of two processes occurring with different rates. Since the formula (10) is symmetric in $\bar{\lambda}_f$ and $\lambda_{pd\mu}$, they could not decide, without additional assumptions, which of the processes is "fast" and which "slow": the nuclear reaction or the formation of the pdμ mesonic molecules.

When processing similar measurements of dn_γ/dt a Columbia group (Bleser *et al.*, 1963) took into account regeneration of μ mesons in the process of catalysis. They also used the theoretical estimates for the relations between the rates of nuclear reactions from different spin states of the pdμ mesonic molecule.‡ The problem of the relationship between the

† Equation (10) is analogous in form to the time dependence of the activity of a daughter product of radioactive decay in the radioactive chain with two different lifetimes. This is quite natural, since in the case under consideration the nuclear reaction $p + d \to {}^3He + \gamma$ is preceded by the formation of pdμ mesonic molecules. In the case $\lambda_0 + \lambda_{pp\mu} \gtrsim \lambda_{pd}C_d$, i.e., under the conditions of very low deuterium concentration, there also arises an expression of the type (10), which, however, contains three exponentials with decay constants $\lambda_0 + \lambda_{pp\mu} + \lambda_{pd}C_d$, $\lambda_0 + \lambda_{pd\mu}$, and $\lambda_0 + \bar{\lambda}_f$, which correspond to the inverse lifetimes of the mesonic atoms pμ and dμ and the mesonic molecule pdμ, respectively (Bleser *et al.*, 1963).

‡ In reality, the time distribution of γ quanta depends weakly on the hfs of the levels of the pdμ mesonic molecule, since the reaction $p + d \to {}^3He + \gamma$ proceeds mainly from two spin states of the pdμ mesonic molecule, which have, moreover, close lifetimes (Section IV, C).

rates $\lambda_{pd\mu}$ and $\bar{\lambda}_f$ was solved by means of a subsidiary experiment in which the mixture of hydrogen and deuterium was enriched in neon (Bleser *et al.*, 1963). Such an addition obviously changes the lifetimes of the dμ mesonic atoms, due to the process $d\mu + Ne \rightarrow Ne\mu + d$, but does not affect the rate of nuclear reaction in the pdμ mesonic molecule. In so doing, one has succeeded in establishing that the fast process is the formation of the pdμ mesonic molecule, and in determining the rate of transfer of the μ meson from the deuterium to the neon (Section II,B,7).

By combining these measurements of dn_γ/dt with those performed under conditions far from saturation (low and different concentrations C_d), it is possible to obtain the constants $\lambda_{pd\mu}$, $\bar{\lambda}_f$, and $\lambda_{pp\mu}$.

4. *Combined Determination of Rates* $\lambda_{pp\mu}$, $\lambda_{pd\mu}$, λ_{pZ}, *and* λ_{dZ} *by the Reactions* $p\mu \rightarrow Z\mu$ *and* $d\mu \rightarrow Z\mu$

A more direct method for determining the quantities $\lambda_{pp\mu}$ and $\lambda_{pd\mu}$ was suggested in 1962 by the Mukhin group (private communication) and applied in experiments of a CERN group (Conforto *et al.*, 1962a, 1964) and a JINR group (Budyashov *et al.*, 1968). This method uses no model theoretical estimates, but proceeds as follows:

In hydrogen with a small admixture of Z atoms ($C_Z \sim 10^{-5}$) the direct landing of μ mesons on the Z atoms is negligible. The subsequent transfer of mesons from the proton to Z nuclei occurs in the highly excited levels of the $Z\mu$ mesonic atom with γ-ray emission. The time distribution of γ quanta from cascade transitions in the $Z\mu$ atom is defined by the lifetime of pμ mesonic atoms, since the time of the radiative transitions is by several orders of magnitude shorter than $\tau_{p\mu} = 1/\lambda_{p\mu}$. Thus, by measuring the time distribution of γ quanta, e.g., from the $2p \rightarrow 1s$ transition in the $Z\mu$ atom,

$$dn_\gamma/dt = \text{const} \exp(-\lambda_{p\mu} t) \tag{11}$$

it is possible to measure the quantity $\lambda_{p\mu} = \lambda_0 + \lambda_{pp\mu} + \lambda_{pZ} C_Z$. The quantity $\lambda_{pZ} C_Z$ can also be determined directly by measuring the decrease in the number of the electrons resulting from the decay (1) in the presence of an admixture of Z atoms when the μ meson capture becomes important. Measuring the quantity

$$B = \frac{N_e^0 - N_e^Z}{N_e^0} = \frac{\lambda_{pZ} C_Z}{\lambda_{p\mu}} \Lambda_Z \tag{12}$$

where N_e^0 and N_e^Z are the numbers of the decay electrons from reaction (1) in pure hydrogen and in the presence of a C_Z impurity, respectively, and Λ_Z is the probability for absorption of the meson by the Z nucleus, it

TABLE 2

RATE OF FORMATION OF ppμ MESONIC MOLECULES

Reference	$\lambda_{pp\mu} \times 10^6$ sec^{-1}
Dzhelopov *et al.* (1962a)	1.5 ± 0.6
Bleser *et al.* (1963)	1.89 ± 0.20
Conforto *et al.* (1964)	2.55 ± 0.18
Budyashov *et al.* (1968)	2.74 ± 0.25
Zeldovich and Gershtein (1958a, 1960)	2.6
Cohen *et al.* (1960)	3.9

is possible to determine the quantity λ_{pZ}. The method proves to be especially effective for large Z for which $\Lambda_Z \approx 1$, since in this case

$$\lambda_0 + \lambda_{pp\mu} \approx \lambda_{p\mu} N_e^Z / N_e^0 \tag{13}$$

If the quantity Λ_Z is unknown, it can be obtained by measuring $\lambda_{p\mu}$ and the ratio B at different concentrations C_Z.

In this way the quantity $\lambda_{pp\mu}$ was measured in liquid hydrogen with a neon admixture (Conforto *et al.*, 1964) and in gaseous hydrogen with a xenon admixture (Budyashov *et al.*, 1968). The results of measurements and the theoretical estimates for $\lambda_{pp\mu}$ are given in Table 2.

If in similar experiments the concentration of deuterium $C_d \sim 1\%$ is sufficient for the fast transfer p$\mu \to$ dμ to occur, but insufficient for the ddμ mesonic molecule formation to be disregarded, as compared with the pdμ mesonic molecule formation, it is possible to determine the quantity

$$\lambda_{d\mu} = \lambda_0 + \lambda_{pd\mu} C_p + \lambda_{dZ} C_Z \tag{14}$$

from the measurements of the time distribution of the γ quanta of the cascade transitions in the $Z\mu$ atom that follow the d$\mu \to Z\mu$ transitions. In combination with the B measurements, the latter allows one to find the rates $\lambda_{pd\mu}$ and λ_{dZ} (Conforto *et al.*, 1964).

At sufficiently high concentration ($C_d \sim 1$), this method can be used to measure the formation rates for the ddμ mesonic molecules (Bystritskij *et al.*, 1974a).

5. *Determination of the Rates* $\lambda_{pd\mu}$ *and* $\lambda_{dd\mu}$ *from the Yields of Nuclear Reactions in* pdμ *and* ddμ *Mesonic Molecules*

Measuring the yield Y_μ of nuclear synthesis in the pdμ mesonic molecule for different hydrogen densities, it is possible to measure the rate $\lambda_{pd\mu}$ and the nuclear reaction probability Λ_μ. At a deuterium concentration that is

TABLE 3

RATE OF FORMATION OF pdμ MESONIC MOLECULES

Reference	$\lambda_{pd\mu}$, $\times 10^6$ sec^{-1}
Bleser *et al.* (1963)	5.8 $\pm$ 0.3
Conforto *et al.* (1964)	6.82 $\pm$ 0.25
Dzhelepov *et al.* (1966)	1.8 $\pm$ 0.6
Belyaev *et al.* (1959), Zeldovich and Gerstein (1960)	1.3
Cohen *et al.* (1960)	3.0

low but sufficient for the p$\mu \to$ dμ transfer to be saturated, the yields Y_μ and Y_μ' of the reaction pd$\mu \to {}^3$He + μ in liquid and gaseous hydrogen are, respectively,

$$Y_\mu = \frac{\lambda_{pd\mu}}{\lambda_0 + \lambda_{pd\mu}} \Lambda_\mu, \qquad Y_\mu' = \frac{\lambda_{pd\mu}\rho/\rho_0}{\lambda_0 + \lambda_{pd\mu}\rho/\rho_0} \Lambda_\mu \tag{15}$$

In liquid hydrogen the yield Y_μ was measured by Alvarez *et al.* (1957), Fetkovich *et al.* (1960), Schiff (1961), and Doede (1963), and in the two latter measurements it was found to be $Y_\mu = (2.84 \pm 0.25) \times 10^{-2}$. In experiments of Dzhelepov *et al.* (1966) the yield Y_μ' was measured in a diffusion chamber, at a pressure of 23 atm. The value of $\lambda_{pd\mu}$ calculated on the basis of these measurements is smaller by a factor of 3 than the value of Bleser *et al.* (1963) and Conforto *et al.* (1964) measured in liquid hydrogen (Table 3). The cause of this difference is unclear. In particular, it may be due to a possible nontrivial dependence of the rate of the pdμ mesonic molecule formation on the hydrogen density or the kinetic energy of the dμ mesonic atoms (cf. Section III,G,3).

The rate $\lambda_{dd\mu}$ can be determined by the yield of nuclear synthesis d + d in a mixture with a high deuterium concentration if, according to the theoretical estimates, the reaction probability in the ddμ mesonic molecule is assumed to be unity ($\lambda_f = \infty$)† and the measured values of $\lambda_{pd\mu}$ are used.

The experimental values of $\lambda_{dd\mu}$ differ strongly from one another and in some cases exceed by an order of magnitude the theoretical estimates (Table 4). In the paper of Dzhelepov *et al.* (1966) it is noted that the

† In the paper of Doede (1963) an attempt has been made to determine the rates $\lambda_{dd\mu}$ and λ_f in liquid deuterium with a 0.96% protium impurity by means of a subsidiary experiment when air is dissolved in deuterium. However, within statistical errors, the values obtained—$\lambda_{dd\mu} \approx 3.6 \times 10^5$ sec^{-1} and $\lambda_f \approx 2.5 \times 10^5$ sec^{-1}—are compatible with the result $\lambda_{dd\mu} \approx 10^5$ sec^{-1} and $\lambda_f = \infty$ (Dzhelepov *et al.*, 1965a, 1966).

values of the rate $\lambda_{dd\mu}$ given in Table 4 are measured for different average kinetic energies $\bar{\mathcal{E}}_{d\mu}$ of dμ mesonic atoms. This dependence is naturally explained by the mechanism of ddμ mesonic molecule formation suggested by Vesman (1967a) (Section III,G,3).

6. *Measurement of Elastic Cross Sections for Mesonic Hydrogen Atoms*

The effective cross sections for scattering of pμ and dμ mesonic atoms by protons, deuterons, and nuclei of heavier elements (carbon, oxygen) can be determined by studying the diffusion of these mesonic atoms. To this end, the Dubna group (Dzhelepov *et al.*, 1962a,b,c, 1964d, 1965b) measured in a diffusion chamber the paths of mesonic atoms directly, from the point of stopping of the μ meson to the point of its decay or capture by a nucleus of the admixture. In these experiments the hydrogen density and the concentration of deuterium and admixtures were varied.

Another method for determining the elastic cross sections for mesonic hydrogen atoms was suggested by Alberigi Quaranta *et al.* (1967a), who used a target filled with gaseous hydrogen or deuterium and divided into thin layers (~1.5 mm) by gold foils. The time of diffusion of mesonic

TABLE 4

RATE OF FORMATION OF ddμ MESONIC MOLECULES

Reference	$\lambda_{dd\mu}$, $\times 10^6$ sec^{-1}	$\bar{\mathcal{E}}_{d\mu}$ (eV)
Dzhelepov *et al.* (1964d)	1.08 ± 0.27	0.043
Dzhelepov *et al.* (1966)	0.62 ± 0.15	0.046
Dzhelepov *et al.* (1966)	0.66 ± 0.19	0.042
Bystritskij *et al.* (1974b)	0.73 ± 0.07	0.046
Dzhelepov *et al.* (1966)	0.53 ± 0.21	0.170
Alvarez *et al.* (1957)[a]	$0.80^{+0.53}_{-0.35}$	0.012
Fetkovich *et al.* (1960)	0.076 ± 0.015	0.0039
Doede (1963)[b]	0.103 ± 0.004	0.0030
Zeldovich and Gershtein (1960)	0.4	independent of $\mathcal{E}_{d\mu}$
Cohen *et al.* (1960)	0.036	independent of $\mathcal{E}_{d\mu}$
Vesman (1967b)	$f(\bar{\mathcal{E}}_{d\mu})$	(Section III, G, 3)

[a] A few reactions d + d → p + t were observed. The data are given under the assumption that "a few" is 3.

[b] The given $\lambda_{dd\mu}$ value is calculated according to the data of this paper under the assumption $(\lambda_f)_{dd\mu} \gg \lambda_{dd\mu}$.

TABLE 5

CROSS SECTIONS FOR ELASTIC SCATTERING OF $p\mu$ MESONIC ATOMS IN THE LOWER HFS STATE

Reference	σ_0, $\times 10^{-21}$ cm^2
Dzhelepov *et al.* (1965b)	167 $\pm$ 30
Alberigi Quaranta *et al.* (1967a)	7.6 $\pm$ 0.7
Zeldovich and Gershtein (1960)	1.2
Cohen *et al.* (1960)	8.2
Matveenko and Ponomarev (1970b)	2.5

atoms from the point of their production to the gold foil was determined by the time distribution of γ quanta from the 2p $\to$ 1s transition in the mesonic atoms of gold produced in the $p\mu \to Z\mu$ transitions on the surface of the foil.

In both cases, the process of diffusion of mesonic atoms was modeled by the Monte-Carlo method under the assumption that their initial energy was ~1 eV and taking into account the molecular structure of hydrogen.† The results of measurements of the effective cross section for $p\mu + p$ scattering and the theoretical estimates are given in Table 5.

It is worth noting (Section III,D) that the fast transition $p\mu$ $(F = 1) \to$ $p\mu$ $(F = 0)$ leads to the fact that in the experiments on scattering of $p\mu$ mesonic atoms by protons one really measures the cross section σ_0, i.e., the elastic scattering cross section of $p\mu$ mesonic atoms in the lower hfs state. The theoretical estimates show that it should be anomalously small (Section III,E,1). The experimental data of both groups, in spite of their obvious disagreement (which is to be further clarified), confirm indirectly the existence of a fast transition $(F = 1) \to (F = 0)$. The experimental data on the cross section for elastic scattering $d\mu + d$ given in Table 6 are in satisfactory agreement with theory.

Compared to the cross section for the $d\mu + d$ process, the measured elastic $d\mu + p$ cross section was found to be anomalously small. This fact has already been mentioned by Alvarez *et al.* (1957). In fact, when natural hydrogen was enriched in deuterium to 0.3% an increase of the number of "gaps" (Fig. 3) was observed due to the path of $d\mu$ in hydrogen after the transition $p\mu \to d\mu$.

† The calculation of these effects is performed in just the same manner as the calculation of the cross section for scattering of slow neutrons on molecules by the Fermi pseudopotential method (Gershtein, 1958b; Matone, 1969).

The fact that by increasing the deuterium concentration to 4.3% the "gap" is not observed may be associated with the increase in importance of the scattering dμ + d, whose effective cross section is much greater than that for dμ + p. The smallness of the elastic dμ + p scattering is due to the Ramsauer–Townsend quantal effect (Section III,E,3).

7. *Determination of the* p$\mu \to Z\mu$ *and* d$\mu \to Z\mu$ *Transfer Rates*

The transfers of μ mesons from protons or deuterons to the nuclei of the Z element can be studied by various methods. To this end, one can use the γ quanta (or Auger electrons) from the cascade transitions in the $Z\mu$ atom, nuclear reactions induced by capture of the μ meson by the Z nucleus, the time distribution or decrease of the number of electrons from the μ-decay, and measurement of the yield of nuclear reactions in hydrogen with an admixture of the Z element. The results of measurements of the transfer rates λ_{pZ} and λ_{dZ} are given in Table 7.

Measurements of the rates of transitions p$\mu \to Z\mu$ to the nuclei of carbon and oxygen in a diffusion chamber were performed by Zaimidoroga *et al.* (1963) and Dzhelepov *et al.* (1962a,b,c, 1964d, 1965b). It was established that this transition proceeds to high orbitals of the $Z\mu$ mesonic atoms, which is confirmed by soft Auger electrons produced in cascade transitions in $Z\mu$ mesonic atoms. The probability for the Auger electron emission was found to be $\approx$80% which is in good agreement with theoretical estimates.

A systematic study of the p$\mu \to Z\mu$ transitions was performed by Basiladze *et al.* (1965), Alberigi Quaranta *et al.* (1967b), and Placci *et al.* (1969), and the d$\mu \to Z\mu$ transitions by Placci *et al.* (1970b).

In the first paper the absolute transfer rate λ_{pZ} was determined by

TABLE 6

Cross Section for Elastic Scattering of dμ Mesonic Atoms in the Lower hfs State[a]

Reference	$\sigma_{1/2}$, $\times 10^{-19}$ cm^2
Dzhelopov *et al.* (1964c)	4.15 ± 0.29
Dzhelepov *et al.* (1966)	1.5 ± 0.5
Alberigi Quaranta *et al.* (1967a)	0.55 ± 0.20
Zeldovich and Gershtein (1960)	3.3
Cohen *et al.* (1960)	3.5
Matveenko and Ponomarev (1970b)	1.8

[a] $\sigma_{1/2} = \frac{1}{3}\sigma_{1/2}(J = \frac{1}{2}) + \frac{2}{3}\sigma_{1/2}(J = \frac{3}{2})$.

TABLE 7

RATES OF TRANSITIONS OF THE μ^- MESON FROM THE PROTON AND DEUTERON TO Z NUCLEI[a]

Z	Element	λ_{pZ}, $\times 10^{11}$ sec^{-1}	λ_{dZ}, $\times 10^{11}$ sec^{-1}
2	He	$\sim 10^{-3}$[b]	$\sim 10^{-2}$[i]
		$\sim 10^{-2}$[i]	$(1.6 \pm 2) \times 10^{-4}$[g]
6	C	0.51 ± 0.10[d]	—
		0.31 ± 0.14[b]	
6, 8	C, O		
		0.26 ± 0.11[c]	0.47 ± 0.14[e]
7	N	—	1.00 ± 0.30[g]
		0.3 ± 0.1[i]	1.42 ± 0.20[g]
10	Ne	1.16 ± 0.28[f]	1.8 ± 0.7[i]
			0.81 ± 0.10[j]
			1.35[k]
		1.21 ± 0.18[d]	
18	Ar	3.48 ± 0.62[f]	0.94 ± 0.08[g]
		1.46 ± 0.14[h]	
36	Kr	6.84 ± 1.26[f]	2.58 ± 0.24[g]
		5.68 ± 0.56[h]	
		4.46 ± 0.36[d]	
54	Xe	9.92 ± 0.92[f]	4.72 ± 0.46[g]
		8.82 ± 0.40[h]	

[a] The quantities λ_{pZ} and λ_{dZ} are normalized by the atomic concentration $N_p{}^0 = 4.2 \times 10^{22}$ cm^{-3} in liquid hydrogen (see the list of notations). Therefore, the values given in Alberigi Quaranta *et al.* (1967b), Placci *et al.* (1967), and Placci *et al.* (1969) are increased by a factor of two. The values of λ_{pHe} and λ_{dNe} (Schiff, 1961) are determined from the measurements of X_{pNe} and X_{dNe} using $\lambda_{pp\mu}$ and $\lambda_{pd\mu}$ from Bleser *et al.* (1963) and Conforto *et al.* (1964).

[b] Zaimidoroga *et al.* (1963).

[c] Dzhelepov *et al.* (1965b).

[d] Basiladze *et al.* (1965).

[e] Dzhelepov *et al.* (1964d).

[f] Alberigi Quaranta *et al.* (1967b).

[g] Placci *et al.* (1967).

[h] Placci *et al.* (1969).

[i] Schiff (1961).

[j] Bleser *et al.* (1963).

[k] Conforto *et al.* (1964).

analyzing the decrease in the number of the electrons from the μ decay (1) with increasing concentration C_Z.

For small Z, this method is ineffective due to the small probability of nuclear capture of mesons by light nuclei. Therefore, an interesting method has been used in this work ("method of three gases"), which consists of the following: Hydrogen is slightly enriched in a heavy gas (argon or xenon) that is, however, sufficient for the transfer $p\mu \rightarrow Z\mu$ to occur and, consequently, for reaction (1) to be almost completely suppressed. An admixture of a light gas Z' to this mixture results in a competitive process $p\mu \rightarrow Z'\mu$, which partially regenerates reaction (1) since the μ capture rate in light nuclei is smaller than the rate λ_0 of the decay of a free μ^- meson. Studying reaction (1) in such a mixture it is possible to determine $\lambda_{pZ'}$.

A modified version of the method of three gases has been used by Alberigi Quaranta *et al.* (1967b), who determined λ_{pXe} by measuring the time distribution of γ quanta from the $2p \rightarrow 1s$ transition in the mesonic atoms of xenon, produced in the transfer $p\mu \rightarrow Xe\mu$. Then, measuring the relative decrease of the yield of such γ quanta with an addition of the light gas Z' to the mixture of hydrogen and xenon, the authors determined the rate $\lambda_{pZ'}$.

In the work of Placci *et al.* (1967, 1969) the transfer rates λ_{pZ} and λ_{dZ} were determined from measurements of the time distribution of the μ-decay electrons in hydrogen (or deuterium) with Z admixture.

As is seen from Table 7, there is some disagreement between the results of different groups. However, this does not prevent us from noting some similarities:

(a) The values of λ_{pZ} increase about proportionally to Z; and

(b) the same method (Placci *et al.*, 1967, 1969) gives $\lambda_{pZ}/\lambda_{dZ} \approx 2$ for the elements with $Z \gtrsim 20$, and for the elements $Z < 20$ the experimental data do not contradict the equality $\lambda_{pZ} \approx \lambda_{dZ}$.

As was observed by Placci *et al.* (1969) these conclusions follow from the theoretical estimates of Gershtein (1962). It is remarkable that the experimental rate λ_{pHe} is small (Section III,H,3).

Note that the transfer $p\mu \rightarrow Z\mu$ can be used, for example, for the study of the μ capture by the nuclei of rare isotopes playing the role of admixtures to hydrogen or deuterium (Backenstoss *et al.*, 1971).

III. Methods and Results of Theoretical Calculations of Mesoatomic and Mesomolecular Processes

A. General Formulation of the Problem

The main mesoatomic processes occur, as a rule, at distances on the order of a few mesoatomic units of length, $a_\mu = 2.56 \times 10^{-11}$ cm, and their course is independent of the electron shells of the atoms. Due to this, the problems of mesoatomic physics mentioned in the previous section allow a strict quantum-mechanical formulation. Most of them reduce to the problem of three bodies that interact according to the Coulomb law. An accurate

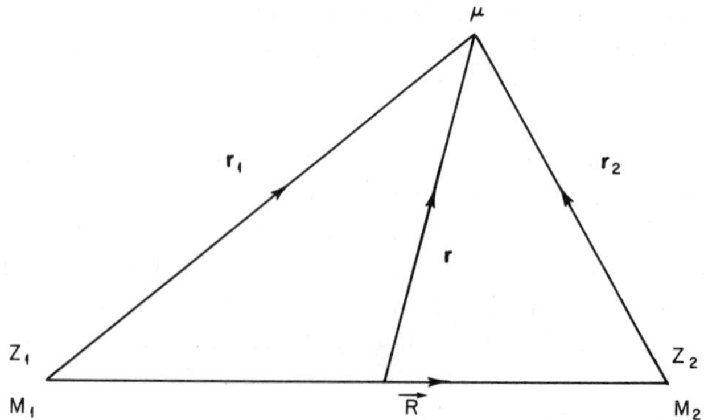

FIG. 4. The Jacobi coordinates in the three-body problem.

solution of three-body problems of such a type requires performing extensive numerical calculations. However, at present, these calculations cause no difficulties in principle, and some of them have already been performed.

The general equation for the problem is, in mesoatomic units $\hbar = e = M\mu = 1$ (Fig. 4),

$$\left(-\frac{m}{2M}\Delta_{\mathbf{R}} + \frac{Z_1Z_2}{R} + H_0\right)\Psi(\mathbf{R}, \mathbf{r}) = E\Psi(\mathbf{R}, \mathbf{r}) \tag{16}$$

where

$$H_0 = -\frac{1}{2m}\Delta\mathbf{r} - \frac{Z_1}{r_1} - \frac{Z_2}{r_2} \tag{17}$$

is the Hamiltonian of the two-center problem, and the following notation is introduced:

$$\frac{1}{m} = 1 + \frac{1}{M_1 + M_2}, \qquad \frac{1}{M} = \frac{1}{M_1} + \frac{1}{M_2} \tag{18}$$

1. *Adiabatic Approximation*

Of special interest in mesoatomic physics are the slow collisions, the energy of which is much smaller than the mesoatomic unit of energy $\mathcal{E}_\mu = 5.63$ KeV. At such collision energies, the adiabatic approximation, according to which the meson in the process of collision follows adiabatically the motion of nuclei, proves to be valid from the physical point of view. Then the wave function $\Psi(\mathbf{R}, \mathbf{r})$ of the three-body system can be written in the form of a product

$$\Psi(\mathbf{R}, \mathbf{r}) = \varphi(\mathbf{r}; R)\chi(\mathbf{R}) \tag{19}$$

where $\chi(\mathbf{R})$ describes the motion of nuclei, and $\varphi(\mathbf{r}; R)$ the motion of the meson for fixed nuclei.

The adiabatic approximation has been used for a long time in atomic physics, in particular, for the calculation of the levels of molecules, where it is known as the Born–Oppenheimer method. In mesoatomic problems, where the ratios of the masses of the meson and the nuclei are not so small as in atomic physics, it is necessary to take into account adiabatic corrections for the motion of nuclei. This is especially important in problems with different M_1 and M_2, since it is precisely these quantities that determine the studied transitions in the system meson + nuclei.

An approximate account of the adiabatic corrections was first made in papers of Gershtein (1957), Belyaev *et al.* (1959), and Cohen *et al.* (1958, 1960). The first-order addiabatic corrections have consistently been taken into account by Matveenko and Ponomarev (1970a,b, 1972a,b).

2. *Two-Center Problem*

The general method of solving the problems of slow collisions in a system of three charged particles is as follows: The wave function $\Psi(\mathbf{R}, \mathbf{r})$ of the three-body problems is sought in the form of the expansion

$$\Psi(\mathbf{R}, \mathbf{r}) = \sum_{\{n\}} \varphi_n(\mathbf{r}; R)\chi_n(\mathbf{R}) \tag{20}$$

in a complete set of the solutions $\varphi_n(\mathbf{r}; R)$ for the two-center problems, i.e., the problem of motion of a negatively charged particle with effective mass m in the field of two fixed nuclei with charges Z_1 and Z_2, masses M_1 and M_2, and at a distance R from each other (Fig. 4):

$$H_0\varphi_n(\mathbf{r}; R) = E_n(R)\varphi_n(\mathbf{r}; R) \tag{21}$$

The eigenfunctions $\varphi_n(\mathbf{r}; R)$ and the eigenvalues $E_n(R)$ for the system $Z_1 = Z_2 = 1$ were first calculated by Bates *et al.* (1953) and Peek (1965a,b), and for the system $Z_1 = 1$, $Z_2 = 2$ by Bates and Carson (1954).

An algorithm for calculating the terms and eigenfunctions of systems with arbitrary Z_1 and Z_2 has been developed by Ponomarev and Puzynina (1967a,b, 1968, 1970).

3. *Method of Perturbed Stationary States*

After inserting the expansion (20) into Eq. (16), averaging over the meson coordinates $\mathbf{r}$, and making a partial wave expansion, there arises a system that is the basis of the method of perturbed stationary states (Mott and Massey, 1965; Halpern, 1968; Vynitskij and Ponomarev, 1974):

$$\frac{m}{2M}\frac{d^2\chi_i}{dR^2} + \left[E - W_i(R) - \frac{m}{2M}K_{ii}(R) - \frac{L(L+1)}{2MR^2}\right]\chi_i$$

$$= \frac{m}{M}Q_{ij}(R)\frac{d\chi_j}{dR} + \frac{m}{2M}K_{ij}(R)\chi_j \tag{22}$$

where

$$W_i(R) = E_i(R) + \frac{Z_1Z_2}{R}$$

is the energy (term) of the three-body system taking into account nuclear repulsion, and

$$Q_{ij}(R) = \frac{\mathbf{R}}{R}\int d\mathbf{r}\varphi_i(\mathbf{r}; R)(-\nabla_{\mathbf{R}})\varphi_j(\mathbf{r}; R)$$

$$K_{ij}(R) = -\int d\mathbf{r}(\nabla_{\mathbf{R}}\varphi_i)(\nabla_{\mathbf{R}}\varphi_j) + \frac{dQ_{ij}(R)}{dR} \tag{23}$$

With such a definition of $K_{ij}(R)$ and $Q_{ij}(R)$, the effect of the meson on the rotation of the internuclear axis in the course of collision is neglected, which is quite justified at slow collisions. The advantage of this system is that in it the small parameter $m/2M$, which allows one to use methods of perturbation theory in the three-body problem, is singled out explicitly.

Some of the matrix elements (23) for the system $Z_1 = Z_2 = 1$ have been computed by Hunter *et al.*, (1966) and Ponomarev and Puzynina (1967c). In the paper by Ponomarev and Puzynina (1970) an algorithm for calculating the functions $W_i(R)$, $Q_{ij}(R)$, and $K_{ij}(R)$ for arbitrary Z_1 and Z_2 is given. The same paper also contains tables of their values for the $Z_1 = 1$ and $Z_2 = 2$ systems, which makes it possible, e.g., to calculate the cross section for the transfer process $\mathrm{p}\mu + \mathrm{He} \rightarrow \mathrm{He}\mu + \mathrm{p}$.

B. Two-Level Approximation

1. *Basic System of Equations*

In describing the wide class of mesoatomic processes, one can restrict oneself to the two-level approximation for which system (22) takes on the following ultimate form:

$$\begin{aligned}\frac{d^2\chi_1}{dR^2} + \left[k_1^2 - \frac{L(L+1)}{R^2}\right]\chi_1 &= K_{11}\chi_1 + K_{12}\chi_2 + 2Q_{12}\frac{d\chi_2}{dR}\\ \frac{d^2\chi_2}{dR^2} + \left[k_2^2 - \frac{L(L+1)}{R^2}\right]\chi_2 &= K_{21}\chi_1 + K_{22}\chi_2 + 2Q_{21}\frac{d\chi_1}{dR}\end{aligned} \tag{24}$$

where

$$K_{ij} = \frac{2M}{m}[W_i(R) - W_i(\infty)] + [K_{ij}(R) - K_{ij}(\infty)], \qquad Q_{ij} = Q_{ij}(R)$$

$$k_1^2 = \frac{2M}{m}\mathcal{E}, \qquad k_2^2 = \frac{2M}{m}\mathcal{E}' = \frac{2M}{m}(\mathcal{E} + \Delta E) = k_1^2 + k_0^2 \tag{25}$$

with $\mathcal{E}$ the collision energy or the binding energy for a three-body system (at $\mathcal{E}' < 0$).

The system of equations (24) underlies all the practical calculations performed to date (Gershtein, 1957; Beljaev *et al.*, 1959; Cohen *et al.* 1958, 1960; Matveenko and Ponomarev, 1969, 1970a,b, 1972c). In such a form it describes, for example, processes of collision of mesonic atoms in a mixture of hydrogen and deuterium isotopes. The scheme of levels at $R \to \infty$ corresponding to this system is given in Fig. 5, the levels E_1 and E_2 corresponding to the systems pμ + d and p + dμ, respectively. Accord-

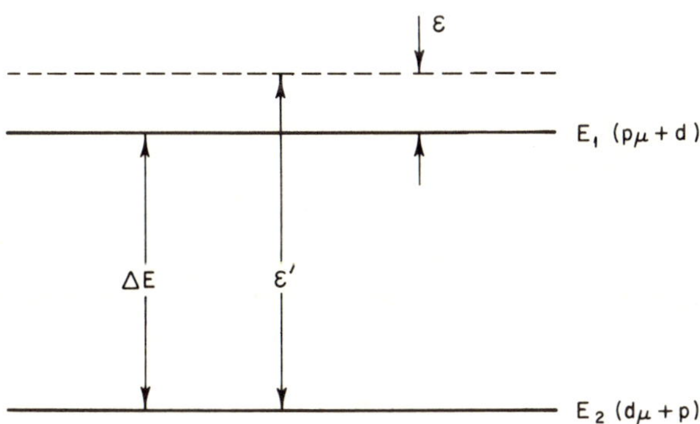

FIG. 5. Scheme of terms of the three-body system at $R \to \infty$. The collision energy $\mathcal{E}$ for the charge exchange process pμ + d → dμ + p is reckoned from the upper level $E_1 = (m/2M)\,K_{11}(\infty)$.

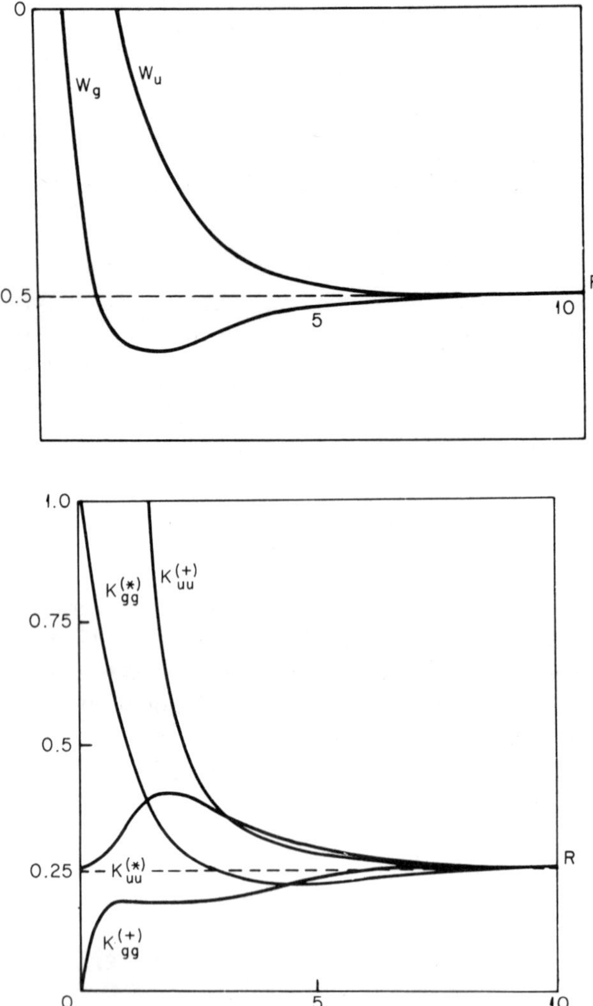

FIG. 6. (a) The symmetric $W_g(R)$ and antisymmetric $W_u(R)$ terms of a system consisting of a meson and two nuclei with charges $Z_1 = Z_2 = 1$. (b, c) Adiabatic corrections for the motion of nuclei with charges $Z_1 = Z_2 = 1$:

$$K_{ij}(R) = K_{ij}^{(+)} + \kappa K_{ij}^{(-)} + \kappa^2 K_{ij}^{(*)}, \qquad Q_{ij}(R) = Q_{ij}^{(+)} + \kappa Q_{ij}^{(-)}$$

$$\kappa = \frac{M_2 - M_1}{M_2 + M_1}, \qquad i, j \equiv \mathrm{g, u}$$

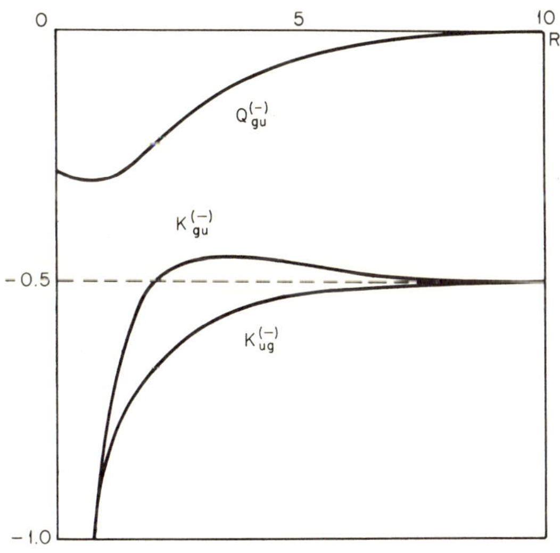

FIG. 6c

ing to the values of $\mathcal{E}$ and $\mathcal{E}'$, processes of three types are possible:

A. $\mathcal{E} > 0$, $\mathcal{E}' > \Delta E$. The system of equations (24) describes the elastic scattering process

$$\mathrm{p}\mu + \mathrm{d} \leftrightarrow \mathrm{p}\mu + \mathrm{d} \tag{26a}$$

and the isotopic exchange reaction

$$\mathrm{p}\mu + \mathrm{d} \leftrightarrow \mathrm{d}\mu + \mathrm{p} \tag{26b}$$

B. $\mathcal{E} < 0$, $0 < \mathcal{E}' < \Delta E$. The system (24) describes the process of elastic scattering of dμ atoms on protons

$$\mathrm{d}\mu + \mathrm{p} \rightarrow \mathrm{d}\mu + \mathrm{p} \tag{26c}$$

C. $\mathcal{E} < 0$, $\mathcal{E}' < 0$. The system allows one to calculate the levels of the mesonic molecule pdμ in different quantum states.

2. *Effective Potentials*

In describing the mesoatomic processes occurring in a mixture of hydrogen isotopes, the effective potentials K_{ij} and Q_{ij} in Eqs. (24) are expressed via the even (*gerade*) $W_g(R)$ and odd (*ungerade*) $W_u(R)$ terms of the system $Z_1 = Z_2 = 1$ and the corresponding matrix elements $K_{gg}(R)$, $K_{uu}(R)$, $K_{gu}(R)$, $K_{ug}(R)$, and $Q_{gu}(R) = -Q_{ug}(R)$. The diagrams of these functions are given in Figs. 6a–c. For processes of the type (26) the corresponding formulas assume the forms

$$K_{11} = \frac{M}{m}(V_g + V_u) - \frac{1}{2}(K_{gu} + K_{ug})$$

$$K_{12} = \frac{M}{m}(V_g - V_u) + \frac{1}{2}(K_{gu} - K_{ug}) \qquad (27)$$

etc., with replacement $1 \leftrightarrow 2$, $K_{ug} \to -K_{ug}$, $K_{gu} \to -K_{ug}$, where

$$V_g = W_g(R) - W_g(\infty) + \frac{m}{2M}[K_{gg}(R) - K_{gg}(\infty)] \qquad (28)$$

and the same for V_u.

3. *Parametrization of the Cross Sections at Low Energies*

It is convenient to calculate the reaction matrix T of the processes under consideration by the phase function method (Calogero, 1967; Babikov, 1968), which allows one to reduce the solution of the system of second-order linear equations (22) for the wave functions $\chi_i(R)$ to the solution of the system of first-order nonlinear equations for the matrix elements $t_{ij}(R)$ of the reaction matrix T.†

At low collision energies ($\mathcal{E} < 0.1$ eV) it is convenient to determine the parameters a_{ij} of the low-energy scattering by the formulas

$$t_{ij} = -a_{ij}(k_i k_j)^{1/2} \qquad (29)$$

TABLE 8

SCATTERING LENGTHS FOR MESONIC ATOMS[a]

Reference \ Scattering lengths	pμ + p		dμ + d		tμ + t	
	a_g	a_u	a_g	a_u	a_g	a_u
Zeldovich and Gershtein (1960)	−17.3	5.3	6.7	5.8	—	—
Cohen *et al.* (1960)	−11.0	5.0	—	—	—	—
Matveenko and Ponomarev (1970a)	−13.3	3.7	5.5	3.1	−6.5	2.4

[a] In units a_μ.

† The reaction matrix T is related to the S-matrix by the equation $S = (1 + iT) \times (1 - iT)^{-1}$.

and then to represent the reaction cross sections in the form

$$\sigma_{ij} = 4\pi \frac{k_j}{k_i} \frac{a_{ij}^2}{1 + k_0^2 a_{22}^2} \tag{30}$$

In particular, the parameter $a_{12} = a_{21}$ is responsible for the isotopic exchange processes (26b), and the parameter a_{11} for the elastic scattering cross section (26a). Formula (30) also gives the threshold behavior of the cross sections σ_{ij} at $\mathcal{E} \to 0$:

$$\sigma_{12} \sim 1/k_1, \qquad \sigma_{11} \sim \text{const}$$

4. *Scattering Lengths for Mesonic-Atom Collisions*

In the case $M_1 = M_2$ the system of equations (24) disintegrates into two independent Schrödinger equations with potentials $V_g(R)$ and $V_u(R)$:

$$\frac{d^2\chi_{g,u}(R)}{dR^2} + \left[k^2 - \frac{2M}{m} V_{g,u}(R)\right] \chi_{g,u}(R) = 0 \tag{31}$$

In the limit $k \to 0$, it is possible to determine from these equations the scattering lengths a_g and a_u by means of which it is possible to describe scattering processes, e.g., in the system $p\mu + p$ at low collision energies. Sometimes, in some physically interesting cases (Section III,E), one can express three parameters a_{11}, a_{22}, $a_{12} = a_{21}$ describing low-energy mesonic-atom scattering, in terms of the two quantities a_g and a_u.

Table 8 gives the values of the lengths a_g and a_u for scattering of mesonic hydrogen atoms on the corresponding nuclei, which have been obtained by a number of authors. The striking difference between the recent values of a_u and those previously known is explained by the fact that earlier authors disregarded the shallow dip of the term $W_u(R)$ for $R = 12.55$ and the long-range asymptotic of the potentials $V_g(R) \approx V_u(R) \approx -9/4R^4$, when $R \to \infty$, the influence of which on the value of a_u was found to be considerable (Matveenko and Ponomarev, 1970a).

C. Isotopic Exchange Reactions

The transfer rates λ_{pd} were first estimated by Zeldovich and Sakharov (1957), Skyrm (1957), Jackson (1957), Gershtein (1957), Burke *et al.* (1959), and Wu *et al.* (1960). More accurate calculations were performed by Belyaev *et al.* (1959), Shimizu *et al.* (1959), and Cohen *et al.* (1958, 1960).

The detailed results of theoretical calculations of the energy dependence of the cross sections σ_{ij} for different reaction channels

$$\mathrm{p}\mu + \mathrm{d} \rightarrow \mathrm{d}\mu + \mathrm{p} \tag{32a}$$

$$\mathrm{p}\mu + \mathrm{t} \rightarrow \mathrm{t}\mu + \mathrm{p} \tag{32b}$$

$$\mathrm{d}\mu + \mathrm{t} \rightarrow \mathrm{t}\mu + \mathrm{d} \tag{32c}$$

are given in the paper by Matveenko and Ponomarev (1970b). The level E_1 of Fig. 5 corresponds to the initial state of the systems (32), and the level E_2 to their final state.

Of special interest is the calculation of the cross sections σ_{12} for transfer of the μ meson from a light hydrogen isotope to a heavy one.

It follows from Eq. (30) that at low initial velocities of the relative motion ($v_1 \ll \alpha c$), the cross section $\sigma_{12} \sim 1/v_1$ because of which the appropriate transfer rate

$$\lambda = \sigma_{12} v_1 N_{\mathrm{p}}{}^0 \ \mathrm{sec}^{-1} \tag{33}$$

is approximately constant.

The results of the calculations show that the condition $\lambda \approx$ const, which is usually used for the analysis of experimental data, is satisfied up to collision energies $\mathcal{E} \sim 1$ eV (with the exception of the system $\mathrm{d}\mu + \mathrm{t}$) although the pure s-scattering approximation is violated at an energy $\mathcal{E}$ as low as about 10^{-2} eV. This fact is due to the increase of the p-wave contribution to the cross section σ_{12}, which compensates for the decrease of the s-wave contribution.

In Table 9 the calculated values of λ for processes (32) are compared with the experimental data.

TABLE 9

TRANSFER RATES FOR ISOTOPIC EXCHANGE PROCESS λ[a]

Literature	λ_{pd}	λ_{pt}	λ_{dt}
Bleser *et al.* (1963)	1.4 ± 0.2	—	—
Dzhelepov *et al.* (1966)	1.2 ± 0.4	—	—
Bertin *et al.* (1972)	0.84 ± 0.13	—	—
Belyaev *et al.* (1959)	1.5	0.65	0.52×10^{-2}
Cohen *et al.* (1960)	1.4	—	—
Matveenko and Ponomarev (1970b)	1.7	0.73	0.77×10^{-2}

[a] $\times 10^{10}$ sec^{-1}.

D. Transitions between the Hyperfine Structure Levels of Mesonic Hydrogen Atoms

Due to the interaction of the spins of the μ meson and nuclei, the ground state of mesonic atoms is split by the value

$$\Delta E_{\text{hfs}} = \frac{8}{3}\frac{\beta_\mu \beta_{\text{N}}}{a_\mu{}^3}\, g_{\text{N}}(2i + 1)\left(1 + \frac{M_\mu}{M_{\text{N}}}\right)^{-3} \tag{34}$$

where β_μ and β_{N} are the meson and nuclear Bohr magnetons, and g_{N} and i the gyromagnetic ratio and the nuclear spin, respectively.

The value of the splitting ΔE_{hfs} (Table 1) exceeds the energy of thermal collisions ($\mathcal{E}_{\text{T}} \approx 0.04$ eV), but it is sufficiently small for the s-scattering approximation to remain valid. In this approximation, the system consisting of two nuclei with spins J_1 and J_2 and a meson with spin $S = \frac{1}{2}$ is characterized by the total momentum $\mathbf{J} = \mathbf{J}_1 + \mathbf{J}_2 + \mathbf{S}$, which is conserved during the scattering process.

Of most interest are the transitions from the state E_1 with total spin of meson and nucleon $F_1 = J_1 + \frac{1}{2}$ to the state E_2 with total spin $F_2 = J_1 - \frac{1}{2}$, which, as in the case of isotopic exchange, may be described by the transfer rate $\lambda \approx$ const. For the case pμ + p, $\sigma_{12} \equiv \sigma_{1\to 0}$, and for the case d$\mu$ + d, $\sigma_{12} \equiv \sigma_{3/2\to 1/2}$.

The probability of transfer $F_1 \to F_2$ depends on the magnitude of the total momentum J of the system, and therefore the λ values calculated by Eq. (33) should be multiplied by the statistical weight of the state with given J and F_1. For the system pμ + p and dμ + d we get, respectively,

$$\begin{aligned} \lambda_{\text{p}} &= \lambda_{1\to 0} = \tfrac{1}{3}\lambda(J = \tfrac{1}{2}) \\ \lambda_{\text{d}} &= \lambda_{3/2\to 1/2} = \tfrac{1}{6}\lambda(J = \tfrac{1}{2}) + \tfrac{1}{3}\lambda(J = \tfrac{3}{2}) \end{aligned} \tag{35}$$

The equations describing the transitions $F_1 \to F_2$ coincide in form with system (24) and were initially solved in the scattering-length approximation by Gershtein (1958a,b, 1961). At collision energies $\mathcal{E} < 10^{-2}$ eV, the cross sections, on account of spin splitting, can be calculated by formulas (30). The values of the low-energy scattering parameters a_{ij} are given by Matveenko and Ponomarev (1970b). They give also the expressions for the parameters a_{ij} in the terms of the scattering lengths a_{g} and a_{u}, which take place at collision energies $\mathcal{E} < 10^{-3}$ eV.

In Table 10 the results of calculations of the constants $\lambda = \lambda\ (F_1 \to F_2)$ performed by a number of authors† are compared. It is remarkable that the

† The experimental estimates for the quantities λ_{p} are not reliable enough, since in determining them from the data on elastic scattering of pμ atoms it was assumed that $a_{\text{u}} = 5$ (Dzhelepov *et al.*, 1965a; Alberigi Quaranta *et al.*, 1967b).

TABLE 10

TRANSFER RATES BETWEEN THE HFS LEVELS OF MESONIC HYDROGEN ATOMS λ^a

Literature	λ_p	λ_d	λ_t
Gershtein (1958a, 1961)	2.2	0.7×10^{-2}	—
Matveenko and Ponomarev (1970b)	5.2	4.7×10^{-2}	0.86
Doede (1963)	—	$\sim 10^{-2}$	—
Bleser *et al.* (1963)	—	$\sim 10^{-2}$	—
Dzhelepov *et al.* (1964d)	4.8 ± 1.0	—	—
Alberigi Quaranta *et al.* (1967a)	2.4 ± 0.3	—	—
	1.1 ± 0.1		

[a] $\times 10^9$ sec^{-1}.

recent value of λ_d exceeds by an order of magnitude the former estimate. This fact is important for the analysis of experiments on μ capture in deuterium and catalysis of the p + d reaction at different deuterium concentrations (Bleser *et al.*, 1963; Doede, 1963; Bystritskij *et al.*, 1974b; Bertin *et al.*, 1973).

The condition $\lambda \approx$ const is fulfilled only in a narrow region of collision energies $\mathcal{E} < 10^{-2}$ eV. In this region, the approximate formulas (Gershtein, 1958a, 1961)

$$\lambda_p \approx \frac{\pi}{4} (a_g - a_u)^2 a_\mu^2 v_0 N_p^0 \text{ sec}^{-1} \tag{36a}$$

$$\lambda_d \approx \frac{\pi}{3} (a_g - a_u)^2 a_\mu^2 v_0 N_p^0 \text{ sec}^{-1} \tag{37b}$$

where

$$v_0 = (2\,\Delta E/M)^{1/2} \text{ cm sec}^{-1}$$

are valid. The transfer rate $\lambda_p \sim 10^9$ sec^{-1} is very large since the quantities a_g and a_u have opposite signs. This fact is of importance because the probability of the weak interaction $\mu^- + p \rightarrow n + \nu_\mu$ from the triplet state of the pμ atom is negligible ($\Lambda_t \approx 12$ sec^{-1}) compared with the reaction from the singlet state ($\Lambda_s \approx 650$ sec^{-1}) (Lee and Wu, 1965, 1966; Primakoff, 1959). In gaseous hydrogen, at a pressure of a few atmospheres, when one can still neglect the production of the ppμ molecules (Section III,G) there occurs an almost complete transition pμ ($F = 1$) $\rightarrow$ pμ ($F = 0$) as a result of which the μ^- capture rate λ increases by about a factor of four compared with the rate in a statistical mixture of the pμ atoms and becomes equal to about Λ_s ($\lambda = 651 \pm 57$, Alberigi Quaranta *et al.*, 1969; $\lambda = 686 \pm 88$, Bystritsky *et al.*, 1973a).

In a similar manner the transition dμ ($F = \frac{3}{2}$) $\rightarrow$ dμ ($F = \frac{1}{2}$) in deuterium leads to an increase of the rate of the process $\mu^- + \mathrm{d} \rightarrow 2\mathrm{n} + \nu_\mu$ by about a factor of three as compared with the μ capture rate for the statistical population of the hfs levels of the dμ atom (Wang *et al.*, 1965; Bertin *et al.*, 1973b). Note that in symmetrical systems the pμ + p and dμ + d transfers to the lower hfs state occur due to exchange effects without meson spin-flip (Section II,A,2). Compared to it, the transition

$$\mathrm{d}\mu\ (F = \tfrac{3}{2}) + \mathrm{p} \rightarrow \mathrm{d}\mu\ (F = \tfrac{1}{2}) + \mathrm{p} \tag{37}$$

should be strongly suppressed because it can occur only due to spin-orbital interaction. From this point of view, the assertion about the large cross section for reaction (37) (Placci *et al.*, 1970b; Bertin *et al.*, 1973b) seems to be surprising.†

E. Elastic Collisions of Mesonic Atoms

1. *Elastic Scattering in the Lower hfs State*

In condensed matter the rates of the transition $F_1 \rightarrow F_2$ are much higher than those of the μ-meson decay; therefore, the μ meson is in the lower hfs state during the main part of its lifetime. At collision energies $\mathcal{E}' < \Delta E$, in the scattering-length approximation, the cross sections σ_{22} for scattering of the mesonic atoms pμ and dμ are, respectively (Gershtein, 1958b, 1961),

$$\sigma_0 \approx 4\pi \left(\frac{a_\mathrm{g} + 3a_\mathrm{u}}{4}\right)^2, \qquad \sigma_{1/2} \approx 4\pi \left[\frac{1}{3}\left(\frac{a_\mathrm{g} + 2a_\mathrm{u}}{3}\right)^2 + \frac{2}{3}\left(\frac{5a_\mathrm{g} + a_\mathrm{u}}{6}\right)^2\right] \tag{38}$$

In Tables 5 and 6 the cross sections calculated by these formulas are compared with the measured ones. It follows from this comparison that there is good agreement between different values of $\sigma_{1/2}$, while the elastic-scattering process in the pμ + p system needs further study. The account

† The estimate $\lambda > 5 \times 10^6\ \mathrm{sec}^{-1}$ at a pressure of 7.6 atm and at the concentration $C_\mathrm{d} = 5\%$ adopted in this chapter means that in liquid hydrogen the rate of reaction (37) is much higher than the rate of formation of pdμ mesonic molecules. However, this conclusion contradicts the experiments by Bleser *et al.* (1963) and Doede (1963), who observed the increase in the yield of nuclear synthesis in the pdμ mesonic molecule at high deuterium concentration ($C_\mathrm{d} \sim 25\%$) due to transitions dμ($F = \frac{3}{2}$) + d $\rightarrow$ dμ($F = \frac{1}{2}$) + d (Section IV, C, D).

of subsequent expansion terms in Eq. (38) leads to the expression

$$\sigma_0 = 4\pi \left[\frac{(a_g + 3a_u) - 4k_0 a_g a_u}{4 - k_0(3a_g + a_u)} \right]^2 \tag{39}$$

and thus also gives an inadequate result $\sigma_0 \approx 5.4 \times 10^{-22}$ cm^2.†

Errors in theoretical calculations can possibly be explained by the resonance character of the process pμ + p, which is seen from an anomalously large scattering length $a_g = -13.3$. In the resonance situation, even small variations of the potential shape can lead to noticeable changes of the cross-section value; therefore, the account of the terms $(m/2M)^2$ in the potential $V_g(R)$ for the process pμ + p may turn out to be essential.

2. *High-Energy Collisions of Mesonic Atoms*

In calculating the cross sections σ_{ij} for various processes with collision energy $\mathcal{E} > 1$ eV, it is necessary to take into account the contribution of the partial waves with orbital angular momenta $L \neq 0$. A strict formulation of the problem requires that the spin–orbital interaction of the meson should also be taken into consideration, which causes considerable difficulties. However, as the practice of calculation shows, at collision energies $\mathcal{E} > 1$ eV, one can neglect both the spin–orbital interaction and the hyperfine splitting of levels, and consider only the identity of nuclei. In this case ($\mathcal{E} \gg \Delta E_{hfs}$), in the system p$\mu$ + p only the symmetrized cross section

$$\sigma_S{}^p = \tfrac{1}{4}\sigma_{S,s} + \tfrac{3}{4}\sigma_{S,a} \tag{40}$$

is meaningful, where $\sigma_{S,s}$ is the scattering cross section in the singlet state of two protons, and $\sigma_{S,a}$ the scattering cross section in the triplet state.

To estimate roughly the cross sections $\sigma_S{}^p$ and $\sigma_S{}^d$ at collision energies $\Delta E_{hfs} \ll \mathcal{E} < 1$ eV, use can be made of the scattering length approximation

$$\begin{aligned} \sigma_S{}^p &\approx 4\pi \left(\frac{1}{4} \frac{a_g{}^2}{1 + k^2 a_g{}^2} + \frac{3}{4} a_u{}^2 \right) \\ \sigma_S{}^d &\approx 4\pi \left(\tfrac{2}{3} a_g{}^2 + \tfrac{1}{3} a_u{}^2 \right) \end{aligned} \tag{41}$$

At an energy $\mathcal{E} > 150$ eV, the account of the proton statistics is also no longer necessary, since in this region of collision energies $\sigma_S \approx \sigma$, where σ is the total cross section disregarding spins. Figure 7 gives the curves of the

† Recently, Matveenko *et al.* (1975) have succeeded in numerically solving Eq. (24) for the case $\mathcal{E} < 0$, $0 < \mathcal{E}' < \Delta E$, and thereby taking into account the influence of the closed channel upon the open one. The value obtained, $\sigma_0 = 2 \times 10^{-22}$ cm^2, is, as before, in disagreement with experiment.

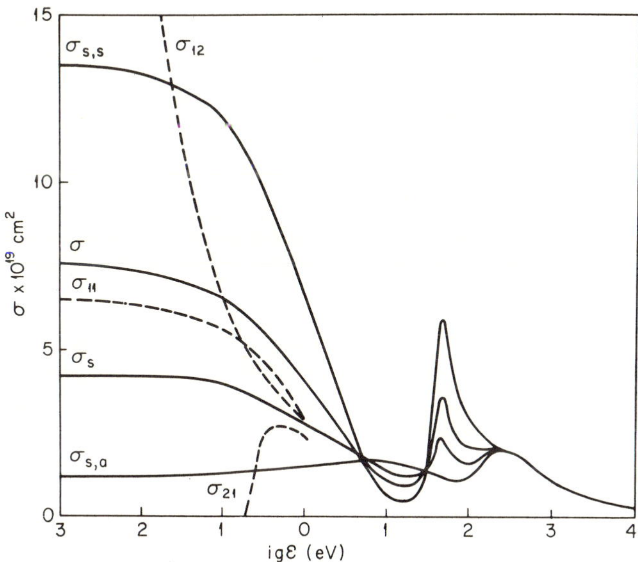

FIG. 7. Cross sections for different processes in the pμ + p system. When $\mathcal{E} < 1$ eV it is necessary to take into account the hfs of the pμ mesonic atom, and therefore only the cross sections σ_{ij} have a physical meaning. When $1 < \mathcal{E} < < 150$ eV it is possible to neglect the mesonic atom hfs, but it is still important to take into account the particle statistics ($\sigma_{S,s}$ scattering cross section for the singlet state of protons, $\sigma_{S,a}$ for the triplet state of protons, σ_S for statistical mixture of both states). When $\mathcal{E} > 150$ eV all these cross sections coincide with the σ cross sections neglecting particle spins (Matveenko and Ponomarev, 1970b).

corresponding cross sections. It is interesting to note that there are resonances in the pμ + p system, at $1 < \mathcal{E} < 100$ eV.

3. *Elastic Scattering in* dμ + p *System*

The cross section of this process at collision energies $\mathcal{E}' < \Delta E$ has been calculated on the basis of the system of Eq. (24) by Belyaev *et al.* (1959) and Cohen *et al.* (1960). It was found that at collision energies $\mathcal{E}' \approx 0.2$ eV, in the cross section for elastic scattering (26c) there is a deep minimum $\sigma_{22} \sim 10^{-21}$ cm^2. This minimum was discovered in the calculations of Cohen *et al.* (1960), when taking into account the long-range asymptotic of the potential $V(R) \sim R^{-4}$. Their calculations explain the large mean free paths of the dμ atoms in hydrogen (Section II,B). In recent calculations (Matveenko *et al.*, 1975) a minimum was found at $\mathcal{E}' \approx 0.6$ eV and, in addition, it was clarified that the scattering in a state with orbital moment $L = 2$

TABLE 11

BINDING ENERGY (eV) OF THE ppμ MESONIC MOLECULE (ADIABATIC CALCULATIONS)

Literature	$L = 0, v = 0$	$L = 1, v = 0$
Cohen *et al.* (1958, 1960)	241	93
Marschall and Schmidt (1958)	224	75
Zeldovich and Gershtein (1958a)	252	106
Mizuno (1961)	241	70
Narumi and Matsuo (1961)	213	43
Joachim and Wantiez (1962)	244	—
Patterson and Becker (1967)	248.6	101.5
Ponomarev *et al.* (1973b)	248	102

is of a resonance character. The corresponding cross section is larger by an order of magnitude than the cross section of the s-scattering and at the maximum (at $\mathcal{E}' \approx 50$ eV) it amounts to $\sigma_{22} \approx 4 \times 10^{-19}$ cm^2.

F. Energy Levels of Mesonic Molecules

The system of equations (24) makes it possible to calculate the energies of the mesonic molecules ppμ, ddμ, ttμ, pdμ, ptμ, and dtμ. We note that in such calculations the account of the adiabatic corrections $(m/2M)K_{ij}(R)$

TABLE 12

BINDING ENERGY (eV) OF THE ppμ MESONIC MOLECULE (VARIATIONAL CALCULATIONS)

Literature	$L = 0, v = 0$	$L = 1, v = 0$
Skyrme (1957)	282	—
Kolos *et al.* (1960)	249	—
Flugge and Schröder (1961)	211	—
Fröman and Kinsey (1961)	230	—
Schröder (1963a)	237	—
Halpern (1964b)	—	107.23
Wessel and Phillipson (1964)	254.3	—
Scherr and Machaček (1965)	—	106.8
Kabir (1965)	254.4	—
Carter (1966)	252.2	—
Delves and Kolotos (1968)	253.14	—
Carter (1968a)	253	—

and $(m/M)Q_{ij}(R)$ is very essential, as far as they are comparable in order of magnitude to the depth of the potential $W_g(R)$ (Fig. 6). As follows from Tables 11–16 the energy levels calculated by Eqs. (24) are in satisfactory agreement with the results of variational calculations.

The variational calculations appear to be preferable for the ground state of mesonic molecules, but the adiabatic calculations are much simpler and more useful visually, especially for the excited states.

1. *Mesonic Molecule* ppμ

The states with orbital angular momenta $L = 0$ and $L = 1$ have one vibrational level each with vibrational quantum number $v = 0$. Tables 11 and 12 give the values of the binding energy of the molecule ppμ calculated by many authors.

2. *Mesonic Molecule* ddμ

In each of the states with $L = 0$ and $L = 1$ there are two vibrational levels: $v = 0$ and $v = 1$; in the state with $L = 2$ there is only the $v = 0$ level. The scheme of the levels of the ddμ mesonic molecule is given in Fig. 8. In Table 13 we give the results of calculations of the binding energy of the mesonic molecule.

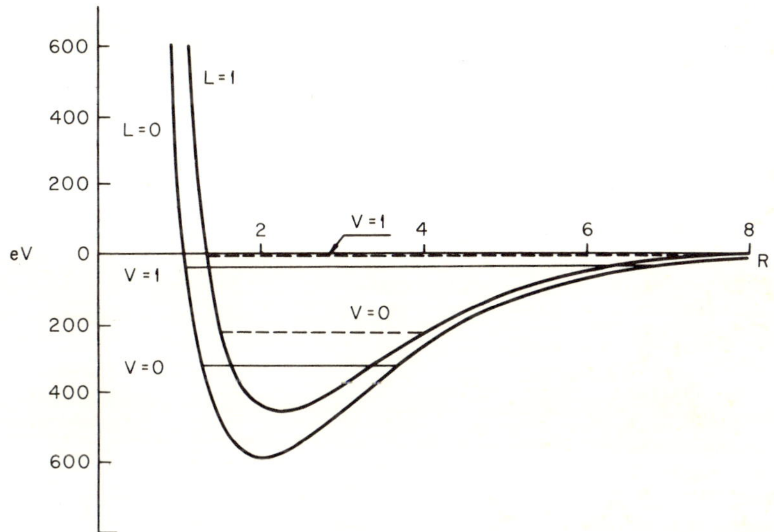

FIG. 8. Scheme of the levels of the ddμ mesonic molecule. In the state with orbital momentum $L = 1$ there exists an excited vibrational level $v = 1$ with small binding energy.

3. *Mesonic Molecule* ttμ

Each state with $L = 0$ and $L = 1$ has two vibrational $v = 0$ and $v = 1$ levels; the states with $L = 2$ and $L = 3$ have one $v = 0$ level each. The results of calculations are given in Table 14.

4. *Mesonic Molecule* pdμ, ptμ, *and* dtμ

Contrary to the molecules with identical nuclei, the energy levels of which can be found from Eq. (31) with potential $V_g(R)$ (Section III,B,2), to calculate the binding energy of the molecules with different nuclei it is necessary to solve the system of Eqs. (24), in which the effective potentials

TABLE 13

Binding Energy (eV) of the ddμ Mesonic Molecule

Literature	$L = 0$		$L = 1$		$L = 2$,	Method of calculation
	$v = 0$	$v = 1$	$v = 0$	$v = 1$	$v = 0$	
Skyrme (1957)	381	—	—	—	—	estimate
Cohen *et al.* (1958, 1960)	322	181	223	—	82	adiabatic
Marschall and Schmidt (1958)	300	—	220	—	25	adiabatic
Belyaev *et al.* (1959), Zeldovich and Gershtein (1960)	330	40	226	7(?)	88	adiabatic
Kolos *et al.* (1960)	318	—	—	—	—	variational
Fröman and Kinsey (1961)	306	—	—	—	—	variational
Narumi and Matsuo (1961)	281	177	190	—	43	adiabatic
Mizuno (1961)	338	9	207	—	59	adiabatic
Joachim and Wantiez (1962)	328	—	—	—	—	adiabatic
Schröder (1963a)	309	—	—	—	—	variational
Halpern (1964b)	—	—	226.55	—	—	variational
Scherr and Machaček (1965)	—	—	226.3	—	—	variational
Carter (1966, 1968a)	324.2	32.7	—	—	—	variational
Patterson and Becker (1967)	323.4	34.1	224.6	—	—	adiabatic
Ponomarev *et al.* (1973b)	323	32.9	224	0.7	83.6	adiabatic

TABLE 14

BINDING ENERGY (eV) OF THE ttμ MESONIC MOLECULE

Literature	$L = 0$		$L = 1$		$L = 2$,	$L = 3$,	Method of calculation
	$v = 0$	$v = 1$	$v = 0$	$v = 1$	$v = 0$	$v = 0$	
Belyaev *et al.* (1959), Zeldovich and Gershtein (1960)	367	86	288	45	170	—	adiabatic
Schröder (1963a)	348	—	—	—	—	—	variational
Halpern (1964b)	—	—	288.72	—	—	—	variational
Scherr and Machaček (1965)	—	—	288.8	—	—	—	variational
Carter (1966)	361.2	75	—	—	—	—	variational
Patterson and Becker (1967)	361.8	82.4	287.7	43.7	—	—	adiabatic
Ponomarev *et al.* (1973b)	361	81.4	288	43.1	171	46.7	adiabatic

TABLE 15

BINDING ENERGY (eV) OF THE pdμ MESONIC MOLECULE[a]

Reference	$L = 0, v = 0$	$L = 1, v = 0$	Method of calculation
Skyrme (1957)	264	—	estimate
Marschall and Schmidt (1958)	250	—	adiabatic
Belyaev *et al.* (1959)	220	90	adiabatic
Cohen *et al.* (1960)	214	90	adiabatic
Zeldovich and Gershtein (1960)	223	95	adiabatic
Fröman and Kinsey (1961)	193	—	variational
Narumi and Matsuo (1961)	223	—	adiabatic
Schröder (1963a)	204	—	variational
Frost *et al.* (1964)	211	—	variational
Carter (1966, 1968a)	221.2	—	variational
Kolos (1968)	220.0	—	variational
Ponomarev *et al.* (1973b)	214	89.7	adiabatic

[a] The binding energy for the pdμ molecule is reckoned from the ground state of the dμ mesonic atom.

TABLE 16

BINDING ENERGY (eV) OF THE ptμ AND dtμ MESONIC MOLECULES[a]

Reference	ptμ		dtμ				Method of calculation
	$L = 0$, $v = 0$	$L = 1$, $v = 0$	$L = 0$, $v = 0$	$L = 0$, $v = 1$	$L = 1$, $v = 0$	$L = 2$, $v = 0$	
Belyaev *et al.* (1959)	213	98	319	32	232	102	adiabatic
Zeldovich and Gershtein (1960)	211	90	323	36	234	103	adiabatic
Schröder (1963a)	195	—	303	—	—	—	variational
Carter (1966)	212.8	—	317	30.4	—	—	variational
Carter (1968a)	—	—	318.1	32.9	—	—	variational
Ponomarev *et al.* (1973b)	206	91.1	317	31.7	230	99.3	adiabatic

[a] The binding energy for the ptμ and dtμ mesonic molecules is reckoned from the ground state of the tμ mesonic atom.

K_{ij} and Q_{ij} are determined by Eq. (25). In Tables 15 and 16 the results of such calculations are compared with variational calculations.

G. Formation of Mesonic Molecules

1. General Picture

In collisions of a mesonic hydrogen atom with hydrogen nuclei there can be produced the binding systems ppμ, pdμ, etc. In this case the binding energy released in the process of the molecular formation is carried, as a rule, away by an atomic electron.† In so far as the dimensions of the mesonic molecules are much smaller than those of the electron shells of the atoms, the probability of formation of mesonic molecules with emission of an atomic electron is calculated in just the same way as the probability of conversion on atomic electrons in nuclear γ transitions. The energy of the conversion electrons accompanying the process of mesonic molecule formation amounts to several tens of hundreds of electron volts. The conversion coefficients for such energies are very large (10^6 to 10^7) and therefore the process of production of mesonic molecules by means of con-

† In the case of ddμ mesonic molecules having a level with very small binding energy, of most importance appears to be the formation mechanism with transfer of the binding energy to the other nucleus of the D_2 molecule. This mechanism is quite ineffective for other molecules.

version on an electron is more likely than by means of radiative transitions, and in the case of E0 transitions, it is the only possible way.

The collisions of mesonic atoms with nuclei resulting in the formation of mesonic molecules occur at very low (most likely thermal) energies. This means that their relative motion is described by the s-wave. The resulting states will be the mesomolecular $L = 0$ and $L = 1$ states, and also $L = 2$ state (for molecules ddμ, dtμ, ttμ) and $L = 3$ state (only for ttμ). Therefore, there can occur electrical dipole transitions E1 $(0 \to 1)$ to the rotational levels with $L = 1$, electrical monopole E0 transitions to the $L = 0$ levels, and E2 and E3 transitions to the $L = 2$ and $L = 3$ levels, respectively. However, the probabilities for the two latter transitions and for magnetic transitions are very small and negligible for the molecule formation.

2. *Molecule Formation by* E1 *Transition with Emission of Conversion Electrons*

The dipole moment of a system consisting of a muon and two nuclei with masses M_1 and M_2 is

$$\mathbf{d} = -\frac{e}{2}[\kappa\mathbf{R} + (\mathbf{r}_1 + \mathbf{r}_2)] \tag{42}$$

where $\kappa = (M_2 - M_1)/(M_2 + M_1)$, e is the electron charge, $\mathbf{R}$ the distance between the nuclei M_1 and M_2, and $\mathbf{r}_1$ and $\mathbf{r}_2$ the distances from nuclei M_1 and M_2 to the meson (Fig. 4).

The first term in expression (42) is connected with the asymmetry of the nucleus charge with respect to the center of mass of the molecule in the case $M_1 \neq M_2$; the second term is related to the muon charge distribution with respect to the center of the molecule, and in the case $M_1 = M_2$ determines the dipole moment of the system completely.

The probability for dipole E1 transition is (Zeldovich and Gershtein, 1958a; Cohen *et al.*, 1960):

$$W_{\mathrm{E1}} = \frac{16}{3}(Na_0)^3\left(\frac{m_e}{m_\mu}\right)^5 \eta\zeta^2\Big(\sum_{\Lambda} |\langle\mathbf{d}\rangle|^2\Big)\frac{e^2}{a_0\hbar}\ \mathrm{sec}^{-1} \tag{43}$$

where $\langle\mathbf{d}\rangle$ is given in mesoatomic units and

$$\eta = e^2/\hbar v_e$$

$$\zeta = \left[\frac{2\pi\eta}{(1+\eta^2)(1-e^{-2\pi\eta})}\right]^{1/2} \exp(-2\eta\cot^{-1}\eta)$$

and a_0 is the Bohr radius of the electron, v_e the velocity of the conversion electron, N the number of nuclei per cubic centimeter, and the summation is performed over all the projections Λ of the orbital momentum L of the mesonic molecule, $\Lambda = 0, \pm 1$.

The first term in the dipole moment (42) has nonzero matrix elements only between the wave functions of the same parity $\varphi_g \rightarrow \varphi_g$ and $\varphi_u \rightarrow \varphi_u$, while the second term contributes to the transitions $\varphi_u \rightleftarrows \varphi_g$. Owing to this, in the process of formation of molecules with identical nuclei there can occur only the transitions $\varphi_u \rightarrow \varphi_g$, i.e., from the state V_u of the continuum. Since the state is antisymmetric with respect to the nucleus interchange it leads, in the s-wave approximation, to the states with odd total spin of both nuclei. Therefore, the formation probability (43) for the case of identical nuclei should be multiplied by the statistical weight of the orthostate $= \frac{3}{4}$ for ppμ and ttμ and $= \frac{1}{3}$ for ddμ molecules.

When mesonic molecules have different nuclei the contribution to the matrix element $\langle \mathbf{d} \rangle$ comes from the both terms in (42). The probabilities of the pdμ mesonic molecule formation calculated by a number of authors are given in Table 3. It should be noted that the molecular structure of hydrogen is neglected in Eq. (43). As the estimates show, accounting for it increases the probability W_{E1} by about a factor of 1.5 (Cohen *et al.*, 1960). The E0 and E2 transition probabilities are suppressed by a factor $(m_e/M_\mu)^2$ as compared with the E1 probability. However, the probability W_{E0} can greatly increase if the mesonic molecule has an excited vibrational level $L = 0$, $v = 1$ with small binding energy (Zeldovich, 1954), since in this case the collision between the mesonic atom and the nucleus occurs under conditions rather close to the resonance. Therefore, in the process of formation of the ddμ and dtμ molecules the E0 transition to the state $L = 0$, $v = 1$ competes with the E1 transitions to the state $L = 1$, $v = 0$, the probability of which is incidentally small.

The E2 transitions can, in principle, go to the rotational level with $L = 2$ of the mesonic molecules ddμ, dtμ, and ttμ. However, contrary to the resonance case, in the E0 transitions to the levels $L = 0$, $v = 1$, the levels with $L = 2$, $v = 0$ are rather deep (Tables 13, 14, and 16) and therefore the probability W_{E2} turns out to be negligible compared with the probability W_{E1}. The probabilities for the ptμ, dtμ, and ttμ molecule formation are $\lambda_{pt\mu} \approx 4 \cdot 10^5$, $\lambda_{dt\mu} \approx 2 \cdot 10^3$, $\lambda_{tt\mu} \approx 7 \times 10^5$ sec^{-1}, respectively, according to the calculations of Belyaev *et al.* (1959).

3. ddμ *Molecule Formation with Excitation of Vibrational Levels of Deuterium Molecule*

The fact that the molecule ddμ has a rotational–vibrational level with $L = 1$, $v = 1$ and very small binding energy may be a cause of a very peculiar mechanism of formation of such mesonic molecules (Vesman, 1967a,b, 1968, 1969); namely, in the E1 transition to this level the binding energy of the molecule ddμ can be transferred to the entire molecule D_2 as a whole. If the binding energy of the ddμ molecule is larger than the dissociation energy of the D_2 molecule ($\approx$4.7 eV) then the probability of such a process is not large: $\lambda_{dd\mu} < 10^2$ sec^{-1}.

If the binding energy of the molecule ddμ in the state with $L = 1, v = 1$ is smaller than the dissociation energy of the molecule D_2 (or DH) then in the collision of the dμ atom with the deuterium nucleus bounded in the molecules D_2 or DH there can occur the formation of the ddμ molecule. Its binding energy will be given up to the excitation of the vibrational levels of a peculiar molecule, one of its nuclei being a system $(dd\mu)^+$ and the other a deuteron or a proton. This process is of resonance character and occurs at the energy $\mathcal{E}_0$ of the relative motion of the dμ atom and the molecule D_2 (or DH) equal to the difference between the excitation energy of a vibrational level n of the molecule $(dd\mu)^+$d and the binding energy of the molecule ddμ, $\mathcal{E}_0 = \mathcal{E}_n - E_{dd\mu}$. The absolute value of the ddμ formation probability depends on the number n of the excited vibrational level of the system $(dd\mu)^+$d or $(dd\mu)^+$p. Good agreement with experiment is achieved under the assumption that the transition goes to the levels with $n = 5$–6. Since the energy difference of the vibrational levels of the system $(dd\mu)^+$d is $\approx$0.336 eV, the binding energy of the ddμ molecule in the state with $L = 1$, $v = 1$ can be inferred to be about 2 eV. Recent calculations performed by Ponomarev *et al.* (1973a,b) confirm the existence of this level (Table 13).†

Figure 9 presents the experimental data on the measurement of the ddμ formation rate and the calculated curve $W_{dd\mu}(\bar{\mathcal{E}}_{d\mu})$ as a function of the average kinetic energy $\bar{\mathcal{E}}_{d\mu}$ of the dμ mesonic atom.

It is seen from the figure that the maximum formation probability for the ddμ mesonic molecule is obtained at an energy $\bar{\mathcal{E}}_{d\mu} = 0.016$ eV. This result can account for the sharp difference in the values of the rates $\lambda_{dd\mu}$ in liquid hydrogen (Fetkovich *et al.*, 1960; Doede, 1963) and in a diffusion chamber (Dzhelepov *et al.*, 1966). Recent measurements (Bystritskij *et al.*, 1973b) are in agreement with this picture.

† The existence of this level was called in question for a long time (Zeldovich and Gershtein, 1960), but this problem has been solved by Matveenko and Ponomarev (1970a). Calculations of the binding energy of the molecule ddμ in the state $L = 1$, $v = 1$ give the value $E \approx 0.7$ eV (Ponomarev *et al.*, 1973a, b). In order to determine reliably the binding energy E, owing to its smallness it is necessary to take into account adiabatic corrections of the second order $\sim(m/M)^2$ to the potentials $K_{ij}(R)$ in Eq. (24). Such corrections are known to increase the E value slightly.

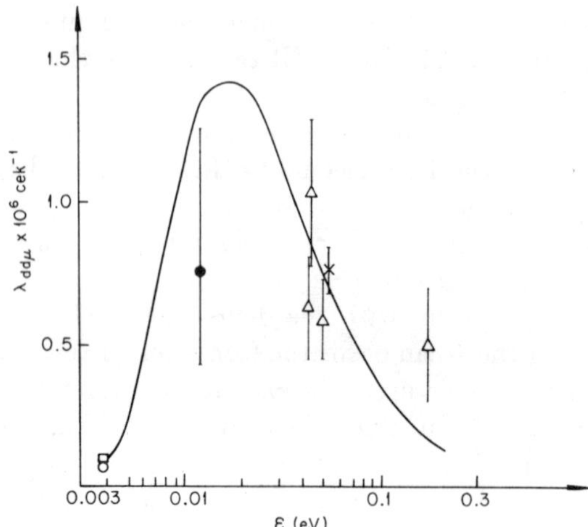

FIG. 9. Probability for ddμ mesonic molecule formation as a function of the mean kinetic energy $\bar{\mathcal{E}}_{d\mu}$ of dμ mesonic atoms, according to the following data: △, Dzhelepov *et al.* (1966); ○, Fetkovich *et al.* (1960); □, Doede (1963); ●, Alvarez *et al.* (1957); ×, Bystritsky *et al.* (1974b). The curve is calculated by Vesman (1967b).

These considerations make it possible to believe that the aforementioned mechanism is predominant in the process of ddμ formation. Nevertheless, direct experimental evidence is desirable, which can be obtained, e.g., in measuring the ddμ formation at different temperatures, as well as from experiments in H_2 with a small admixture of D_2, since this mechanism is very sensitive as to whether the deuterium is present in the form of D_2 molecules or DH.

4. *Transitions between Levels of Mesonic Molecules*

The estimates given in the previous sections show that the μ mesonic molecules are, as a rule, formed in excited states. Mesonic molecules with different nuclei (pdμ, ptμ, and dtμ) can go over from the excited $L = 1$ states to the ground ($L = 0$) state by radiative E1 transitions during a time of 10^{-5} to 10^{-6} sec. It is more likely, however, that during a time of 10^{-11} sec a mesonic molecule in liquid hydrogen captures an electron and passes to the ground state with conversion on this electron during a time of 10^{-10} to 10^{-11} sec (Cohen *et al.*, 1960; Zeldovich and Gershtein, 1960).†

† Another possible mechanism leading to a conversion transition consists in the fact that the $(pd\mu)^+$ mesonic molecule plays the role of the nucleus of the molecular ion $[(pd\mu)^+p]e$ immediately after formation of pdμ with conversion on the electron of the H_2 molecule.

Due to this mechanism, the mesonic molecules pdμ and ptμ reach the ground state in a rather short time. In the dtμ mesonic molecule, there occurs a competition between the $(L = 1) \rightarrow (L = 0)$ transition and the nuclear reaction in the excited state of dtμ.

The excited states of molecules with identical nuclei (ppμ, ddμ, and ttμ) are metastable since for them the transition $(L = 1) \rightarrow (L = 0)$ is accompanied by a change in the total spin of the nuclei. According to the estimates of Vesman (1967), the probability for the conversion transition of the ddμ mesonic molecule from the $L = 1$, $v = 1$ state to the $L = 0$, $v = 0$ and $L = 1$, $v = 0$ states is much less than the probability for the nuclear reaction in the excited state.

The ppμ mesonic molecule can be formed in the $L = 1$ excited state only when the total spin of the nuclei is $I = 1$. The transition $(L = 1) \rightarrow (L = 0)$ must be accompanied by a change in the total spin $(I = 1) \rightarrow (I = 0)$.

According to the estimates of Weinberg (1960), the probability for such an orthoparatransition is essentially smaller than the probability for μ decay. This means that the nuclear capture of a meson by a proton occurs from the excited rotational state of the ppμ mesonic molecule. The fast transition pμ $(F = 1) \rightarrow$ pμ $(F = 0)$ precedes the ppμ molecule formation with the result that the ppμ mesonic molecules with total nuclear spin $I = 0$ are formed in a state with total spin $J = \frac{1}{2}$.

In this state, the probability for the μ meson to occur in the $F = 0$ state of the pμ atom is $\frac{3}{4}$ and that for the $F = 1$ state is $\frac{1}{4}$. Since the nuclear capture of a μ meson by a proton occurs mainly from the $F = 0$ state (Primakoff, 1959), the two subsequent processes, transition of the pμ mesonic atom to the lower hfs state and the ppμ mesonic molecule formation in the state with $L = 1$ and $J = \frac{1}{2}$, increase the probability for μ capture by the proton by a factor of three compared with the isolated pμ atom (Zeldovich and Gershtein, 1958b; Rothberg *et al.*, 1963; Lobov, 1963; Ohtsubo and Fujii, 1966).† This conclusion has an experimental confirmation (Bleser *et al.*, 1962; Bertolini *et al.*, 1962; Rothberg *et al.* 1963).

The interaction of spins with rotation in the $L = 1$ state may, in principle, transform the ppμ system to the state with total spin $J = \frac{3}{2}$, the nuclear μ capture from which is negligible (Zeldovich and Gershtein, 1958;

† To determine the probability for capture of a μ meson by a proton in the ppμ mesonic molecule in the $L = 1$ state, it is essential to know the quantity γ_0 that is equal to the ratio of the probabilities of the μ meson location near protons in the ppμ mesonic molecule and the isolated pμ atom, respectively. According to calculations of Halpern (1964b) and Kabir (1966), $2\gamma_0 = 1.01 \pm 0.01$.

Weinberg, 1960). However, according to the calculations by Halpern (1964a, 1968) the probability for such a transition is very small.

H. Transfer of μ Mesons from Nuclei of Hydrogen Isotopes to Z Nuclei of Heavier Atoms

1. *General Picture*

The system of Eqs. (22) makes it possible, in principle, to calculate the cross section for the transfer process

$$\mathrm{p}\mu + Z \rightarrow Z\mu + \mathrm{p} \tag{44}$$

In slow collisions, the nuclear motion is adiabatic (Section III,A,1). Therefore, it is convenient to use the notion of terms $W_i(R)$, which are the effective potential energy of nuclei for a given quantum state i of the meson,

$$W_i(R) = E_i(R) + (Z/R)$$

As an example, Fig. 10 gives the picture of terms for the case $Z = 5$. At $R \rightarrow \infty$ the term 5gσ $[E_1 = W_1(\infty) = -\frac{1}{2}]$ describes the system p$\mu + Z$ with a μ meson on the K-orbit of the pμ mesonic atom, while, e.g.,

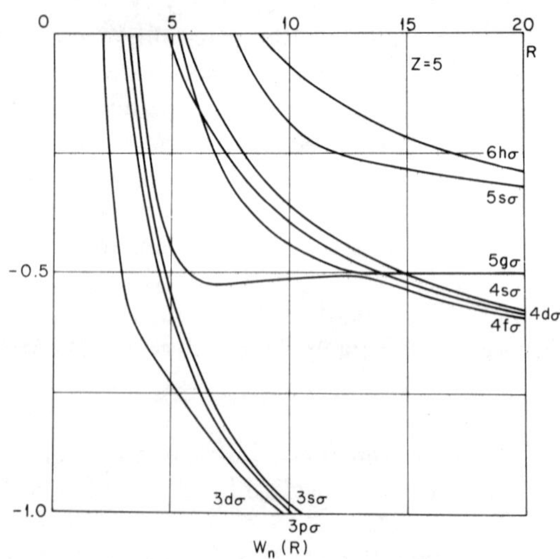

FIG. 10. Terms of the p$\mu + Z$ and $Z\mu$ + p systems for $Z = 5$. At $R \rightarrow \infty$, the term 5gσ corresponds to the level 1sσ of the pμ mesonic atom; the terms 4fσ, 4dσ, and 4sσ to the levels of the $Z\mu$ mesonic atom. For $R \approx 13$ there occurs an intense transfer according to the reaction p$\mu + Z \rightarrow Z\mu$ + p (Ponomarev and Puzynina, 1967a).

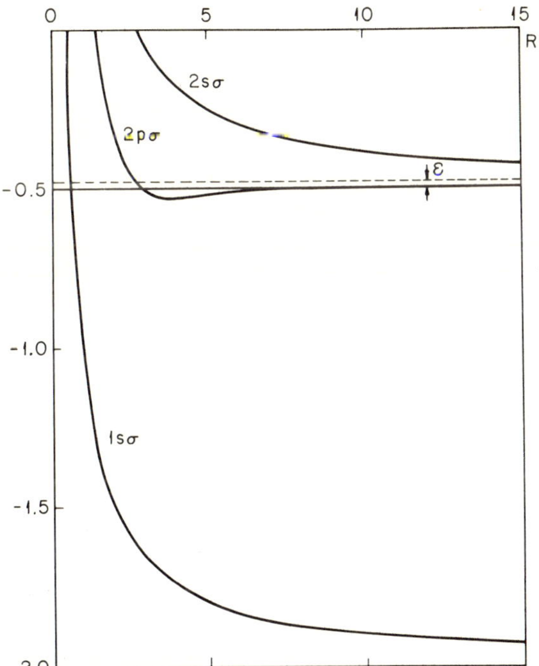

FIG. 11. Terms of the systems pμ + He (term 2pσ) and Heμ + p (terms 1sσ and 2sσ). Intersections and pseudointersections of the terms are absent, and the transfer pμ + He → Heμ + p is suppressed.

terms 4sσ, 4pσ, 4dσ correspond to the system $Z\mu$ + p in different $Z\mu$ atomic states. To the transfer reaction (44) there corresponds a transition from the term 5gσ to one of the terms of the $Z\mu$ + p system.

Because of the large mass of the nuclei, their motion is quasi-classic and thus the probability for transition (44) is exponentially small for the exception of the cases of intersection of terms (see, e.g., Landau and Lifshitz, 1966). The existence of such intersections in the pμZ system was indicated by Gershtein and Krivchenkov (1961) and used by Gershtein (1962) for the calculation of the cross sections for μ-meson transfer from protons to oxygen and carbon nuclei.

The absence of such intersections in the system pμHe (Fig. 11) accounts for the small cross section for transfer of μ mesons from protons to helium.

2. *Effect of Pseudointersection of Terms on Charge-Exchange Processes*

More recent calculations of the terms of systems pμZ (Ponomarev and Puzynina, 1967a) have shown that some of the earlier considered cases of intersections of terms are really the ones of pseudointersections (terms $5g\sigma$ and $4f\sigma$, Fig. 10, for $R \approx 13$). Since in reaction (44) the most intense transfer occurs at points of intersection or pseudointersection, the system (22) again reduces to a two-equation approximation that takes into account only the interaction of intersecting or pseudointersecting terms.

The location R_0 of the pseudointersection points, the splitting δE, and the average value E_0 of terms at these points as well as the total number z_0 of pseudointersections in system (44) can be expressed analytically as follows (Ponomarev, 1968; Komarov and Slavyanov, 1968, 1969):

$$R_0 \approx 2Z\left[\left(\frac{Z-1}{n'-n}\right)^2 - 1\right]^{-1}, \qquad 2p = R_0(-2E_0)^{1/2}$$

$$E_0 \approx -\frac{1}{2}\left(\frac{Z-1}{n'-n}\right)^2 \tag{45}$$

$$\delta E \approx -\frac{4E_0}{n'-n}\frac{(4p)^{n_2+n_2'+m+1}e^{-2p}}{[n_2!n_2'!(n_2+m)!(n_2'+m)!]^{1/2}}$$

$$z_0 = \mathcal{E}nt[n(\sqrt{Z}-1)(\sqrt{Z}+1-\sqrt{1+2\sqrt{Z}})]$$

where (n, n_2, m) and (n', n_2', m) are the parabolic quantum numbers of the levels of the $p\mu$ and $Z\mu$ mesonic atoms, respectively. The first pseudointersection of such a kind ($\bar{\bar{z}}_0 = 1$) appears in the $p\mu + Z$ system only for $Z = 5$ ($n = 1, n' = 4$) and the first intersection for $Z = 4$ ($n = 1, n' = 3$). Therefore, for $Z = 2$ and 3 reaction (44) must be strongly suppressed.

3. *μ-Meson Transfer from Proton to Helium*

The suppression of the probability of the reaction

$$p\mu + He \rightarrow He\mu + p \tag{46}$$

is verified experimentally (Schiff, 1961; Zaimidoroga *et al.*, 1963; Placci *et al.*, 1967; see Table 7). The rates of the processes $p\mu \rightarrow {}^3He\mu$, $p\mu \rightarrow {}^4He\mu$, and $d\mu \rightarrow {}^3He\mu$ calculated by Eqs. (24) are, respectively, $\lambda = 6.4 \times 10^6$, 5.5×10^6, and 1.3×10^6 sec^{-1} (Matveenko and Ponomarev, 1972c).

This computation is made for the transfer of a μ meson from the ground state of the $p\mu$ mesonic atom. It follows from Eqs. (45); however, for $n = 5$ in the $p\mu$ + He system there also occurs a pseudointersection. This means that the probability for the $p\mu \rightarrow He\mu$ reaction from the excited states of $p\mu$ mesonic atoms may turn out to be significant. This conclusion is confirmed in experiments on the study of the reaction $p\pi + Z \rightarrow Z\pi^- + p$ (Petrukhin *et al.*, 1968).

4. *Radiative Transfer of* μ *Mesons*

It follows from the above considerations that at the intersection and pseudointersection points there occurs a nonradiative transition of a μ meson from the proton to the Z nuclei. For small Z this process seems to be predominant. However, for large Z, it comes into competition with the radiative transition

$$\mathrm{p}\mu + Z \rightarrow Z\mu + \mathrm{p} + \gamma \tag{47}$$

The experimental evidence for the occurrence of this reaction has been obtained by Budjashov *et al.* (1967c), who have measured the K series of μ mesonic X rays in pure Ar and in a mixture 0.5% Ar + 99.5% H_2, i.e., under the conditions when all the μ mesons are initially captured by hydrogen forming $\mathrm{p}\mu$ mesonic atoms. Figure 12 shows that, due to the $\mathrm{p}\mu \rightarrow \mathrm{Ar}\mu$ transfer, there appear noticeable changes in the K series structure of the $\mathrm{Ar}\mu$ mesonic atom. One of the possible mechanisms of this phenomenon can be explained by means of the scheme of terms, Fig. 10. In the process of collision (57) for each value of R there exists a nonzero

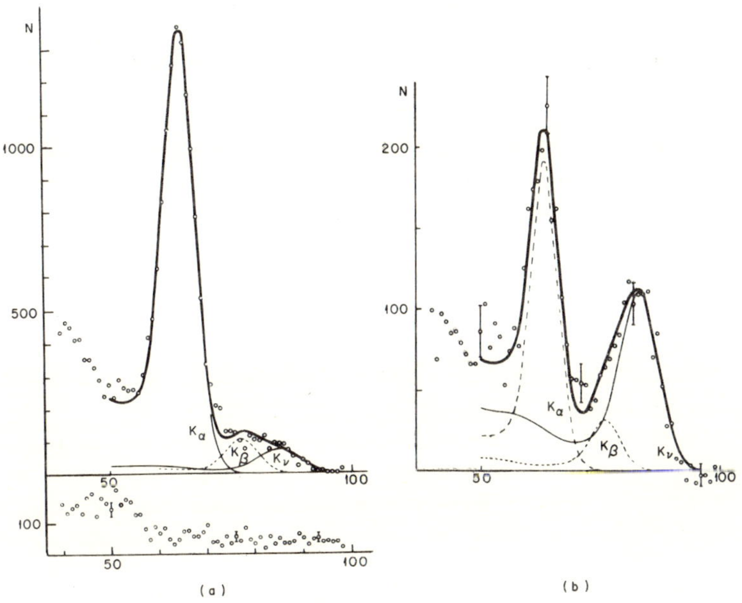

FIG. 12. The structure of the Ar K-series in pure argon (a) and in a mixture 0.5% Ar + 99.5% H_2 (b). In the latter case the contribution of the K_ν series is much larger than in the former (Budyashov *et al.*, 1967c).

probability for the radiative transition from the term $n = 1$ of the pμ mesonic atom to the terms $n' \leq (Z - 1)$ of the $Z\mu$ mesonic atom. The energy of the $n = 1$ level of the pμ mesonic atom is equal to the energy of the $Z\mu$ atom in the $n' = Z$ state; therefore, the energy of γ quanta from the direct transition pμ $(n = 1) \rightarrow Z\mu$ $(n' \sim 1)$ corresponds to high transitions of the X series of the $Z\mu$ mesonic atom. In a free $Z\mu$ atom such transitions are hindered due to the selection rules with respect to the orbital momentum. However in the process of collision with a proton, the central symmetry of the field in the $Z\mu$ mesonic atom is violated, which must lead to an increase of the intensity in the $Z\mu$ mesonic atom spectrum.

Another possible cause of the observed change in the K-series structure of Arμ consists in the following. The pμ $\rightarrow$ Arμ transfer occurs only to some levels selected by pairs of interacting terms. This fact alters the initial population of the levels of the free atom. This in turn effects the K series of the Arμ mesonic atom.

Recent studies of the reaction pμ $\rightarrow$ Arμ in a high-pressure gas target performed by Backenstoss *et al.* (1971) are expected to make a contribution to the understanding of the true mechanism of the indicated phenomenon.

IV. Catalysis of Nuclear Reactions by μ^- Mesons in Hydrogen

It follows from simple considerations that nuclear reactions induced by μ^- mesons occur from mesomolecular states of nuclei rather than "in flight," i.e., in collisions of mesonic atoms with the nuclei. In fact, we may picture the hydrogen nucleus in the mesomolecule as being bound in a potential well with dimensions of the order of a_μ, whereas the mean distances between the nucleus of a free mesonic atom and the nuclei of the atoms of a substance are equal in order of magnitude to the Bohr radius a_0. This implies that nuclear reaction in flight is suppressed by about a factor of $(a_0/a_\mu)^3 \approx 10^7$, compared with the same reaction in a mesonic molecule.†

When the formation of mesonic hydrogen molecules occurs, the nuclei of hydrogen isotopes come to one another within a distance a_μ and their fusion becomes possible due to quantum-mechanical tunneling through the Coulomb barrier. Frank (1947)‡ was the first to point to such a possibility.

† The probability for nuclear reaction "in flight" can noticeably increase in the resonance situation, for example, when the mesonic molecule has a level with small binding energy (Zeldovich, 1954).

‡ It is worth noting that this idea was suggested by Frank in connection with the experiments of Lattes *et al.* (1947) in which the π mesons were first identified.

Later, nuclear reactions in mesonic molecules were independently considered by Sakharov (1948) and Zeldovich (1954). The catalysis of nuclear reactions $p + d \rightarrow {}^3He$ and $d + d \rightarrow t + p$ by μ mesons in hydrogen (with conversion of muons) was experimentally discovered in Berkeley (Alvarez *et al.*, 1957).† The reaction $p + d \rightarrow {}^3He + \gamma$ was observed and studied by a Liverpool group (Ashmore *et al.*, 1958). Later, nuclear reactions in the $pd\mu$ and $dd\mu$ molecules were investigated in detail at Columbia University (Bleser *et al.*, 1962, 1963), Chicago (Schiff, 1961; Doede, 1963), Carnegie Institute (Fetkovich *et al.*, 1960), CERN (Conforto *et al.*, 1962, 1964), and Dubna (Dzhelepov *et al.*, 1962a,b,c, 1964a,b, 1966). The experimental investigations have stimulated a large number of theoretical studies (Skyrme, 1957; Zeldovich and Sakharov, 1957; Zeldovich, 1957; Hayashi *et al.*, 1957; Jackson, 1957; Cohen *et al.*, 1958, 1960; Belyaev *et al.* 1959; Mizuno, 1959, 1961; Schröder 1963; Vesman, 1967–1969; Kolos, 1968).

A. *Estimation of the Probability for Nuclear Reaction in Mesonic Molecules*

The rates of nuclear reactions in mesonic molecules can be estimated by using the experimental data on the cross sections for synthesis at small colliding energies. The main channels of nuclear reactions are the following:

$$\begin{aligned}
&p + d \rightarrow {}^3He + \gamma + 5.4\ \text{MeV} && \text{(I)}\\
&d + d \rightarrow \begin{cases} t + p + 4\ \text{MeV} \\ {}^3He + n + 3.3\ \text{MeV} \end{cases} && \text{(II)}\\
&d + t \rightarrow {}^4He + n + 17.6\ \text{MeV} && \text{(III)}\\
&p + t \rightarrow {}^4He + \gamma + 20\ \text{MeV} && \text{(IV)}\\
&t + t \rightarrow {}^4He + 2n + 10\ \text{MeV} && \text{(V)}
\end{aligned} \tag{48}$$

At small colliding energies the cross sections for reactions I–V can be

† It may be concluded from these first experiments that nuclear reactions of synthesis proceed mainly from the mesomolecular state rather than "in flight." In fact, since the constant of the $d + d$ reaction is larger by a factor of 10^6 than the corresponding constant for the $p + d$ reaction (see Section IV, A), then under experimental conditions ($C_d \approx 4\%$) the $d + d$ reaction probability would exceed the $p + d$ reaction probability by several times. The predominance of the latter means that the reaction of the $pd\mu$ molecule formation anticipates nuclear synthesis and its probability much exceeds the probability of the nuclear reaction "in flight."

represented in the form

$$\sigma = C\,|\,\psi(0)\,|^2/v \tag{49}$$

where $\psi(0)$ is the value of the wave function of the relative motion of nuclei normalized to the unit flux, at distances on the order of the range of nuclear forces, v the relative velocity of nuclei, and C the reaction constant. For the value $\psi(0)$ the following expression holds:

$$|\,\psi(0)\,|^2 = 2\pi\eta(e^{2\pi\eta} - 1)^{-1} \approx 2\pi\eta e^{-2\pi\eta}, \qquad \eta = e^2/\hbar v \gg 1 \tag{50}$$

The energy dependences (49) and (50) are confirmed experimentally at energies 10–30 KeV and make it possible to determine the constants C from the measurements of the cross sections for reactions I–V:

$$C_{\mathrm{I}} = (2.5 \pm 0.6) \times 10^{-22}\ \mathrm{cm^3\,sec^{-1}} \qquad \text{(Griffits } et\ al.\text{, 1963)}$$
$$C_{\mathrm{II}} = 2 \times 10^{-16}\ \mathrm{cm^3\,sec^{-1}} \qquad \text{(Arnold } et\ al.\text{, 1954)} \tag{51}$$
$$C_{\mathrm{III}} = 2 \times 10^{-14}\ \mathrm{cm^3\,sec^{-1}}$$

The rates of nuclear reactions in the pdμ, ddμ, and dtμ mesonic molecules in the $L = 0$ states can then be estimated from the relation

$$\lambda_f = C\,|\,\chi(0)\,|^2 \tag{52}$$

where $\chi(0)$ is the wave function describing the motion of the nuclei of the mesonic molecule at distances of the order of range of nuclear forces. The quantity $\chi(0)$ can be calculated from Eqs. (24). For the pdμ molecule this yields $\chi_{\mathrm{pd}\mu}(0) \approx 0.0146(4\pi)^{-1/2}a_\mu^{-3/2}$, i.e., $|\,\chi(0)\,|^2_{\mathrm{pd}\mu} \approx 10^{27}\ \mathrm{sm^{-3}}$ (Cohen *et al.*, 1960). A close value $\approx 1.4 \times 10^{27}\ \mathrm{sm^{-3}}$ is obtained by the WKB method (Jackson, 1957; Zeldovich and Sakharov, 1957; Skyrme, 1957). According to these calculations the rate of reaction I for the pdμ molecule is estimated as

$$(\lambda_f)_{\mathrm{pd}\mu} = (0.25 \pm 0.06) \times 10^6\ \mathrm{sec^{-1}} \tag{53}$$

An analogous calculation of the same authors for the ddμ and dtμ molecules leads to the results

$$(\lambda_f)_{\mathrm{dd}\mu} \sim 10^{11}\ \mathrm{sec^{-1}}, \qquad (\lambda_f)_{\mathrm{dt}\mu} \sim 10^{12}\ \mathrm{sec^{-1}} \tag{54}$$

i.e., the ddμ and dtμ mesonic molecule formation leads in almost all the cases to nuclear reactions.

The given estimates should be essentially improved. First of all, in the ddμ mesonic molecule formed in a rotational metastable state with $L = 1$, nuclear reaction must occur not from the s- but from the p-state. We should, however, bear in mind that the p-wave contribution to the cross section for reaction II is not small down to the lowest energies, which is seen from the angular distribution of its products.† Using the data on the p-wave contribution in the (d + d) reaction, Vesman (1967c) has found that the rate for nuclear reaction II from the $L = 1$, $v = 1$ state of the ddμ mesonic molecule does not practically differ from the rate of the same reaction from the s-state.

In addition, the estimates indicated do not take into account the fact that reactions I–V can proceed with different probabilities from different spin states. The account of this fact can change these estimates several times and leads to very peculiar phenomena when the reaction occurs in the ddμ mesonic molecule.

B. Dependence of the Nuclear Reaction Rate on the H.F.S. of the Mesonic Molecule pdμ

In the $L = 0$ state, the total nuclear spin of the system can assume the values $I = \frac{1}{2}$ and $I = \frac{3}{2}$, the rates of the nuclear reaction

$$p + d \rightarrow {}^3He + \gamma \tag{55}$$

from these states being strongly different. The model estimates (Cohen *et al.*, 1960) show that the probabilities for M1- and E2-transitions from the $I = \frac{3}{2}$ state of the pdμ molecule can be disregarded compared to that for M1-transitions from the $I = \frac{1}{2}$ state and to the rate λ_0 of the free μ-meson decay. One succeeds in explaining the magnitude of the relatively large contribution of the reaction pd$\mu \rightarrow {}^3$He + μ by an E0-transition from the $I = \frac{1}{2}$ state with conversion on the μ meson‡ (Zeldovich and Sakharov, 1957; Skyrme, 1957; Jackson, 1957).

Thus, the fact that in the pdμ mesonic molecule nuclear reaction occurs only from the state with the total proton and deuteron spin $I = \frac{1}{2}$ can be considered to be established. However, it should be pointed out that, in the pdμ mesonic molecule, states with a definite I are not always eigenstates of the system, since in the three-body system their total momentum J is just the conserved quantity. The interaction between the spins of mesons and nuclei splits the ground state of the pdμ mesonic molecule

† As is known, for slow particles the reactions from the p-wave are usually suppressed due to the centrifugal barrier. However, in subbarrier reactions with charged particles, the ratio of the squared modules of the Coulomb s- and p-wave functions is independent of the collision energy and equals the ratio $(R_N/a_N)^2$, where R_N is the effective radius of nuclear forces and $a_N = \hbar^2/Me^2$ is the Bohr radius of a particle with reduced nuclear mass M (see, e.g., Landau and Lifshitz, 1966).

‡ For the M1 transition the conversion coefficient is very small ($\sim 4 \times 10^{-4}$).

with orbital momentum $L = 0$ into four hfs levels with total momenta $J = 2, 1, 1', 0$. The total spin of the nuclei has the definite values $I = \frac{3}{2}$ and $I = \frac{1}{2}$ only in the states with total momentum $J = 2$ and $J = 0$, respectively. The $J = 1$ and $J = 1'$ states are a superposition of the $I = \frac{3}{2}$ and $I = \frac{1}{2}$ states with weights $|C_1|^2$ and $|C_2|^2$, respectively (Fig. 13). Following the estimates of Zeldovich and Gershtein (1960), $|C_1|^2 = (0.41)^2 = 0.17$ and $|C_2|^2 = (0.91)^2 = 0.83$. Then the rates of the nuclear

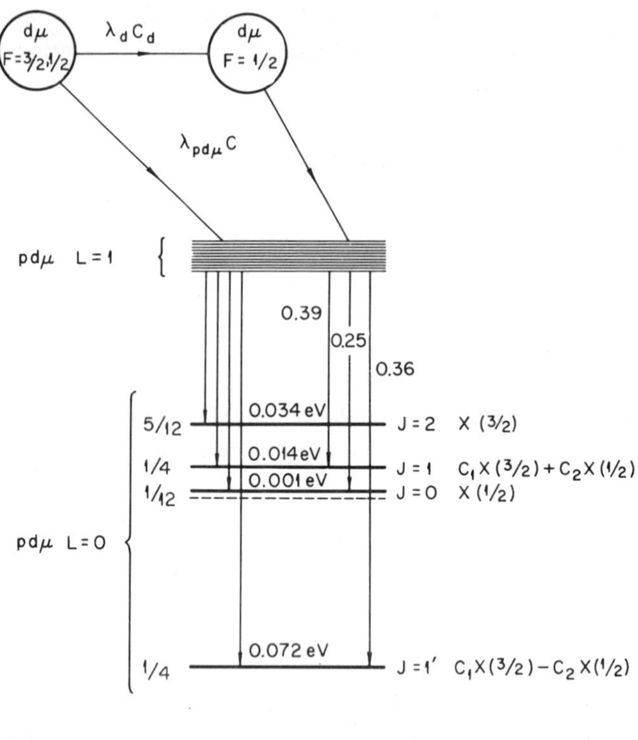

FIG. 13. Scheme of formation of pdμ mesonic molecules and transitions occurring in them. At low deuterium concentration pdμ mesonic molecules are formed in the $L = 1$ state with statistical distribution over the hfs levels. (Statistical weights of the levels are shown on the left.) When $\lambda_d C_d > \lambda_{pd\mu}(1 - C_d)$, there initially occurs the transition dμ $(F = \frac{3}{2}) \rightarrow$ dμ $(F = \frac{1}{2})$ and only after it the reaction dμ $(F = \frac{1}{2})$ + p $\rightarrow$ pdμ proceeds. In this case, after the transitions $(L = 1) \rightarrow (L = 0)$ the $J = 2$ level of the $L = 0$ state is not populated, and the values of population of the levels with momenta 1, 0, 1′ are 0.39, 0.25, and 0.36, respectively (provided that $C_1 \approx 0.41$ and $C_2 \approx 0.91$). The sequence of the above processes results in the increase of the yield of nuclear reaction pd$\mu \rightarrow {}^3$He + μ^-.

reaction from different hfs states of the pdμ mesonic molecule are

$$\lambda_f^2 = 0, \qquad \lambda_f^1 = |C_1|^2 \lambda_f = 0.17\lambda_f$$
$$\lambda_f^0 = \lambda_f, \qquad \lambda_f^{1'} = |C_2|^2 \lambda_f = 0.83\lambda_f \tag{56}$$

where $\lambda_f = \lambda_f^0$ is the rate of the nuclear reaction from the state with total spin $I = \frac{1}{2}$.

The total yield of the reaction (55) under conditions of saturation of the p$\mu \to$ dμ transfer and the assumption of the statistical population of the hfs levels of the pdμ mesonic molecule is

$$Y_\gamma = \frac{\lambda_{pd\mu}}{\lambda_0 + \lambda_{pd\mu}} \Lambda_\gamma$$
$$\Lambda_\gamma = \frac{1}{4}\frac{\lambda_\gamma C_1^2}{\lambda_0 + \lambda_f C_1^2} + \frac{1}{4}\frac{\lambda_\gamma C_2^2}{\lambda_0 + \lambda_f C_2^2} + \frac{1}{12}\frac{\lambda_\gamma}{\lambda_0 + \lambda_f} \tag{57}$$

where for brevity one puts $\lambda_\gamma = \lambda_\gamma^0$.

Since the rates (56) of nuclear reaction (55) from the $J = 0$ and $J = 1'$ states are close to each other and both exceed the rate of the reaction from the $J = 1$ state, the contribution of which to the total yield is small, the time distribution of the γ quantum yield is insensitive to the hfs effects. The rate of reaction (55) measured by Bleser *et al.* (1963) is† (Section II,B,3)

$$\bar{\lambda}_f = (0.305 \pm 0.010) \times 10^6 \text{ sec}^{-1} \tag{58}$$

The total yield of reaction (55) measured in this work under the condition of saturation of the transfer p$\mu \to$ dμ is

$$Y_\gamma = 0.140 \pm 0.024 \tag{59}$$

Comparing it with the conversion μ-meson yield measured by Schiff (1961) and Doede (1963)

$$Y_\mu = 0.028 \pm 0.003 \tag{60}$$

it is possible to find the ratio of the reaction channels

$$\alpha = Y_\mu / Y_\gamma = 0.20 \pm 0.04$$

and from the $\bar{\lambda}_f$ measurements to calculate the absolute rates of both

† For the statistical population of the hfs levels, the $J = 1'$ state gives the largest contribution to the total yield of nuclear reaction; therefore, it may be assumed that the quantity $\bar{\lambda}_f$ is the rate $\lambda_f^{1'}$. In this case $\lambda_f = (0.367 \pm 0.012) \times 10^6 \text{ sec}^{-1}$.

channels:

$$\bar{\lambda}_\gamma = (0.255 \pm 0.012) \times 10^6 \text{ sec}^{-1}, \qquad \bar{\lambda}_\mu = (0.050 \pm 0.010) \times 10^6 \text{ sec}^{-1} \tag{61}$$

Inserting these values of Eq. (57) we find that the value Y_γ calculated in such a way fits well with the measured one (59). This confirms the assumption that the mesonic molecule hfs influences the course of the nuclear reaction.†

The account of the spin dependence of nuclear reaction (55) changes the previous estimates. Since the statistical weight of the states with total spin $I = \frac{1}{2}$ is $\frac{1}{3}$, then the constant C of the nuclear reaction from this state is three times the value (51) determined from relations (49) and (50). Therefore, the rate $\lambda_\gamma{}^0$ is three times the value (61) and is $\lambda_\gamma{}^0 = (0.75 \pm 0.10) \times 10^6 \text{ sec}^{-1}$. This value is different from the experimental one. However, to draw more definite conclusions, more accurate calculations of $|\chi(0)|^2$ should be made.

C. Influence of Transitions $d\mu(F = \frac{3}{2}) \rightarrow d\mu(F = \frac{1}{2})$ on Catalysis of Nuclear Reaction in the $pd\mu$ Mesonic Molecule

At low deuterium concentration, the statistical distribution of the $d\mu$ mesonic atoms over the $F = \frac{3}{2}$ and $F = \frac{1}{2}$ states is conserved, which leads to the statistical population of the levels of the $pd\mu$ mesonic molecules formed. However, as was noticed by Wolfenstein, at high deuterium concentrations the transitions $(F = \frac{3}{2}) \rightarrow (F = \frac{1}{2})$ can essentially affect the yield of the nuclear reaction in the $pd\mu$ mesonic molecule, since in this case the $J = 2$ level giving no contribution to the reaction is not populated, whereas the population of the $J = 0$ and $J = 1'$ levels increases. According to calculations (Gershtein, 1960) under these conditions the populations for $J = 1, 1'$, and 0 are 0.39, 0.36, and 0.25, respectively, which corresponds to the following yield of reaction (55):

$$\begin{gathered} Y_\gamma \approx \frac{\lambda_{pd\mu}(1 - C_d)}{\lambda_0 + \lambda_{pd\mu}(1 - C_d) + \lambda_{dd\mu}C_d} \Lambda_\gamma{}' \\ \Lambda_\gamma{}' = 0.39 \frac{0.17\lambda_\gamma}{\lambda_0 + 0.17\lambda_f} + 0.36 \frac{0.83\lambda_\gamma}{\lambda_0 + 0.83\lambda_f} + 0.25 \frac{\lambda_\gamma}{\lambda_0 + \lambda_f} \end{gathered} \tag{62}$$

† Doede (1963) remarked that earlier measurements $Y_\nu = 0.34 \pm 0.06$ (Ashmore *et al.*, 1958) and $Y_\nu = 0.33 \pm 0.08$ (Bleser *et al.*, 1962) are not compatible with the hfs effects.

and exceeds by about a factor of 1.7 the yield for the statistical population of levels.†

Wang *et al.* (1965) and Bleser *et al.* (1963) confirmed this effect experimentally: the observed increase of the γ quantum yield at a concentration $C_d = 0.25$ was in agreement with the one calculated. Doede (1963) observed the corresponding increase of the yield of the conversion μ meson at $C_d = 0.99$.

D. Nuclear Reaction in the ddμ Mesonic Molecule

The nuclear synthesis in the ddμ mesonic molecule was first observed by Alvarez *et al.* (1957) in a liquid-hydrogen chamber at a concentration $C_d \approx 0.04$. It was further studied by Fetkovich *et al.* (1960) and more completely by Doede (1963) in a liquid-hydrogen chamber with a protium admixture $C_p = 0.96\%$, as well as in a diffusion chamber by the Dubna group (Dzhelepov *et al.*, 1964a,b,d, 1966), who observed, in addition to the reaction $d + d \rightarrow t + p$ (Fig. 14a), the reaction $d + d \rightarrow {}^3He + n$ (Fig. 14b). Recently these reactions have been studied using counter methods in a superpure deuterium at 41 atm pressure (Bystritsky *et al.*, 1974b).

As follows from the estimates (Section III,G), the formation of the ddμ mesonic molecules in the excited $L = 1$ state actually always leads to the reaction $(\lambda_f)_{dd\mu} \sim 10^{11}$ sec^{-1}. Doede's (1963) conclusion concerning the value $(\lambda_f)_{dd\mu} = (2.5 \pm 0.4) \times 10^5$ sec^{-1} seems to have no statistical grounds (Dzhelepov *et al.*, 1965a) and should be studied in more detail.

Under the conditions of a liquid-deuterium chamber (Doede, 1963) a few cases of multiple nuclear catalysis of the reaction $d + d \rightarrow t + p$ by the same μ meson were observed. The probability of "sticking" of a μ meson to tritium was thus estimated to be 0.24 ± 0.11 (Dzhelepov *et al.*, 1965a). The direct measurement (Dzhelepov *et al.*, 1964a,b, 1966) gives for this probability the upper estimate 0.14 (the theoretical estimate, $\sim$0.006) and for the probability of "sticking" to He^3 in the reaction $d + d \rightarrow {}^3He + n$ the value $\leq$0.13 (the theoretical estimate, $\sim$0.13).

E. Nuclear Reaction in the ptμ Mesonic Molecule

The ptμ system is interesting from the point of view of observation of "unusual" channels of nuclear reactions. For the nuclear total spin $I = 1$,

† Wolfenstein noticed that as far as the pdμ mesonic molecules are produced in the $L = 1$ state, the spin–rotation interaction can transfer them to the state with total $J = 2$ momentum. The appropriate calculations have not been performed yet; however, similar calculations for the ppμ mesonic molecule (Halpern, 1964a) allow one to think that this transfer is small.

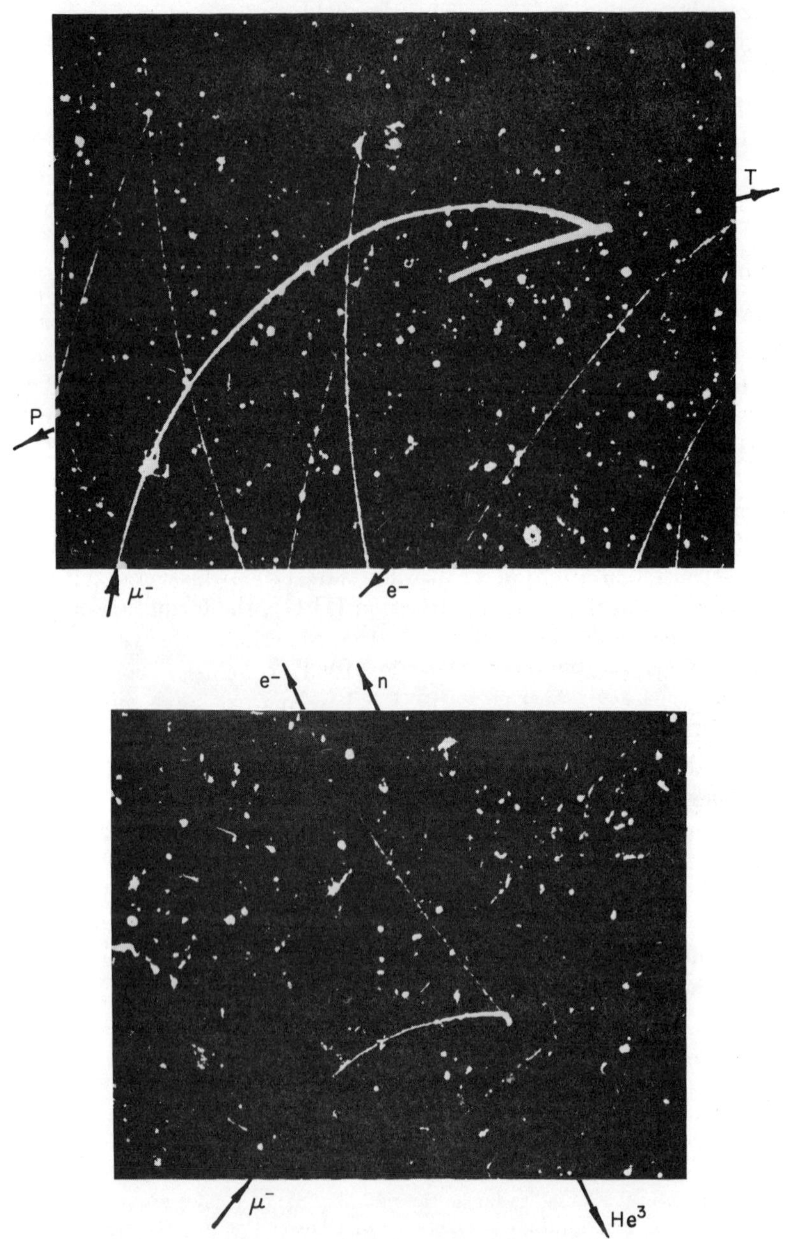

FIG. 14. μ^- catalysis of nuclear reaction in deuterium. (a) $dd\mu \to t + p + \mu^-$, (b) $dd\mu \to {}^3He + n + \mu^-$. The figures are taken from the paper of Dzhelepov *et al.* (1966).

the reaction $p + t \rightarrow {}^4He + \gamma$ from the $L = 0$ state can occur only via an M1-transition, but its probability is larger by a factor of 60 than the probability for the $p + d \rightarrow {}^3He$ reaction. The probability for transfer of the energy to a μ meson in this state is negligible ($\sim 10^{-5}$).

In a state with spin $I = 0$ the emission of a γ quantum is impossible, while the probability for a reaction with conversion on a meson in an E0 transition is on the order of 10^5 to 10^6 sec^{-1}. The hfs of the ground $L = 0$ state of the ptμ mesonic molecule consists of a level with total momentum $J = \frac{3}{2}$ and two levels with $J = \frac{1}{2}$. Due to the similarity of the values of the magnetic moments p and t one of the $J = \frac{1}{2}$ states is an almost pure state of the nuclei with total spin $I = 0$. Under these conditions the formation of electron–positron pairs (Zeldovich and Gershtein, 1960) competes successfully with conversion of the μ meson. At an energy of 20 MeV both channels have approximately the same probability. For the p + d reaction the formation probability for the e^+e^- pair is $\sim 10^{-3}$.

F. *Estimation of the Efficiency of μ^- Catalysis*

The rate of formation of mesonic molecules is, as a rule, comparable with the μ-meson decay rate. Therefore, the probability for a series of nuclear reactions induced by the same meson is not large. However, even in the case of the stable meson, there is a reason for which it is impossible to realize a sustained nuclear reaction (Skyrme, 1957; Jackson, 1957; Zeldovich, 1957), which is just that one of the channels of all the reactions is the formation of He nuclei. There exists a nonzero probability of "sticking" of the μ meson to helium in the course of the reaction, as a result of which it is eliminated from catalysis. For the p + d, d + d, and d + t reactions these probabilities are 0.006, 0.13, and 0.01, respectively.† Thus, even the stable meson can induce not more than 8 reactions in deuterium and not more than 100 reactions in the mixture d + t (in reality this number is still smaller, since in the mixture d + t there occur the d + d and t + t reactions, which decrease the μ-meson regeneration).

V. Influence of the Molecular Structure of Substances on the Course of Mesoatomic Processes

The majority of mesoatomic processes studied to date is independent of the physical and chemical structure of the substances in which they occur.

† The probability of pulling out of the μ meson from the Heμ mesonic atom in the collision with the nuclei of matter is not large, ~ 0.04 and ~ 0.2 for the d + d and d + t reactions, respectively (Jackson, 1957).

However, a number of experiments has been performed recently that show convincingly that the mesoatomic processes and the probability of the nuclear capture of mesons depend on the chemical structure of substances in which these phenomena occur. From the point of view of earlier notions concerning the mechanism of mesoatomic processes, observation of effects of this kind was somewhat unexpected, since the chemical properties of substances are determined by the outer electronic shells the dimension of which ($\sim 10^{-8}$ cm) greatly exceed the effective radius of the nuclear forces ($\sim 10^{-13}$ cm) or the size of the mesonic atoms ($\sim 10^{-11}$ to 10^{-10} cm) from the levels of which the nuclear capture of the mesons takes place.

Attempts to interpret these facts led to an appreciable refinement of the picture of the capture of negative mesons in matter. In particular, it was established that in molecular compounds an appreciable fraction of the mesons is first captured not at the levels of isolated mesonic atoms but at the levels belonging to the molecule as a whole. Such complexes, in which the dimensions of the meson orbits are comparable to the dimension of the molecules of the substance (and exceed by hundreds of times the characteristic mesoatomic distances) were called "large mesonic molecules" (Ponomarev, 1965, 1967). Within the framework of this model, it was possible to explain consistently a number of phenomena, which we shall discuss below. A more detailed consideration of these problems is given in the review papers by Gershtein *et al.* (1969) and Ponomarev (1973).

A. Nuclear Capture of μ^- Mesons in Chemical Compounds

According to the Fermi–Teller Z-law (1947), when negative mesons are stopped in substances of the type $Z_k Z_m'$ the ratio of the probabilities of capture of mesons by nuclei Z and Z' is, respectively,

$$\frac{W(Z)}{W(Z')} = \frac{k}{m}\frac{Z}{Z'} \tag{63}$$

Some indications of Z-law violation were early obtained in papers by Sens *et al.* (1958), Astruby (1960), Lathrop *et al.* (1960, 1961), Camal (1961), Cohen *et al.* (1964), Blair *et al.* (1962), and Eckhause *et al.* (1962). Data on the Z-law verification were systematized in the paper of Baijal *et al.* (1963). Figure 15 taken from it shows the results of this verification. The straight lines correspond to the dependence

$$A(Z/Z') = \frac{m}{k}\frac{W(Z)}{W(Z')} = \left(\frac{Z}{Z'}\right)^s \tag{64}$$

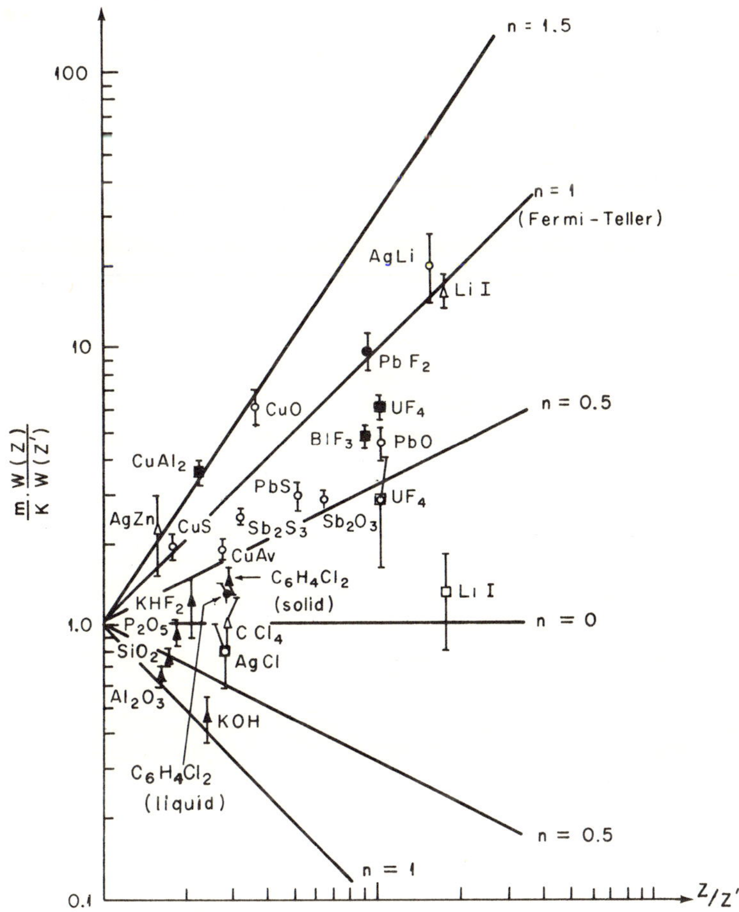

FIG. 15. Results of verification of the Fermi–Teller Z-law (Baijal *et al.*, 1963).

for different values of the power s, the value $s = 1$ corresponding to the Z-law. It is obvious from the figure that the experimental data testify against the Z-law and there arises the problem of what causes its violation.

One of the possible causes of the Z-law violations was discussed by Bobrov *et al.* (1965), who studied the capture of μ mesons by atomic nuclei making up various chemical compounds. It turned out that in 26 out of 31 cases mesons are captured by the nuclei of the atoms with greater affinity to the electron.

B. Study of the μ-Meson X-Ray Series

1. Verification of the Z-Law

In the majority of the experiments mentioned, the indicator of the atomic capture of a μ^- meson by the element with atomic number Z was the reaction $\mu^- Z \to (Z - 1) + \nu_\mu$. A shortcoming of this procedure is that for heavy atoms, the rates λ_Z of the capture of mesons by nuclei depend little on the values of the charge Z, and therefore, in compounds of the type $Z_k Z_m'$ with close values of Z and Z' it is difficult to extract the corresponding decay exponentials. Experiments on mesonic X radiation are free of this shortcoming since the characteristic series of the Z atoms of the different elements are distributed over the energy scale as Z^2, and can be easily separated from one another. This method has been used by Zinov *et al.* (1965a,b, 1967) for the study of μ-meson X rays in chemical compounds.

The investigations have shown that in mixtures of inert gases the Z-law holds satisfactorily (Budyashov *et al.*, 1967b), although small systematic deviations are observed (the probability of the μ-meson capture calculated per unit charge decreases weakly with increasing Z). At the same time, the Z-law is violated in the $Ar + CO_2$ mixture: the capture probability in CO_2 is approximately double the calculated value. Studies of muonic X-ray spectra in metal–halogen compounds (Zinov *et al.*, 1965a) showed that the Z-law is satisfactorily fulfilled, though the measured dependence

$$A(Z/Z') = 0.66Z/Z' \tag{64a}$$

differs from the theoretical one (64) by a constant factor.

2. Periodic Dependence of Relative Intensity of the K Series in Oxides of Metals

In experiments by Zinov *et al.* (1965a) the total intensity $J(Z)$ of the K series of the μ-mesonic X radiation of Z elements in oxides of the type $Z_k O_m$ has been studied. The relative intensity of the K series of metals in these compounds is found to be

$$A(Z) = \frac{m}{k}\frac{J(Z)}{J(0)} \neq \frac{Z}{8} \tag{65}$$

[$J(0)$ is the total intensity of the oxygen K series] i.e., it is not equal to the ratio of the charges of the nuclei of oxygen and an element bound with it as should be expected according to the Z-law. Measurements in a wide interval of Z-values have shown that with increasing Z the quantity $A(Z)$

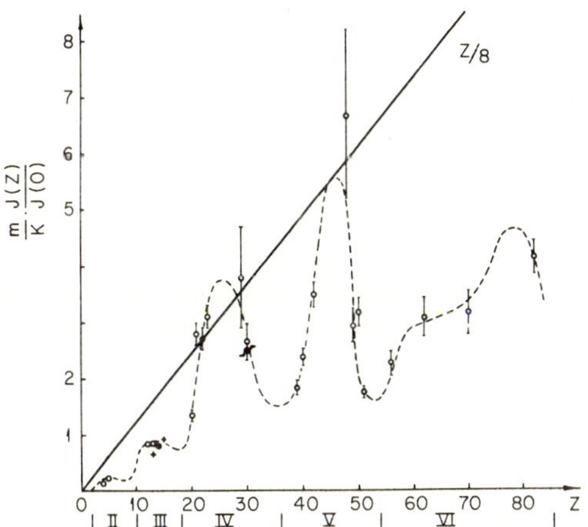

FIG. 16. The relative intensity $A(Z) = (m/k)\, J(Z)/J(0)$ of the K series of different elements in oxides $Z_k\, O_m$ (Zinov *et al.*, 1965a).

changes periodically, according to the periods of the Mendeleev table, the dips of the function $A(Z)$ coinciding with the position of the alkali metals.

Figure 16 gives the function $A(Z)$ measured in the paper cited. The straight line $A(Z) = Z/8$ corresponds to the Z-law. It should be noted that the entire curve measured lies below the straight line $Z/8$. This means that the atomic capture of μ mesons occurs predominantly to oxygen independently of the kind of Z element making a compound with it. This conclusion is also drawn from the results of the paper of Goldanskii *et al.* (1973d).

The same group of authors (Zinov *et al.*, 1965a) has established that, even for the same metal, the quantity $A(Z)$ depends on the kind of oxide in which it is measured. In particular,

$$\frac{[A(Z)]_{Sb_2O_3}}{[A(Z)]_{Sb_2O_5}} = \frac{3.5}{1.7} \qquad (Z/8 = 6.4) \tag{66}$$

i.e., in this case the values of $A(Z)$ for different antimony oxides differ by a factor of two.

3. *Structure of the K Series in Chemical Compounds*

More detailed information on the μ^--meson cascade transitions in various substances is extracted from investigations of the structure of the muon X-ray series in metals and chemical compounds. In particular, an anomalously large contribution of the K_ν series (high transitions np → 1s) to the total intensity of the K series for the elements with $Z > 20$ has been observed (Stearns and Stears, 1957; Quitman *et al.*, 1964). This fact contradicts the calculations by Eisenberg and Kessler (1961a,b, 1963a,b) and cannot be understood in the framework of the theory of cascade transitions in the isolated atom $Z\mu$.

The influence of the structure of the external electron shells forming a chemical bond between the atoms on the structure of the mesoatomic K series is clearly demonstrated by parallel measurements of the structure of the K series in pure metals and their oxides. Figure 17a gives the results of measurements for Ti and TiO_2. It is seen that the contribution of the K_ν series of titanium in the pure metal is larger by a factor of 1.5 than in the oxide of this metal (Zinov *et al.*, 1965b, 1967).

The use of germanium–lithium detectors for registering μ-mesonic X rays made it possible to advance these investigations and, in particular, to measure the intensities of individual lines of the K_ν series in Ti and TiO_2 (Kessler *et al.*, 1967). It is seen from Fig. 17b, where the results of these measurements are given, that when going over from the metal to its oxide, not only the total intensity of the K_ν series, but also the intensity of all the lines decreases: the higher the transitions $\mathrm{np} \to 1\mathrm{s}$, the stronger this decrease.

Experiments of such a kind have also been performed by Daniel *et al.* (1968) on a μ^- meson beam and by Grin and Kunselman (1970) on a π^- meson beam. In the former, the ratio of the intensities of separate lines of the K_ν series of carbon in (CH_2) and in pure graphite has been measured. It turned out that the intensity of muonic X-ray lines $\mathrm{np} \to 1\mathrm{s}$ in CH_2 is higher than in graphite: the higher the transitions, the greater the intensity:†

$$(3\mathrm{p} \to 1\mathrm{s}) \left[\frac{\mathrm{CH_2}}{\mathrm{C}}\right] = 1.25 \pm 0.05, \qquad (5\mathrm{p} \to 1\mathrm{s}) \left[\frac{\mathrm{CH_2}}{\mathrm{C}}\right] = 1.99 \pm 0.15 \tag{67}$$

In the latter experiments, the intensities of high transitions of π mesons

† In experiments on the study of the μ X-ray series, one fails to understand the particular features of the atomic capture of μ mesons in chemical substances containing hydrogen. This is explained by the experimental conditions, which do not permit soft γ quanta from μ^- meson transitions in the $\mathrm{p}\mu$ mesonic atom to be registered, as well as by the intense transfer $\mathrm{p}\mu \to Z\mu$, which violates the initial distribution of μ mesons over the levels of isolated mesonic atoms. Recently, detecting and studying the μ and π series in hydrogen has been achieved (Placci *et al.*, 1970a; Bajley *et al.*, 1970; Budick *et al.*, 1971). The appropriate cascade calculations have been performed by Leon (1971).

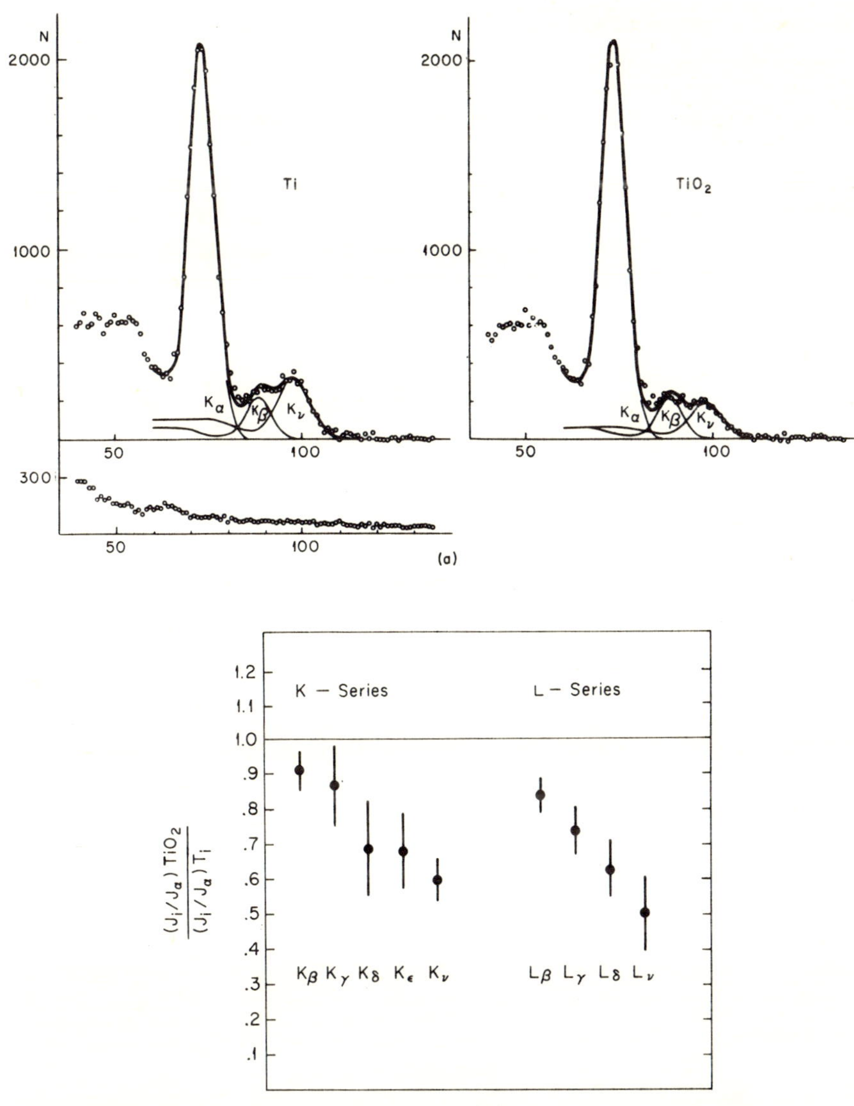

FIG. 17. The intensity of the Ti K_ν series in pure metal and in the oxide TiO_2. (a) Change in the total K_ν series intensity (Zinov *et al.*, 1967); (b) changes in the intensities of individual K_ν series lines (Kessler *et al.*, 1967).

in chemical compounds and mechanical mixtures equivalent to them have been compared: ZnSe and Zn + Se, ZnS and Zn + S, FeS and Fe + S, etc. In all the cases studied the contribution of the high transitions of the pionic X-ray series of elements in chemical compounds was found to be larger than in the corresponding mixtures. This effect was observed to be especially great for the pair ZnSe and Zn + S, for which the ratio of the transition intensities

$$\frac{(4\mathrm{f} - 3\mathrm{d})_{\mathrm{Zn}}}{(4\mathrm{f} - 3\mathrm{d})_{\mathrm{Se}}} = R$$

is, respectively,

$$R(\mathrm{ZnSe}) = 4.1 \pm 0.5, \qquad R(\mathrm{Zn} + \mathrm{Se}) = 0.74 \pm 0.09 \tag{68}$$

i.e., differs by about a factor of 6.

Finally, Tauscher *et al.* (1968) have established that the intensity of high mesonic X-ray transitions of elements depends not only on the chemicals but also on their physical state. By measuring the ratio $R = (5\mathrm{f} - 3\mathrm{d})/(4\mathrm{f} - 3\mathrm{d})$ they have found that it is far smaller for amorphous than for metallic selenium:

$$\frac{R(\mathrm{Se}_{\mathrm{amorph}})}{R(\mathrm{Se}_{\mathrm{met}})} = 0.74 \pm 0.04 \tag{69}$$

All these experiments prove with certainty that the chemical and physical structure of substances affects the course of mesoatomic processes.

C. *Capture of π^- Mesons by Nuclei of Chemically Bound Hydrogen*

Especially great and characteristic deviations from the Z-law were observed in experiments on the capture of π^- mesons in hydrogeneous substances, namely, in the measurements of the charge-exchange reaction probability

$$\pi^- + \mathrm{p} \rightarrow \mathrm{n} + \pi^0 \tag{70}$$

on the nuclei of chemically bound hydrogen.

It is known that this reaction proceeds only on hydrogen nuclei from the low s-states on the $\mathrm{p}\pi$ mesonic atom. For other nuclei instead of reaction (70) there occurs either nucleus fission or the emission of a few fast nucleons (Petrukhin and Prokoshkin, 1963). Thus, observation of reaction (70) means that the π^- meson was captured by the proton. According to this, the probability for reaction (70) in pure hydrogen may be normalized to unity†

$$W(H_2) = 1 \tag{71}$$

When hydrogen is bound in the chemical compound Z_mH_n, one should expect a suppression of the reaction (70) in so far as on slowing down a fraction of mesons is captured at the levels of the Z atoms and gives no contribution to reaction (70). The Z-law predicts for the probability of reaction (70) the following expression:

$$W(Z_mH_n) \approx \frac{n}{m} Z^{-1} \tag{72}$$

However, as experiments (Dunaitsev *et al.*, 1962, 1964; Charbe *et al.*, 1963; Bartlett *et al.*, 1964; Krumshtein *et al.*, 1968a,b) have shown, this probability is suppressed more strongly, namely, according to the law

$$W(Z_mH_n) \approx a_L \frac{n}{m} Z^{-3} \tag{73}$$

where a_L is a certain coefficient whose value depends on the period L of the Mendeleev table to which the atom Z belongs.

Such a sharp suppression of reaction (70) (for example, $W(\text{LiH}) = 3.5 \times 10^{-2}$, $W(H_2O) = 3.5 \times 10^{-3}$) cannot be explained by the π^- meson transfer processes in the reaction

$$p\pi^- + Z \rightarrow Z\pi^- + p \tag{74}$$

as was suggested by Panofsky *et al.* (1951). Special experiments on the study of reaction (70) in a mixture of gases H_2 + He, Ne, Ar, Kr, Xe, H_2 (Petrukhin *et al.*, 1972) have shown that the reaction (74) depends weakly on the charges Z of the atoms of heavy gases and, therefore, cannot be a cause of dependence (73).‡

Test measurements of the probability of reaction (70) in hydrazine, N_2H_4, and in the equivalent mechanical mixture $N_2 + 2H_2$ have given the following ratio:

$$W(N_2H_4)/W(N_2 + 2H_2) \approx 1/30 \tag{75}$$

† In reality, reaction (70) proceeds in parallel with the other reaction $\pi^- + p \rightarrow n + \gamma$, the ratio of the probabilities of these channels being about 0.6. However, this fact is nonessential for further considerations. The probability for reaction (70) on ^{3}He is $W(\text{He}) = 0.155$ and on all the remaining nuclei it is suppressed down to 10^{-4} to 10^{-5}. For deuterium $W(D_2) < 10^{-3}$.

‡ The measured probability $W(H_2 + Z) \sim (I + \Lambda C)^{-1}$, where $\Lambda = f(Z, C)$ is a weakly changing function $\Lambda = 5 \div 10$ in the limits $Z \leq 36$ and $C \lesssim 1$ ($C = n_Z/n_H$ is the relative concentration of Z nuclei).

i.e., under identical conditions the chemically bound hydrogen enters into reaction (70) 30 times more rarely than the free one.

The results of these measurements for compounds Z_mH_n with the elements Z of period II of Mendeleev's table are given in Fig. 18, where the function

$$P = \left(\frac{m}{n} + \frac{1}{Z}\right) W(Z_mH_n) = a_L Z^{-3} \tag{76}$$

is presented. It is seen from the figure that all the data ln P fit well with the straight line for $a_L \approx 1.3$.

D. Model of Large Mesonic Molecules

1. *Assumptions of the Model*

All the indicated phenomena provide evidence of the influence of chemical forces on the processes of nuclear capture of mesons and cannot be explained by the mechanisms of the capture and slowing down of mesons in free atoms. In this connection a model of "large mesonic molecules" has been suggested (Ponomarev, 1965, 1967, 1973; Ponomarev and Prokoshkin, 1968; Gershtein, *et al.*, 1969).

The main assumption of the model is that when the mesons are slowing down, some of them are captured by very high mesomolecular orbitals located in the region of the valence electrons of the molecule and only these mesons are capable further of inducing reaction (70). Such an assumption is equivalent to asserting the possibility of existence of stable complexes of the type $Z_m\pi^-H_n$, the dimensions of which exceed by hundreds of times the characteristic mesoatomic distances and the lifetime of which is rather long.

It is easily seen that this hypothesis alone can explain qualitatively many indicated phenomena. However, to make any quantitative predictions more concrete assumptions are needed.

First of all it should be noted that the height of the potential barrier separating the potential wells of the proton and the nucleus Z corresponds to the energy of the level of the pπ atom with quantum number

$$n_0 \approx [R/2(1 + 2\sqrt{Z})]^{1/2} \tag{77}$$

where R is the distance between the nuclei in mesoatomic units (Ponomarev, 1965). Since for all the compounds R = 400 to 600, then for $I \leq Z \leq 10$, one gets n_0 = 5 to 7 (for hydrogen R = 371, n_0 = 8). This implies that the mesonic pπ atom cannot be produced in states with

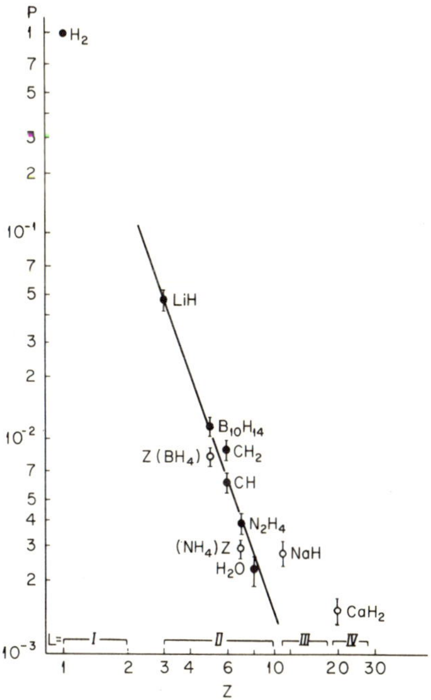

FIG. 18. The reduced probabilities

$$P = \left(\frac{m}{n} + \frac{1}{Z}\right) W(Z_m H_n)$$

for reaction (70) in hydrogenous substances $Z_m H_n$. The straight line is the dependence $P = a_L Z^{-3}$, which results from Eq. (70) (Krumshtein *et al.*, 1968a).

quantum numbers larger than $n_0 = 8$, and the levels $n > n_0$ of the pπ atom belong to the entire molecule $Z_m\pi^-H_n$ as a whole.† The value of n_0 obtained is much smaller than the previous estimate of Wightman (1950), $n \approx 15$ for the number of the pπ atom level starting with which an atomic capture of π^- mesons by hydrogen is possible.

Thus, all the levels of the $Z_m\pi^-H_n$ system are divided into n levels of the isolated pπ atom, n' levels of the $Z\pi$ atom, and N "molecular" levels of the large mesonic molecule $Z_m\pi^+H_n$ (see Fig. 19). In order to occupy one of these levels a meson should give up its energy to one of the electrons of the Z_mH_n molecule. As in the case of hydrogen (Wightman, 1950), an adiabatic capture is here the most likely mechanism. A characteristic feature of this

† The account of screening of the Z nucleus by electrons does not essentially change the given estimates due to the weak dependence of n_0 on Z.

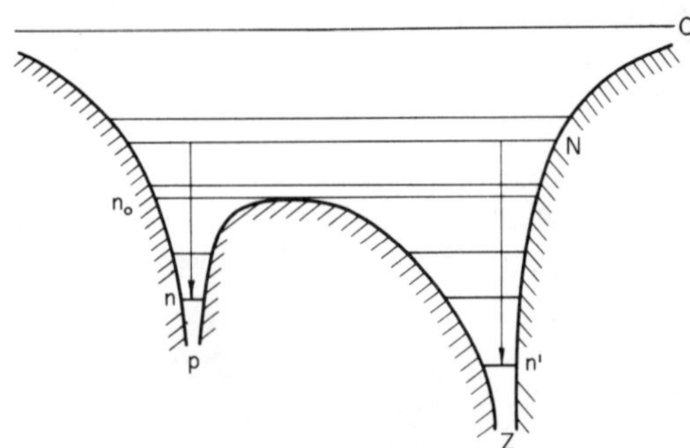

FIG. 19. Scheme of the levels of a "large mesonic molecule" pπZ. N are the mesomolecular levels located in the region of the valence electrons of the pZ molecule; n and n' are the levels of the isolated pπ and $Z\pi$ mesonic atoms, respectively; n_0 is the level separating the levels N and n.

mechanism is a small transfer of energy to electrons, as a result of which the orbit of the captured meson must be geometrically similar to the orbit of the electron knocked out. Hence, this confirms once more that in the process of initial capture the π^- meson cannot be captured into the n levels of an isolated pπ atom [from which nuclear capture (70) occurs], since the only electron of the hydrogen atom is spent for the formation of the chemical bond. Therefore, the charge exchange reaction (70) can be induced only by the mesons that first found themselves at the levels N.†

It is rather hard to determine what the fraction W_1 of all the mesons captured by the Z_mH_n molecule is. In the first studies (Ponomarev, 1965, 1967) it was assumed to be

$$W_1 = a_L n/(mZ + n) \tag{78}$$

where a_L is a certain coefficient depending on the L row of the Mendeleev table, to which the Z atom belongs, and accounting for the particular features of the chemical bond Z—H.‡

† The fraction of the mesons that fell initially on the n' levels of the $Z\pi$ atom does not contribute to reaction (70) and therefore does not take part in further considerations.

‡ This expression for W_1 is equivalent to the assertion of the validity of the Z-law at the stage of atomic capture of π^- mesons. The experimental grounds for this supposition are the experiments on muon absorption in a mixture of inert gases (Budyashov *et al.*, 1967b) and in halogen salts (Zinov *et al.*, 1965b), as well as experiments on the study of reaction (70) in compounds like $(NH_4)Z$ (Krumshtein *et al.*, 1968b). It is, however, known that in oxygen-containing substances the Z-law is strongly violated (Zinov *et al.*, 1965a; Goldanskii *et al.*, 1973d), which appears to be due to the nonadiabatic mechanism of capture of mesons by oxygen (Au-Yang and Cohen, 1968).

The fraction of the mesons that fell from the very beginning on the common molecular levels N of the $Z_m\pi^-H_n$ system can pass to the separated levels n and n' of the mesonic atoms $p\pi$ and $Z\pi$ with the probabilities W_{Nn} and $W_{Nn'}$, either by ejecting Auger electrons or by emitting gamma quanta. The former mechanism is effective only for transitions between high levels ($n' > 7$) and the latter for the transitions $n'\mathrm{p} \to 1\mathrm{s}$ from the $n' \sim Z$ levels. Such a two-stage deexcitation mechanism of large mesonic molecules

$$N \xrightarrow{\text{Auger}} n' \xrightarrow{\text{rad}} 1\mathrm{s}$$

appears to be the most probable one. Some experimental grounds for this assumption are given in a recent paper (Goldanskii *et al.*, 1973c) in which high transitions from the levels $n' \gtrsim Z$ of the carbon and oxygen mesonic atoms have not been observed.

In the $Z_m\pi^-H_n$ system, as compared with the isolated mesonic atoms $p\pi$ and $Z\pi^-$, the probability of Auger and radiative transitions should appreciably increase. Indeed, in this system the field is no longer central and therefore the selection rules with respect to the orbital angular momentum $\Delta l = 1$, which greatly increases the cascade time in isolated mesonic atoms, are weakened.

The estimate for the probabilities w_{Nn} and $w_{Nn'}$ of such radiative transitions (taking into account the screening of the nucleus Z by outer electrons) yields

$$w_{Nn'}/w_{Nn} \approx Z^2 \tag{79}$$

Taking into account relation (78), it follows from this formula that the probability of the capture of a π^- meson by a proton in chemical compounds Z_mH_n is (Ponomarev, 1967)

$$W(Z_mH_n) = a(Z)\,\frac{w_{Nn}}{w_{Nn} + w_{Nn'}} \approx a_L\,\frac{nZ^{-2}}{mZ + n} \tag{80}$$

It is easily seen that this formula is in agreement with the experimental dependence (73).

2. *Consequences of the Model*

The model in question allows one to make some qualitative predictions:

(a) All the processes, which in the framework of the model result in the absorption of the π^- meson by the proton, proceed completely inside the molecule Z_mH_n and therefore are independent of the collisions with other atoms. Thus, the probability of reaction (70) should not depend on the density and aggregate state of substances. This conclusion has been confirmed experimentally by Petrukhin and Prokoshkin (1965), who measured the probability of reaction (70) in ethane, C_2H_6. The density of ethane changed by a factor of 110, from gas to liquid, but $W(C_2H_6)$ remained constant.

(b) In strong ionic compounds Z_mH_n, when the density of the valence cloud in the vicinity of the proton is insignificant, and consequently the coefficient a_L is anomalously small, the reaction probability is expected to be still smaller than in covalent compounds (80). Measurements of the probability for reaction (70) in strong acids HCl, HNO_3, and H_2SO_4 (Krumshtein *et al.*, 1970) have shown that the suppression of the charge exchange reaction is indeed much larger than what follows from Eq. (80) with coefficient $a_{II} \approx 1$. (For example, for HNO_3 and H_2SO_4 $a < 0.1$.)

(c) In chemical compounds, the intensity of the K_ν series should increase compared with that of the K_ν series in isolated atoms. This statement is in agreement with the measurements by Kessler *et al.* (1967), who have established that the intensities of the K_ν series of the elements in chemical compounds are approximately twice as large as the calculated intensities for the cascades in the isolated atom.

Next, it follows from the model that the larger the effective valence of an element in the compound the higher the K_ν series intensity of this element. By the example of the pairs of substances indicated above (Section V,B,3)—Ti and TiO_2, ZnSe and Zn + Se, Se_{met} and Se_{amorph}—it can be seen that this rule holds, since the effective valence of atoms of metals is maximum in pure metals.

(d) The structure of the muonic X-ray series in ionic compounds should not differ strongly from the structure of the spectra of noble gases. This conclusion is confirmed experimentally. Indeed, the K_ν series intensities reveal a dip in the region of potassium for which $J_\nu \approx 0.07$. This value is approximately equal to the K_ν series intensity of argon and coincides with the one calculated in the isolated atom. At the same time in metallic chrome $J_\nu \approx 0.22$ (Zinov *et al.*, 1967).

3. *Systematic Investigations*

The probability of the charge exchange reaction (70) in the compounds Z_mH_n can be written in the general form as follows:

$$W(Z_mH_n) = aW_1W_2W_3 \tag{81}$$

where W_1 is the π^- meson capture probability into the highly excited levels N of large mesonic molecules $Z_m\pi^-H_n$, W_2 the π^- meson transition probability from the molecular levels N to the n levels of an isolated pπ mesonic atom, and W_3 the coefficient accounting for the decrease of the reaction (70) probability in the mesonic atom $p\pi$ due to transfer to the Z nuclei according to the reaction $p\pi + Z \rightarrow Z\pi + p$. The coefficient a accounts phenomenologically for the difference in the structure of the valence shells of the molecules of different chemical compounds, and its measurements are of most interest for "meson chemistry."

At present a total of about one hundred measurements of reaction (70) have been performed in various substances or mixtures of substances: acids, alkalis, salts, metal hydrides, and various organic compounds (Petrukhin and Prokoshkin, 1963; Krumshtein *et al.*, 1968a,b; Vilgel'mova *et al.*, 1973; Goldanskii *et al.*, 1974a,b). Some of these measurements are devoted to checking the assumptions on the magnitude of the probabilities W_i; in the others Eq. (80) is used to determine the coefficient a with the aim of establishing correlations between its magnitude and the particular features of the hydrogen chemical bond in various compounds.

The transfer stage. In the previous considerations the transfer stage was not taken into account, i.e., $W_3 \approx 1$ was assumed. Special experiments on the study of reaction (70) in gas mixtures $H_2 + Z$, where Z implies He, Ne, Ar, Kr, or Xe, undertaken by Petrukhin *et al.* (1968) showed that with increasing Z and a relative concentration $C = n_Z/n_H$ the quantity W_3 decreases essentially (see Fig. 20). However, similar measurements in mixtures $CH_4 + Z$ (Petrukhin *et al.*, 1974) showed that in this case the π^- meson transfer from protons to nuclei Z is smaller by about a factor of 3 to 5 than in the case of the mixture $H_2 + Z$. Therefore, in the first approximation we may assume $W_3 \approx 1$ and include the resulting uncertainty in the coefficient a in Eq. (80).

Hydrides of the second row. Table 17 gives the results of measurements of the probability $W(Z_mH_n)$ in simple hydrides of Z elements of the second row of the Mendeleev table. The coefficient a calculated by formula (80) has its peak in the middle of the row for covalent compounds (CH_4 and NH_3) and falls to the row boundary for ionic compounds (LiH and HF) (see Fig. 21). This fact is in agreement with the basic assumptions of the model of the process under discussion.

Bond hybridization effect. The influence of the particular features of the chemical bond Z—H on the quantity $W(Z_mH_n)$ is established from the measurement of the reaction (70) probability in substances that are constructed out of the same atoms but differ by the type of the chemical bond. Table 18 gives the reduced probabilities $P(Z_mH_n) = mW(Z_mH_n)/n$, i.e., the reaction (70) probabilities per bond Z—H.

It is seen from the table that the $P(Z_mH_n)$ values differ noticeably for substances with different types of bond (H_2O and H_2O_2, NH_3 and N_2H_4,

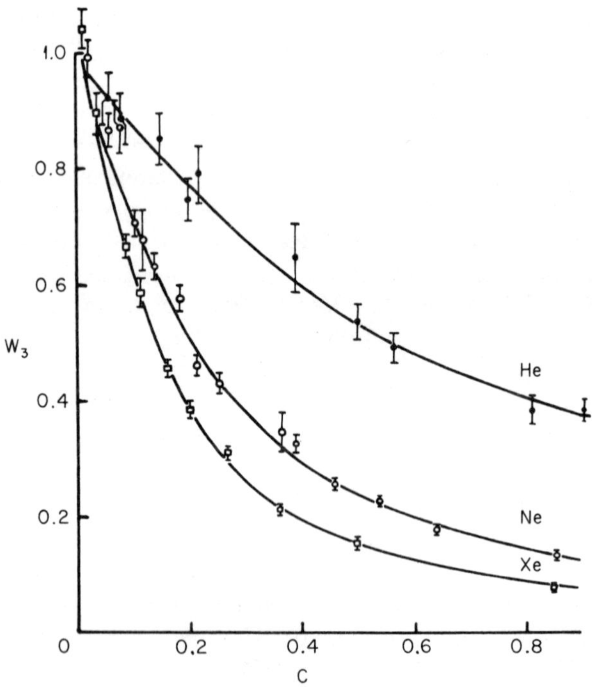

FIG. 20. Suppression of the reaction $p\pi \to n + \pi^0$ due to the transfer $p\pi + Z \to Z\pi + p$ in the gas mixtures $H_2 + Z$ on the basis of the work of Petrukhin *et al.* (1968) (see footnote p. 211), $C = n_Z/n_H$.

TABLE 17

HYDRIDES Z_mH_n OF THE SECOND ROW OF THE MENDELEEV TABLE[a]

Substance	Z	$W \times 10^{-3}$	a
LiH	3	35 ± 4	1.26 ± 0.14
$B_{10}H_{14}$	5	12.6 ± 1.4	1.44 ± 0.16
CH_4	6	26.7 ± 0.5	2.40 ± 0.05
NH_3	7	14.2 ± 0.9	2.32 ± 0.14
H_2O	8	3.9 ± 0.3	1.24 ± 0.08
HF	9	0.66 ± 0.08	0.53 ± 0.06

[a] The table is compiled on the basis of the data from the papers of Petrukhin and Prokoshkin (1965), Krumshtein *et al.* (1968a, 1973), Goldanskii *et al.* (1973a), and Petrukhin (1972).

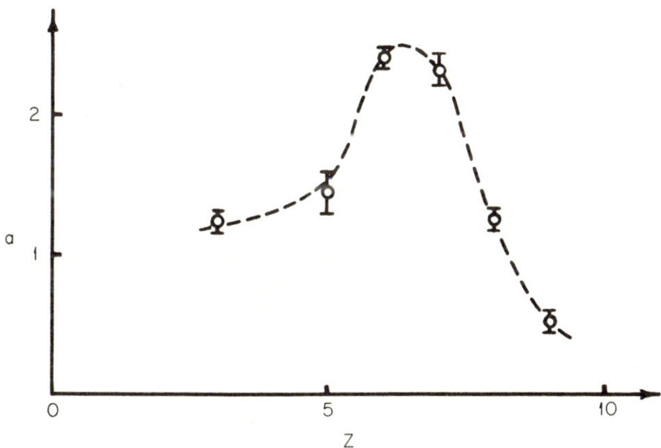

FIG. 21. Coefficient a in Eq. (80) as a function of the atomic number Z of the elements of the second period in simple hydrides Z_mH_n (see Table 17) following Petrukhin and Prokoshkin (1965), Krumshtein *et al.* (1968), Goldanskii *et al.* (1973b), and Petrukhin *et al.* (1973).

CH_4 and C_2H_2, C_6H_{12} and C_6H_6). At the same time, the P values are practically identical for substances with different stoichiometric compounds: CH_4 (methane), $(CH_2)_n$ (polyethylene), and C_6H_{12} (cyclohexane)—sp^3 hydrization of molecular orbitals; C_2H_4 (ethylene), $(CH)_n$(polystirol), and C_6H_6 (benzene)—sp^2 hydridization.

TABLE 18

COMPOUNDS WITH VARIOUS TYPES OF CHEMICAL BOND[a]

Substance	$W \times 10^{-3}$	$P \times 10^{-3}$
CH_4	26.7 ± 0.5	6.7 ± 0.1
$(CH_2)_n$	13.7 ± 0.7	6.8 ± 0.3
C_6H_{12}	14.3 ± 0.6	7.1 ± 0.3
C_2H_4	10.0 ± 0.6	5.0 ± 0.3
$(CH)_n$	5.1 ± 0.6	5.1 ± 0.6
C_6H_6	4.5 ± 0.2	4.5 ± 0.2
NH_3	14.2 ± 0.9	4.7 ± 0.5
N_2H_4	6.5 ± 0.5	3.3 ± 0.3
H_2O	3.9 ± 0.3	1.9 ± 0.1
H_2O_2	0.62 ± 0.13	0.62 ± 0.13

[a] See the footnote to Table 17.

Inductive effect. The study of the reactivity of radicals as a function of the type of substituent (e.g., the CH_3 radical in the row CH_3X or the C_6H_5 radical in the row C_6H_5X, where $X = J$, CN, NO_2, COCl, CH_3CO, etc.) is based on the correlation analysis widely used in chemistry (see, e.g., Hammett, 1970). The reactivity of the radicals is quantitatively described by the Hammett–Taft induction constants σ_I, the determination of which by chemical methods is rather difficult and not always possible.

In Fig. 22 the coefficients a calculated by the formula

$$W(Z_k'Z_m\mathrm{H}_n) = a \times nZ^{-2}/(n + mZ + kZ') \qquad (82)$$

are given as functions of the constants σ_I. The coefficients a are determined on the basis of the measured values for the reaction (70) probability in the row CH_3X (Vil'gelmova *et al.*, 1973; Goldanskii *et al.*, 1973a). There is observed a definite correlation, which can be used for determining the unknown constants σ_I.

Deuterated compounds. Reaction (70) occurs on the hydrogen nuclei alone, and its probability on deuterium nuclei is practically zero (see footnote, p. 211) though hydrogen and deuterium do not differ in their chemical properties. This fact can be employed in studying an isolated bond Z—H in complex molecules by subjecting their retaining bonds to deuteration.

Such measurements were performed by Goldanskii *et al.* (1973b) for the following sets of substances: CH_3OH, CH_3OD, and CD_3OH; for the pairs C_2H_5OH and C_2H_5OD, CH_3NH_2 and CH_3ND_2, etc.; as well as for the benzene derivatives C_6H_5OH and C_6H_5OD, $C_6H_5ND_2$, $C_6H_5CH_3$, and $C_6H_5CD_3$, etc. In all the cases a noticeable change of the coefficients a_{C-H}, a_{O-H}, and a_{N-H} was observed as compared with their values in the compounds H_2O, C_mH_{2m+2}, and NH_3. For example,

$$a_{C-H}(C_mH_{2m+2}) = 2.05 \pm 0.04, \qquad a_{O-H}(H_2O) = 1.24 \pm 0.08$$

while in the compound CH_3OH,

$$a_{C-H}(CH_3OH) = 1.32 \pm 0.06, \qquad a_{O-H}(CH_3OH) = 2.58 \pm 0.48$$

In a similar manner, in the compound CH_3NH_2,

$$a_{C-H}(CH_3NH_2) = 1.34 \pm 0.07, \qquad a_{N-H}(CH_3NH_2) = 3.11 \pm 0.22$$

while in hydrazine N_2H_4,

$$a_{N-H}(N_2H_4) = 1.30 \pm 0.15$$

Thus, an increase of the coefficient a_{Z-H} in some group is accompanied by a simultaneous decrease of the coefficient $a_{Z'-H}$ in some other group. This

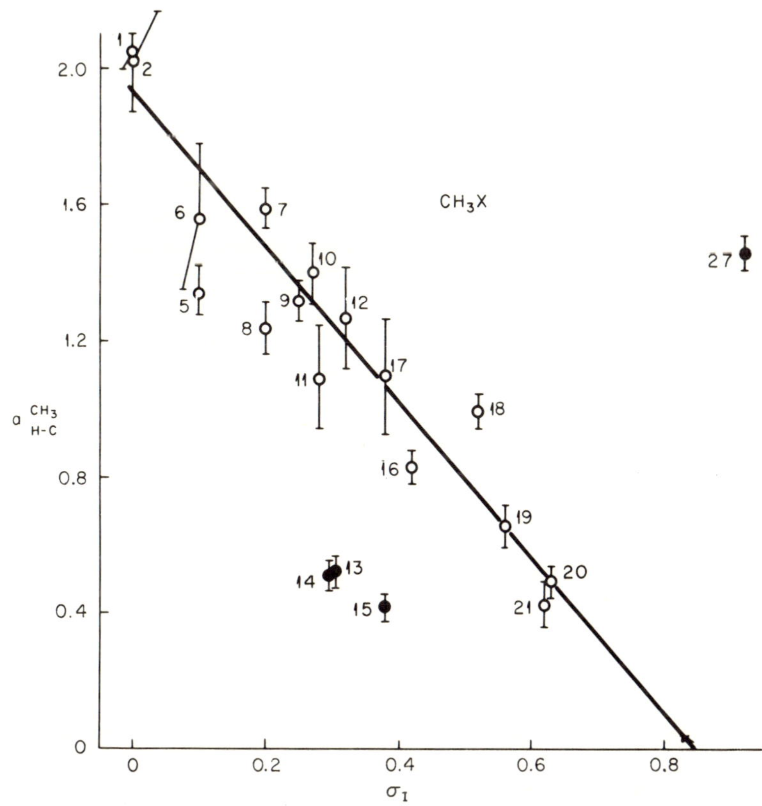

FIG. 22. Correlation between the values of the coefficients $a_{C-H}{}^{CH_3}$ of methyl groups and the values of the induction constants σ_I in methane substituents CH_3X, following Vilgelmova *et al.* (1973) and Goldanskii *et al.* (1973a). The numbering of the points corresponds to the following series of substituents X: 1, H; 2, $(CH_3)_3Sn$; 5, NH_2; 6, C_6H_5; 7, C_2H_5S; 8, CH_3S; 9, OH; 10, CH_3COO; 11, C_6H_5CO; 12, $COCH_3$; 13, CH_3OOC; 14, HOOC; 15, COCl; 16, J; 17, C_6H_5O; 18, $SOCH_3$; 19, CN; 20, NO_2; 21, OSO_2OCH_3. The straight line is described by the equation

$$a_{C-H}{}^{CH_3} = (1.93 \pm 0.03) - (2.26 \pm 0.07)\ \sigma_I$$

fact may be interpreted as a displacement of the electronic cloud from the radical CH_3 to the electronegative substituents OH and NH_2.

Meson charge exchange in organic compounds. In organic compounds the conditions of the transfer (probability W_3) and the transition from the levels N to the levels n (probability W_2) are practically identical, which allows studying directly the dependence of the coefficient a_{C-H} on the type of the compound. As the analysis of the experimental data shows (Ponomarev, 1973), in this case the expression

$$W_1 = n/m\nu \tag{78a}$$

where ν is equal to the number of valence electrons of the carbon atom in the compound C_mH_n, is the most preferable one for the probability of the initial capture W_1. From the physical viewpoint this means that the electrons of the internal shells of the carbon atom little affect the process of atomic capture of mesons. The general expression for the probability $W(Z_m\text{H}_n)$ for the charge exchange reaction (70) then takes the form

$$W(Z_m\text{H}_n) = \alpha n Z^{-2}/m\nu \tag{83}$$

E. *Possible Applications*

The phenomena, methods, and problems described in this section may be united under the common name "meson chemistry." The aim of this field of investigation may be formulated as follows: use of specific properties of the mesons and nuclear interactions for the study of the structure of matter.

The main goal of meson chemistry is to find a tool for a more detailed study of the electronic structure of molecules and, consequently, their chemical properties. This fact is especially attractive because almost all the available approaches make it possible to measure only the integral characteristics of molecules, i.e., sizes, moments of inertia, level structure, etc.

In addition to this main problem it is worth discussing some practical applications of meson chemistry, in particular, for the qualitative and quantitative analysis of matter (Petrukhin *et al.*, 1967a). Using the charge-exchange reaction for π^- mesons by protons, we are able even now to distinguish reliably between bound and free hydrogen. New methods of meson production will provide us with intense meson beams and allow us to perform precise measurements of such a kind.

At present there exists a wide program of studies of μ X-ray spectra of elements in chemical compounds (Zinov *et al.*, 1972; Knight *et al.*, 1972) with the aim of qualitative and quantitative "nondestructive chemical analysis." In particular, varying the incident beam energy and consequently, the point of its stopping in the target it is possible to identify step by step the composition of the target material even in a closed container.

On the basis of recent results of Varlamov *et al.* (1971) we can in principle investigate the kinetics of chemical reactions using the polarization properties of $Z\mu$ mesonic atoms. They have established that when Neμ mesonic atoms of neon are formed, its electron shell becomes identical with the electron shell of fluorine, the μ^- meson possessing rather large residual polarization and its spin being connected with the spin of the electron shell.†

In the interaction with the μ meson spin the two fine structure levels of the ground state of the F atom with $J = \frac{3}{2}$ and $J = \frac{1}{2}$ momentum split into four "hfs" levels, two of which have the total momentum 1 and the two others 2 and 0, respectively.

In a weak magnetic field the Neμ mesonic atom in the state with total momentum 2 precesses with muonium frequency ω_{Mu}, and to the two states with momentum 1 there correspond the precession frequencies $\frac{1}{3}\omega_{\mathrm{Mu}}$ and $\frac{5}{3}\omega_{\mathrm{Mu}}$.

The $Z\mu$ mesonic atoms are thus "labeled" atoms, which possess intrinsic standard frequencies. The latter provide us with a time scale allowing us to investigate the rates of chemical reactions of elements with atomic number $Z - 1$.

Continued work along these lines seems to be very promising. This method, in its conception, is analogous to the use of muonium (μ^+e^- system) for the study of the kinetics of chemical reactions involving hydrogen (see the review articles of Hughes, 1966; Goldanskii and Firsov, 1971; Gurevich, 1972 and Brewer *et al.*, Chapter VII,1 in this volume).

VI. Conclusion

At present the general picture of mesoatomic and mesomolecular processes occurring in a mixture of hydrogen isotopes with small impurities of heavier elements can be considered as established. Calculations of various processes, such as the formation of mesonic molecules, isotopic exchange reactions, transitions between hfs levels, etc. are in satisfactory agreement with experiment.

It should be noted that the existing picture of phenomena is self-consistent and is confirmed by measurements of some effects being a result of a large number of consecutive mesoatomic phenomena. This self-consistency is important not only by itself, but mainly for a reliable separation of meso-

† In experiments of Varlamov *et al.* (1971), μ^- mesons are stopped in gaseous neon with a xenon impurity. Under these conditions, the electron shell of the $Z\mu$ mesonic atom, which has lost some of the electrons in the cascade Auger transitions of the μ meson, can be reconstructed at the expense of the electrons of xenon since the ionization potential of the latter is less than that of fluorine.

atomic and mesomolecular effects in the study of the weak interaction of μ^- mesons with the hydrogen nuclei.

As an example we may indicate the measurements of the probability for capture of μ^- mesons by protons. This probability is found to be four times larger than should be expected from the statistical distribution of pμ mesonic atoms over hfs levels. This effect is known to be due to the fast transition pμ $(F = 1) \rightarrow$ pμ $(F = 0)$ (Section III,D), and the very effect of its observation with the account of the measured capture rate gives evidence for the negative relative sign of the (V − A) weak interaction. (For more details, see Zavattini, Chapter V5, Volume II.)

In liquid hydrogen the transition pμ $(F = 1) \rightarrow$ p $(F = 0)$ is followed by the ppμ mesonic molecule formation. According to theoretical estimates, the ppμ mesonic molecules are formed in the orthostate with the orbital moment $L = 1$, and during the lifetime of the μ^- meson, transitions to the parastate and the spin–flip of the meson and proton are negligible (Section III,G). Under these conditions the μ capture probability in the molecule must increase by about a factor of three compared with the capture in an isolated pμ atom for statistical population of the hfs levels. Experiment confirms this prediction and thereby points to the validity of this mechanism of the mesonic molecule formation.

Another example is the reaction of synthesis in the mesonic molecules that are formed in hydrogen with deuterium impurities according to the reactions p$\mu \rightarrow d\mu \rightarrow$ pd$\mu \rightarrow {}^3$He $+ \gamma$ (or ^{3}He $+ \mu^-$). The account of the hfs levels of the molecule pdμ allows us to bring the observed rates of nuclear reactions (λ_γ and λ_μ) into agreement with the magnitude of the yield of the reaction products (Y_γ and Y_μ) and predict the increase of the yield with increasing deuterium concentration, since in this case the pdμ mesonic molecule formation is preceded by the transition dμ $(F = \frac{3}{2}) \rightarrow$ dμ $(F = \frac{1}{2})$, which changes the statistical population of the levels of the molecule (Sections IV,C and D). These predictions are also verified and confirmed by experiment.

It should, however, be noted that recent measurements of the capture rate for d$\mu \rightarrow 2$n $+ \nu_\mu$ in a gaseous mixture of hydrogen and deuterium (Bertin *et al.*, 1973; Placci *et al.*, 1971) are not yet interpretable, since the mechanism of spin–flip according to the reaction dμ $(F = \frac{3}{2}) +$ p $\rightarrow$ dμ $(F = 0) +$ p is still unknown (see Section III,D). Direct measurements of the rates of the processes dμ $(F = \frac{3}{2}) \rightarrow$ dμ $(F = \frac{1}{2})$ and pμ $(F = 1) \rightarrow$ pμ $(F = \frac{1}{2})$ would therefore be desirable, which can, in principle, be realized by observing the μ^- meson residual polarization in gaseous hydrogen or deuterium.

The problems that must be solved are also discrepancies between different measurements of the rates of the $pd\mu$ and $dd\mu$ mesonic molecule formation, the rates of the $p\mu \rightarrow Z\mu$ transfer, the cross sections for elastic $p\mu + p$ scattering, etc. These processes (especially $p\mu + p$ scattering and $dd\mu$ molecule formation at various temperatures) need further experimental and theoretical study. It is also very interesting to study experimentally the catalysis of the synthesis reactions in the mesonic molecules $pt\mu$ and $dt\mu$.

According to the experiments of the past years there are no grounds to doubt that the molecular structure of substances essentially affects the course of mesoatomic processes and the probability of some nuclear reactions (e.g., charge-exchange reaction of π^- mesons on protons of hydrogeneous substances, μ^- X-ray series structure in chemical compounds, etc.). The qualitative picture suggested for describing in outline the discovered phenomena appears to be valid. However, some important details need further improvement and development. To this end, it is necessary to perform calculations of the atomic and molecular systems in the presence of the negative meson, and, in particular, calculations of the π^- and μ^- meson capture probabilities to highly excited mesoatomic and mesomolecular levels, distribution of mesons over levels with different orbital moments, and mesoatomic cascade calculations in complex atoms. It may be hoped that such calculations performed at the present-day level of rigor will allow connecting measured quantities with the parameters characterizing the density distribution for the valence electronic cloud.

Independently of the success of this program, the discovered phenomena are so characteristic that their use for studying the properties of matter seems to be reasonable and promising. To realize this program it is necessary to improve the accuracy of the former measurements and set up new experiments that will establish the correspondence between the mesonic chemistry methods and other hitherto developed methods of studying the structure of matter and will define the range of their applicability.

References

Adair, R. K. (1959). *Phys. Rev. Lett.* **3,** 438.
Alberigi Quaranta, A., *et al.* (1967a). *Nuovo Cimento* **47B,** 72.
Alberigi Quaranta, A., *et al.* (1967b). *Nuovo Cimento* **47B,** 92.
Alberigi Quaranta, A., *et al.* (1967c). *Phys. Lett.* **25B,** 429.
Alberigi Quaranta, A., *et al.* (1969). *Phys. Rev.* **177,** 2118.
Alvarez, L. W., *et al.* (1957). *Phys. Rev.* **105,** 1127.
Ammiraju, A., and Lederman, L. M. (1956). *Nuovo Cimento* **4,** 283.
Arnold, A. R., *et al.* (1954). *Phys. Rev.* **93,** 483.
Ashkin, J. (1960). *Nuovo. Cimento* **16,** 490.

Ashmore, A., Nordhagen, R., Strauch, K., and Townes, B. M. (1958). *Proc. Phys. Soc. (London)* **71**, 161.
Astruby, A., Hattersley, P. M., Hussain, M., Kemp, M. A. R., Muirhead, H. (1960). *Nuovo Cimento* **18**, 1267.
Atarashi, M., and Narumi, H. (1971). *Progr. Theor. Phys.* **45**, 1779.
Au Yang, M. Y., and Cohen, M. L. (1968). *Phys. Rev.* **174**, 468.
Babikov, V. V. (1968). "Metod Fasovikh Funkzij v Kvantovoj Mekhanike." Nauka, Moscow.
Backenstoss, G. (1970). *Ann. Rev. Nucl. Sci.* **20**, 467.
Backenstoss, G., *et al.* (1971). *Phys. Lett.* **36B**, 422.
Baijal, J. S., Diaz, J. A., Kaplan, S. N., and Pyle, R. V. (1963). *Nuovo Cimento* **30**, 711.
Bajley, J., *et al.* (1970). *Phys. Lett.* **33B**, 369.
Baker, G. A. (1960). *Phys. Rev.* **117**, 1130.
Balats, M. Ya., Kondrat'ev, L. N., Landsberg, L. G., Lebedev, P. I., Obukhov, B. V., and Pontecorvo, B. (1960). *Zh. Eksper. Teor. Fiz.* **39**, 1168 [*English transl.: Sov. Phys.-JETP* **12**, 813 (1961)].
Barkas, W. H., *et al.* (1958). *Phys. Rev.* **112**, 622.
Bartlett, D., Devons, S., Meyer, S. L., and Rosen, J. L. (1964). *Phys. Rev.* **B136**, 1452.
Basiladze, S. G., Yermolov, P. F. and Oganesian, K. O. (1965). *Zh. Eksper. Teor. Fiz.* **49**, 1042 [*English transl.: Sov. Phys.-JETP* **22**, 725 (1966)].
Bates, D. R., and Carson, T. R. (1954). *Proc. Roy. Soc.* **A234**, 207.
Bates, D. R., Ledsham, K., and Stewart, G. L. (1963). *Phil. Trans. Roy. Soc.* **A246**, 215.
Belyaev, V. B., Gershtein, S. S., Zakhar'ev, B. N., and Lomnev, S. P. (1959). *Zh. Eksp. Teor. Fiz.* **37**, 1652 [*English transl.: Sov. Phys.-JETP* **10**, 1171 (1960)].
Berezin, S., Burleson, G., Eartly, D., Roberts, A. R., and White, T. O. (1969). *Phys. Lett.* **30B**, 27.
Berezin, S., Burleson, G., Eartly, D., Roberts, A. R., and White, T. O. (1970a). *Nucl. Phys.* **B16**, 389.
Berezin, S., Burleson, G., Eartly, D., Roberts, A. R., and White, T. O. (1970b). *Phys. Rev.* **A2**, 1630.
Bertin, A., Bruno, M., Vitale, A., Placci, A., and Zavattini, E. (1972). *Nuovo Cimento Lett.* **4**, 449.
Bertin, A., Bruno, M., Vitale, A., Placci, A., and Zavattini, E. (1973a). *Phys. Rev.* **7A**, 462.
Bertin, A., Vitale, A., Placci, A., and Zavattini, E. (1973b). *Phys. Rev.* **8D**, 3774.
Bierman, E., Taylor, S., Koller, E. L., Stamer, P., and Huetter, T. (1963). *Phys. Lett.* **4**, 351.
Bjorkland, J. A., Raboy, S., Trail, C. C., Ehrlich, R. D., and Powers, R. J. (1965). *Nucl. Phys.* **69**, 161.
Blair, J. M., Muirhead, H., and Woodhead, T. (1962). *Proc. Phys. Soc.* **80**, 945.
Bleser, E., Lederman, L., Rosen, J., Rothberg, J., and Zavattini, E. (1962). *Phys. Rev. Lett.* **8**, 128.
Bleser, E., *et al.* (1963). *Phys. Rev.* **132**, 2679.
Block, M. M., *et al.* (1963). *Phys. Rev. Lett.* **11**, 301.
Block, M. M., Kopelman, J. B., and Sun, C. R. (1965). *Phys. Rev.* **B140**, 143.
Bobrov, V. D., *et al.* (1965). *Zh. Eksper. Teor. Fiz.* **48**, 1197 [*English transl.: Sov. Phys.-JETP* **21**, 798 (1965)].
Budick, B., Toroskar, J. R., and Yaghoobia, I. (1971). *Phys. Lett.* **B34**, 539.

Budyashov, Yu. G., Zinov, V. G., Konin, A. D., and Mukhin, A. I. (1967a). *Yad. Fiz.* **5,** 134 [*English transl.: Sov. J. Nucl. Phys.* **5,** 93 (1967)].
Budyashov, Yu. G., Zinov, V. G., Konin, A. D., and Mukhin, A. I. (1967b). *Yad. Fiz.* **5,** 830 [*English transl.: Sov. J. Nucl. Phys.* **5,** 589 (1967)].
Budyashov, Yu. G., Ermolov, P. F., Zinov, V. G., Konin, A. D., and Mukhin, A. I. (1967c). *Yad. Fiz.* **5,** 599 [*English transl.: Sov. J. Nucl. Phys.* **5,** 426 (1967)].
Budyashov, Yu. G., Ermolov, P. F., Zinov, V. G., Konin, A. D., Mukhin, A. I., and Oganesyan, K. O. (1968). Preprint JINR P15-3964, Dubna.
Burbidge, G. R., and de Borde, A. H. (1953). *Phys. Rev.* **89,** 189.
Burhop, E. H. S. (1969). "High Energy Physics" (D. R. Bates, ed.), Vol. 3, p. 109. Academic Press, New York.
Burke, P. G., Haas, F., and Percival, I. C. (1959). *Proc. Phys. Soc.* **73,** 912.
Bystritsky, V. M., *et al.* (1974a). *Zh. Eksper. Teor. Fiz.* **66,** 43.
Bystritsky, V. M., *et al.* (1974b). *Zh. Eksper. Teor. Fiz.* **66,** 61.
Calogero, F. (1967). "Variable Phase Approach to Potential Scattering." Academic Press, New York.
Camal, A. A. (1961). *Nuovo Cimento* **19,** 738.
Carter, B. P. (1966). *Phys. Rev.* **141,** 863.
Carter, B. P. (1967). *Phys. Rev.* **153,** 1358.
Carter, B. P. (1968a). *Phys. Rev.* **165,** 139.
Carter, B. P. (1968b). *Phys. Rev.* **173,** 55.
Carter, B. P. (1969). *J. Comput. Phys.* **4,** 54.
Charbe, M., Depommier, P., Heintze, J., and Soegel, V. (1963). *Phys. Lett.* **5,** 67.
Cohen, S., Judd, D. L., and Riddel, R. J. (1958). *Phys. Rev.* **110,** 1471.
Cohen, S., Judd, D. L., and Riddel, R. J. (1960). *Phys. Rev.* **119,** 384.
Cohen, R. C., Devons, S., Kanaris, A. D., and Nissim-Sabat, C. (1964). *Bull. Amer. Phys. Soc.* **9,** 652.
Comac, M., McCuire, A. D., Platt, J. B., and Schulte, H. J. (1955). *Phys. Rev.* **99,** 897.
Condo, G. T. (1964). *Phys. Lett.* **9,** 65.
Condo, G. T. (1974). *Phys. Rev. Lett.* **33,** 126.
Condo, G. T., Hill, R. D., and Martin, A. D. (1964). *Phys. Rev.* **133A,** 1280.
Conforto, G., Focardi, S., Rubbia, C., and Zavattini, E. (1962a). *Phys. Rev. Lett.* **9,** 432.
Conforto, G., Focardi, S., Rubbia, C., and Zavattini, E. (1962b). *Phys. Rev.* **9,** 525.
Conforto, G., Rubbia, C., Zavattini, E., and Focardi, S. (1964). *Nuovo Cimento* **33,** 1001.
Conversi, M., Pancini, E., and Piccioni, O. (1947). *Phys. Rev.* **71,** 209.
Cosyns, M. G. E., Dilworth, C. C., Occhialini, G. P. S., Schoenberg, M., and Page, N. (1949). *Proc. Phys. Soc. (London)* **A62,** 801.
Cresti, M., Gottstein, K., Rosenfeld, A. H., and Ticho, H. K. (1957). Preprint UCRL-3782.
Cresti, M., Limentani, S., Loria, A., Peruzzo, L., and Santangelo, R. (1965). *Phys. Rev. Lett.* **14,** 847.
Daniel, H., *et al.* (1968). *Phys. Lett.* **26B,** 281.
Day, T. B. (1960). *Nuovo Cimento* **18,** 381.
Day, T. B., and Snow, G. A. (1960). *Phys. Rev. Lett.* **5,** 112.
Day, T. B., Snow, G. A., and Sucher, J. (1959). *Phys. Rev. Lett.* **3,** 61.
de Benedetti, M. S. (1956). *Nuovo Cimento* **4,** 1209.
de Borde, A. H. (1954). *Proc. Phys. Soc. (London).* **67A,** 57.
Delves, L. M., and Kolotas, T. (1968). *Aust. J. Phys.* **21,** 1.

Demuer, M. (1956). *Nucl. Phys.* **1,** 516.
Derrick, K., Derrick, M., Fetkovich, J. C., Fields, T. H., Pewitt, E. G., and Yodch, G. B. (1966). *Phys. Rev.* **151,** 82.
Devons, S., and Duerdoth, I. (1969). *Advan. Nucl. Phys.* **2,** 295.
Doede, J. H. (1963). *Phys. Rev.* **132,** 1782.
Doede, J. H., Hildebrand, R. H., and Israel, M. H. (1963). *Phys. Rev.* **129,** 2808.
Doede, J. H., Hildebrand, R. H., and Israel, M. H. (1964). *Phys. Rev.* **136B,** 1609.
Dunaitsev, A. F., Petrukhin, V. I., Prokoshkin, Yu. D., and Rykalin, V. I. (1962). *Zh. Eksper. Teor. Fiz.* **42,** 1680 [*English transl.: Sov. Phys.-JETP* **15,** 1167 (1962)].
Dunaitsev, A. F., Petrukhin, V. I., and Prokoshkin, Yu. D. (1964). *Nuovo Cimento* **34,** 521.
Dzhelepov, V. P. (1963). *At. Energija* **14,** 27.
Dzhelepov, V. P., and Yermolov, P. F. (1964). *C. R. Congr. Int. Phys. Nucl. Paris* **2,** 1063.
Dzhelepov, V. P., Ermolov, P. F., Kushnirenko, E. A., Moskalev, V. I., and Gershtein, S. S. (1962a). *Zh. Eksper. Teor. Fiz.* **42,** 439 [*English transl.: Sov. Phys.-JETP* **15,** 306 (1962)].
Dzhelepov, V. P., Gershtein, S. S., Kushnirenko, E. A., Moskalev, V. I., and Yermolov, P. F. (1962b). *Nucl. Phys.* **34,** 424.
Dzhelepov, V. P., Friml, M., Gershtein, S. S., Katyshev, Yu. V., and Moskalev, V. I. (1962c). *Proc. Int. Conf. High Energy Phys.* p. 484. CERN, Geneva.
Dzhelepov, V. P., Filchenkov, V. V., Friml, M., Katyshev, Yu. V., Moskalev, V. I., and Yermolov, P. F. (1964a). *Nuovo Cimento* **33,** 40.
Dzhelepov, V. P., Yermolov, P. F., Katyshev, Yu. V., Moskalev, V. I., and Filchenkov, V. V. (1964b). *Zh. Eksper. Teor. Fiz.* **46,** 2042 [*English transl.: Sov. Phys.-JETP* **19,** 1376 (1964)].
Dzhelepov, V. P., Yermolov, P. F., Moskalev, V. I., Filchenkov, V. V., and Friml, M. (1964c). *Zh. Eksper. Teor. Fiz.* **47,** 1243 [*English transl.: Sov. Phys.-JETP* **20,** 841 (1965)].
Dzhelepov, V. P., Yermolov, P. F., Moskalev, V. I., Filchenkov, V. V., and Friml, M. (1964d). *Proc. Int. Conf. High Energy Phys. Dubna.*
Dzhelepov, V. P., Yermolov, P. F., Moskalev, V. I., and Filchenkov, V. V. (1965a). Preprint JINR P-2356, Dubna.
Dzhelepov, V. P., Yermolov, P. F., and Filchenkov, V. V. (1965b). *Zh. Eksper. Teor. Fiz.* **49,** 393 [*English transl.: Sov. Phys.-JETP* **22,** 275 (1966)].
Dzhelepov, V. P., Yermolov, P. F., Moskalev, V. I., and Filchenkov, V. V. (1966). *Zh. Eksper. Teor. Fiz.* **50,** 1235 [*English transl.: Sov. Phys.-JETP* **23,** 820 (1966)].
Eckhause, M., Filippas, T. A., Sutton, R. B., Welsh, R. E., and Romanowski, A. (1962). *Nuovo Cimento* **24,** 666.
Eisenberg, Y., and Kessler, D. (1961a). *Nuovo Cimento* **19,** 1195.
Eisenberg, Y., and Kessler, D. (1961b). *Phys. Rev.* **123,** 1472.
Eisenberg, Y., and Kessler, D. (1963a). *Phys. Rev.* **130,** 2349.
Eisenberg, Y., and Kessler, D. (1963b). *Phys. Rev.* **130,** 2352.
Fafarman, A., and Shamos, M. H. (1955). *Phys. Rev.* **100,** 874.
Fetkovich, J. G., and Pewitt, E. G. (1963). *Phys. Rev. Lett.* **11,** 289.
Fetkovich, J. G., Fields, T. H., Yodh, G. B., and Derrick, M. (1960). *Phys. Rev. Lett.* **4,** 570.
Fetkovich, J. G., *et al.* (1970). *Phys. Rev.* **D2,** 1803.
Fetkovich, I. G., Riley, B. R., and Wang, I.-T. (1971). *Phys. Lett.* **35B,** 178.

Fermi, E., and Teller, E. (1947). *Phys. Rev.* **72,** 399.
Fermi, E., Teller, E., and Weisskopf, V. (1947). *Phys. Rev.* **71,** 314.
Ferrel, R. A. (1960). *Phys. Rev. Lett.* **4,** 425.
Fields, T. H., Yodh, G. B., Derrick, M., and Fetkovich, I. G. (1960). *Phys. Rev. Lett.* **5,** 69.
Firsov, V. G., and Ponomarev, L. I. (in print). *In* "Modern Physics in Chemistry" (E. Flack and V. I. Goldanskii, eds.). Academic Press, New York.
Flugge, S., and Schröder, V. (1961). *Z. Phys.* **162,** 28.
Frank, F. C. (1947). *Nature (London)* **160,** 525.
Fried, Z., and Martin, A. D. (1963). *Nuovo Cimento* **29,** 574.
Fröhlich, H., Hubby, R., Kolodzievski, R., Rosenberg, R. L., and Wills, H. H. (1948). *Nature (London)* **162,** 450.
Fröman, A., and Kinsey, J. L. (1961). *Phys. Rev.* **123,** 2077.
Frost, A. A., Inokuti, M., and Lowe, J. P. (1964). *J. Chem. Phys.* **41,** 482.
Gershtein, S. S. (1957). *Dokl. Acad. Nauk USSR* **117,** 956.
Gershtein, S. S. (1958a). *Zh. Eksper. Teor. Fiz.* **34,** 463 [*English transl.: Sov. Phys.-JETP* **7,** 318 (1958)].
Gershtein, S. S. (1958b). *Zh. Eksper. Teor. Fiz.* **34,** 993 [*English transl.: Sov. Phys.-JETP* **7,** 685 (1958)].
Gershtein, S. S. (1959). *Zh. Eksper. Teor. Fiz.* **36,** 1309 [*English transl.: Sov. Phys.-JETP* **9,** 927 (1959)].
Gershtein, S. S. (1960). *Zh. Eksper. Teor. Fiz.* **39,** 1170 [*English transl.: Sov. Phys.-JETP* **12,** 815 (1961)].
Gershtein, S. S. (1961). *Zh. Eksper. Teor. Fiz.* **40,** 698 [*English transl.: Sov. Phys.-JETP* **13,** 488 (1961)].
Gershtein, S. S. (1962). *Zh. Eksper. Teor. Fiz.* **43,** 706 [*English transl.: Sov. Phys.-JETP* **16,** 501 (1962)].
Gershtein, S. S., and Krivchenkov, V. D. (1961). *Zh. Eksper. Teor. Fiz.* **40,** 1491 [*English transl.: Sov. Phys.-JETP* **13,** 1044 (1962)].
Gershtein, S. S., Petrukhin, V. I., Ponomarev, L. I., and Prokoshkin, Yu. D. (1969). *Usp. Fiz. Nauk* **97,** 1 [*English transl.: Sov. Phys.-Usp.* **12,** 1 (1970)].
Goldanskii, V. I., and Firsov, V. G. (1971). *Sov. Usp. Khim.* **40,** 1353.
Goldanskii, V. I., *et al.* (1973a). *Dokl. Acad. Nauk SSSR* **211,** 60.
Goldanskii, V. I., *et al.* (1973b). *Dokl. Akad. Nauk SSSR* **211,** 316.
Goldanskii, V. I., *et al.* (1974a). *Dokl. Acad. Nauk SSSR* **214,** 1337.
Goldanskii, V. I., *et al.* (1974b). *Dokl. Acad. Nauk SSSR* **214,** 1105.
Griffith, J. E., Haff, P. K., and Tombrello, T. A. (1974). *Ann. Phys.* **87,** 1.
Griffiths, G. M., Lal, M., and Scarfe, C. D. (1963). *Can. J. Phys.* **41,** 724.
Grin, G. A., and Kunselman, R. (1970). *Phys. Lett.* **31B,** 116.
Gurevich, I. S. (1972). *Proc. Int. Conf. High Energy Phys. Nucl. Structure, 4th, Dubna, 1971.*
Haff, P. K., and Tombrello, T. A. (1974). *Ann. Phys.* **86,** 178.
Halpern, A. (1964a). *Phys. Rev.* **135A,** 34.
Halpern, A. (1964b). *Phys. Rev. Lett.* **13,** 660.
Halpern, A. (1968). *Phys. Rev.* **174,** 62.
Hammett, L. P. (1970). "Physical Organic Chemistry." McGraw-Hill, New York.
Hayashi, C., Nakano, T., Nishida, M., Suekane, S., and Yamaguchi, Y. (1957). *Progr. Teor. Phys.* **17,** 615.
Hubby, R. (1949). *Phil. Mag.* **40,** 685.

Hughs, V. W. (1966). *Ann. Rev. Nucl. Sci.* **16,** 445.
Hunter, G., and Prichard, H. O. (1967a). *J. Chem. Phys.* **46,** 2146.
Hunter, G., and Prichard, H. O. (1967b). *J. Chem. Phys.* **46,** 2153.
Hunter, G., Gray, B. F., and Prichard, H. O. (1966). *J. Chem. Phys.* **45,** 3806.
Jackson, J. D. (1957). *Phys. Rev.* **106,** 330.
Joachain, C., and Wantiez, N. (1962). *Bull. Acad. Roy. Belg. Cl. Sci.* **48,** 147.
Kabir, P. K. (1965). *Phys. Lett.* **14,** 257.
Kabir, P. K. (1966). *Z. Phys.* **191,** 447.
Kessler, D., *et al.* (1967). *Phys. Rev. Lett.* **18,** 1179.
Kim, Y. N. (1962). *Phys. Lett.* **3,** 33.
Klem, R. D. (1967). *Nuovo Cimento* **48A,** 743.
Knight, J. D., Schillaci, M. E., and Naumann, R. A. (1972). *Proc. Int. Conf. High Energy Phys. Nucl. Structure 4th, Dubna 1971.*
Knop, R., Burnstein, R. A., and Snow, G. A. (1965). *Phys. Rev. Lett.* **14,** 767.
Kolos, W. (1968). *Phys. Rev.* **165,** 165.
Kolos, W. (1969). *Acta Phys. Hung.* **27,** 241.
Kolos, W., Roothaan, C. C. J., and Sack, R. A. (1960). *Rev. Mod. Phys.* **32,** 178.
Komarov, I. V., and Slavyanov, S. Yu. (1968). *J. Phys.* **B1,** 1066.
Komarov, I. V., and Slavyanov, S. Yu. (1969). *Vestnik Leningradskogo Univ.* (*Fiz. Khim.*). **16** (3), 30.
Konin, A. D., *et al.* (1974). *Phys. Lett.* **50A,** 57.
Krumshtein, Z. V., Petrukhin, V. I., Ponomarev, L. I., and Prokoshkin, Yu. D. (1968a). *Zh. Eksper. Teor. Fiz.* **54,** 1690 [*English transl.: Sov. Phys.-JETP*, **27,** 906 (1968)].
Krumshtein, Z. V., Petrukhin, V. I., Ponomarev, L. I., and Prokoshkin, Yu. D. (1968b). *Zh. Eksper. Teor. Fiz.* **55,** 1640 [*English transl.: Sov. Phys.-JETP* **28,** 860 (1969)].
Krumshtein, Z. V., Petrukhin, V. I., Smirnova, L. I., Suvorov, V. M., and Yutlandov, I. A. (1970). Preprint JINR P12-5224, Dubna.
Krumshtein, Z. V., Petrukhin, V. I., Risin, V. E., Smirnova, L. M., Suvorov, V. M., and Yutlandov, I. A. (1973). *Zh. Eksp. Teor. Fiz.* **65,** 455.
Landau, L. D., and Lifshitz, E. M. (1966). "Quantum Mechanics." Pergamon, Oxford.
Lathrop, J. L., Lundy, R. A., Swanson, R. A., Telegdi, V. L., and Yovanovitch, D. D. (1960). *Nuovo Cimento* **15,** 831.
Lathrop, J. L., Lundy, R. A., Telegdi, V. L., Winston, R., and Yovanovitch, D. D. (1961a). *Phys. Rev. Lett.* **7,** 107.
Lathrop, J. L., Lundy, R. A., Telegdi, V. L., and Winston, R. (1961b). *Phys. Rev. Lett.* **7,** 147.
Lattes, C. M., Muirhead, H., Occhialini, G. P. S., and Powell, C. F. (1947). *Nature* (*London*) **159,** 694.
Lee, T. D., and Wu, C. S. (1965). *Ann. Rev. Nucl. Sci.* **15,** 381.
Lee, T. D., and Wu, C. S. (1966). *Ann. Rev. Nucl. Sci.* **16,** 471.
Leon, M. (1971a). *Phys. Lett.* **35B,** 413.
Leon, M. (1971b). *Phys. Lett.* **37B,** 87.
Leon, M., and Bethe, H. A. (1962). *Phys. Rev.* **127,** 636.
Lobov, G. A. (1963). *Zh. Eksper. Teor. Fiz.* **45,** 713 [*English transl.: Sov. Phys.-JETP* **18,** 479 (1964)].
Mann, R. A., and Rose, M. E. (1961). *Phys. Rev.* **121,** 293.
Marschall, H., and Schmidt, T. (1958). *Z. Phys.* **150,** 293.
Martin, A. D. (1963). *Nuovo Cimento* **27,** 1359.

Massey, H. S. W., Burhop, E. H. S., and Gilbody, H. B. (1974). "Electronic and Ionic Impact Phenomena," p. 3276. Oxford (Clarendon), London.
Matone, G. (1969). Preprint LNF-69/5, Frascati, Rome.
Matone, G. (1971). *Lett. Nuovo Cimento* **2,** 151.
Matveenko, A. V., and Ponomarev, L. I. (1969). *Zh. Eksp. Teor. Fiz.* **57,** 2084 [*English transl.: Sov. Phys.-JETP* **30,** 1131 (1970)].
Matveenko, A. V., and Ponomarev, L. I. (1970a). *Zh. Eksp. Teor. Fiz.* **58,** 1640 [*English transl.: Sov. Phys.-JETP* **31,** 880 (1970)].
Matveenko, A. V., and Ponomarev, L. I. (1970b). *Zh. Eksp. Teor. Fiz.* **59,** 1593 [*English transl.: Sov. Phys.-JETP* **32,** 871 (1971)].
Matveenko, A. V., and Ponomarev, L. I. (1972a). *Zh. Teor. Mat. Fiz.* **12,** 64.
Matveenko, A. V., and Ponomarev, L. I. (1972b). *Jadern. Fiz.* **16,** 620.
Matveenko, A. V., Ponomarev, L. I., and Faifman, M. P. (1975). *Zh. Eksper. Teor. Fiz.* **68,** 437.
Mizuno, Y. (1959). *Progr. Teor. Phys.* **21,** 479.
Mizuno, Y. (1961). *J. Phys. Soc. Japan* **16,** 1043.
Mott, N. F., and Massey, H. S. W. (1965). "The Theory of Atomic Collisions," 3rd ed. Oxford Univ. Press, London and New York.
Narumi, H., and Matsuo, S. (1961). *Progr. Teor. Phys.* **25,** 290.
Ohtsubo, H., and Fujii, A. (1966). *Nuovo Cimento* **42,** 109.
Panofsky, W. K. F., Aamodt, R., and York, H. F. (1950). *Phys. Rev.* **78,** 825.
Panofsky, W. K. F., Aamodt, R., and Hadley, J. (1951). *Phys. Rev.* **81,** 565.
Patterson, M. R., and Becker, R. L. (1967). Thesis, Preprint ORNL-TM-1850, Oak Ridge, Tennessee.
Pauling, L. C. (1960). "The Nature of the Chemical Bound and the Structure of Molecules and Crystals." Cornell Univ. Press, Ithacca, New York.
Pawlewich, W. T., Murphy, C. T., Fetkovich, J. G., Dombeck, T., Derrick, M., and Wangler, T. (1970). *Phys. Rev.* **2D,** 2538.
Peek, T. M. (1965a). *J. Chem. Phys.* **43,** 3004.
Peek, T. M. (1965b). Sandia Corp. Rep., SC-RR-65-67.
Petrukhin, V. I. (1972). *Proc. Int. Conf. High Energy Phys. Nucl. Structure 4th, Dubna 1971.*
Petrukhin, V. I., and Prokoshkin, Yu. D. (1963). *Nuovo Cimento* **28,** 99.
Petrukhin, V. I., and Prokoshkin, Yu. D. (1965). *Dokl. Akad. Nauk USSR* **160,** 71.
Petrukhin, V. I., Prokoshkin, Yu. D., and Suvorov, V. M. (1968). *Zh. Eksper. Teor. Fiz.* **55,** 2173.
Petrukhin, V. I., Ponomarev, L. I., and Prokoshkin, Yu. D. (1967b). *Khimiya vysokih energij* (*High Energy Chem.*) **1,** 283.
Petrukhin, V. I., Rysin, V. E., and Suvorov, V. M. (1974). *Yad. Fiz.* **19,** 626.
Pevsner, A., Strand, R., Madansky, L., and Toohig, T. (1961). *Nuovo Cimento* **19,** 409.
Placci, A., Zavattini, E., Bertin, A., and Vitale, A. (1967). *Nuovo Cimento* **52A,** 1274.
Placci, A., Zavattini, E., Bertin, A., and Vitale, A. (1969). *Nuovo Cimento* **64A,** 1053.
Placci, A., *et al.* (1970a). *Phys. Lett.* **32B,** 413.
Placci, A., Zavattini, E., Bertin, A., and Vitale, A. (1970b). *Phys. Rev. Lett.* **25,** 475.
Podgoretskii, M. I. (1953). *Usp. Fiz. Nauk* **51,** 253.
Ponomarev, L. I. (1965). *Yad. Fiz.* **2,** 223 [*English transl.: Sov. J. Nucl. Phys.* **2,** 160 (1966)].
Ponomarev, L. I. (1967). *Yad. Fiz.* **6,** 389 [*English transl.: Sov. J. Nucl. Phys.* **6,** 281 (1968)].

Ponomarev, L. I. (1968). *Zh. Eksper. Teor. Fiz.* **55,** 1836 [*English transl.: Sov. Phys.-JETP* **28,** 971].
Ponomarev, L. I. (1973). *Ann. Rev. Nucl. Sci.* **23,** 395.
Ponomarev, L. I., and Prokoshkin, Yu. D. (1968). *Comments Nucl. Particle Phys.* **2,** 176.
Ponomarev, L. I., and Puzynina, T. P. (1967a). *Zh. Eksp. Teor. Fiz.* **52,** 1273 [*English transl.: Sov. Phys.-JETP* **25,** 846 (1967)].
Ponomarev, L. I., and Puzynina, T. P. (1967b). Preprint JINR, R4-3175, Dubna.
Ponomarev, L. I., and Puzynina, T. P. (1967c). Preprint JINR P4-3405, Dubna.
Ponomarev, L. I., and Puzynina, T. P. (1968). *Zh. Vyc. Mat. Mat. Fiz.* **8,** 1256.
Ponomarev, L. I., and Puzynina, T. P. (1970). Preprint JINR P4-5040, Dubna.
Ponomarev, L. I., Puzynin, I. V., and Puzynina, T. P. (1973a). *J. Comp. Phys.* **13,** 1.
Ponomarev, L. I., Puzynin, I. V., and Puzynina, T. P. (1973b). *Zh. Eksp. Teor. Fiz.* **65,** 28.
Primakoff, M. (1959). *Rev. Mod. Phys.* **31,** 802.
Quitman, D., *et al.* (1964). *Nucl. Phys.* **51,** 609.
Rook, J. P. (1963). *Nucl. Phys.* **43,** 363.
Rook, J. R. (1969). *Nucl. Phys.* **9B,** 441.
Rook, J. P. (1970). *Nucl. Phys.* **20B,** 14.
Rosenberg, R. L. (1949). *Phil. Mag.* **40,** 759.
Rothberg, J. E. (1963). Thesis, Columbia Univ.
Rothberg, J. E., *et al.* (1963). *Phys. Rev.* **132,** 2664.
Rudermann, M. (1960). *Phys. Rev.* **118,** 1632.
Russel, J. E. (1965). *Proc. Phys. Soc.* **85,** 245.
Russel, J. E. (1970). *Phys. Rev.* **1A,** 721.
Russel, J. E., and Shaw, G. L. (1960). *Phys. Rev. Lett.* **4,** 769.
Sakharov, A. D. (1948). Rep. Lebedev Phys. Inst., Acad. Sci. USSR.
Scherr, C. W., and Machaček Milos (1965). *Phys. Rev.* **138A,** 371.
Schiff, M. (1961). *Nuovo Cimento* **22,** 66.
Schröder, U. (1963a). *Z. Phys.* **173,** 221.
Schröder, U. (1963b). *Z. Phys.* **173,** 432.
Selinov, I. P. (1970). "Izotopy," Vol. 3. Nauka, Moscow.
Sens, J. C., Swanson, R. A., Telegdi, V. L., and Yovanovich, D. D. (1958). *Nuovo Cimento* **7,** 4.
Shimizu, M., Mizuno, Y., and Izuyama, T. (1959). *Progr. Theor. Phys.* **20,** 777.
Skyrme, T. H. R. (1957). *Phil. Mag.* **2,** 910.
Srinivasan, V., and Sundaresan, M. K. (1968). *Nuovo Cimento* **57B,** 235.
Stearns, M. B. (1957). *Progr. Nucl. Phys.* **6,** 108.
Stearns, M. B., and Stearns, M. (1957). *Phys. Rev.* **105,** 1573.
Stearns, M. B., Culligan, G., Sherwood, B., Telegdi, V. L., and Stearns, M. (1969). *Phys. Rev.* **184,** 22.
Suzuki, A. (1967). *Phys. Rev. Lett.* **19,** 1005.
Tauscher, L., *et al.* (1968). *Phys. Lett.* **27A,** 581.
Taylor, B. N., Parker, W. H., and Langenberg, D. N. (1969a). *Rev. Mod. Phys.* **41,** 375.
Taylor, B. N., Parker, W. H., and Langenberg, D. N. (1969b). "The Fundamental Constants and Quantum Electrodynamics." Academic Press, New York.
Varlamov, V., Dobretzov, U., Dolgoshein, B., Kirillov-Ugryumov, V., and Rogozhin, A. (1971). *Proc. Int. Conf. High Energy Phys. Nucl. Structure, 4th, Dubna.*
Vermeuler, T. L. (1969). *Nucl. Phys.* **12B,** 506.
Vesman, E. A. (1967a). *Zh. Eksper. Teor. Fiz. Pisma* **5,** 113.

Vesman, E. A. (1967b). Preprint JINR P4-3256, Dubna.

Vesman, E. A. (1967c). Preprint JINR P4-3384, Dubna.

Vesman, E. A. (1968). *Toimet. Eesti. NSV Teaduste Acad.* (*USSR*) **17**, 65.

Vesman, E. A. (1969). *Eesti NSV Teaduste Akad. Fuusika Astron.-Inst. Uurimused* (*USSR*) **18**, 429.

Vilgelmova, L., *et al.* (1973). *Zh. Eksp. Teor. Fiz.* **65**, 24.

Vynitskij, S. I., and Ponomarev, L. I. (1974). *Yad. Fiz.* **20**, 576.

Wang, I.-T., *et al.* (1965). *Phys. Rev.* **139B**, 1528.

Weinberg, S. (1960). *Phys. Rev. Lett.* **4**, 575.

Wessel, W. R., and Phillipson, P. (1964). *Phys. Rev. Lett.* **13**, 23.

West, D. (1958). *Rep. Progr. Phys.* **21**, 271.

Wheeler, J. A. (1947). *Phys. Rev.* **71**, 320.

Wightman, A. S. (1950). *Phys. Rev.* **77**, 521.

Wolfenstein, L. (1960). *Proc. Conf. High Energy Physics, Rochester*, p. 529. Univ. of Rochester Press, Rochester, New York.

Wu, C. S., and Wilets, L. (1969). *Ann. Rev. Nucl. Sci.* **19**, 527.

Wu, T. Y., Rosenberg, R. L., and Sandström, H. (1960). *Nucl. Phys.* **16**, 433.

Zaimidoroga, O. A., Kulyukin, M. M., Sulyaev, R. M., Filippov, A. I., Tsupko-Sitnikov, V. M., and Shcherbakov, Yu. A. (1963). *Zh. Eksp. Teor. Fiz.* **44**, 1852 [*English transl.: Sov. Phys.-JETP* **17**, 1246 (1963)].

Zeldovich, Ya. B. (1954). *Dokl. Akad. Nauk USSR* **95**, 493.

Zeldovich, Ya. B. (1957). *Zh. Eksp. Teor. Fiz.* **33**, 310 [*English transl.: Sov. Phys.-JETP* **6**, 242 (1958)].

Zeldovich, Ya. B., and Gershtein, S. S. (1958a). *Zh. Eksp. Teor. Fiz.* **35**, 649 [*English transl.: Sov. Phys.-JETP* **8**, 453 (1959)].

Zeldovich, Ya. B., and Gershtein, S. S. (1958b). *Zh. Eksp. Teor. Fiz.* **35**, 649 [*English transl.: Sov. Phys.-JETP* **8**, 453 (1959)].

Zeldovich, Ya. B., and Gershtein, S. S. (1960). *Usp. Fiz. Nauk* **71**, 581 [*English transl.: Sov. Phys.-Usp.* **3**, 593 (1961)].

Zeldovich, Ya. B., and Sakharov, A. D. (1957). *Zh. Eksp. Teor. Fiz.* **32**, 947 [*English transl.: Sov. Phys.-JETP* **5**, 775 (1957)].

Zinov, V. G., Konin, A. D., and Mukhin, A. I. (1964). *Zh. Eksp. Teor. Fiz.* **46**, 1919 [*English transl.: Sov. Phys.-JETP* **19**, 1292 (1964)].

Zinov, V. G., Konin, A. D., and Mukhin, A. I. (1965a). *Yad. Fiz.* **2**, 859 [*English transl.: Sov. J. Nucl. Phys.* **2**, 613 (1965)].

Zinov, V. G., Konin, A. D., Mukhin, A. I. and Polyakova, P. V. (1965b). Preprint JINR P-2039, Dubna.

Zinov, V. G., Konin, A. D., Mukhin, A. I., and Polyakova, P. V. (1967). *Yad. Fiz.* **5**, 591 [*English transl.: Sov. J. Nucl. Phys.* **5**, 420 (1967)].

Zinov, V. G., Konin, A. D., and Mukhin, A. I. (1972). Preprint JINR P14-6407, Dubna.

Section 3

DEPOLARIZATION OF NEGATIVE MUONS AND INTERACTION OF MESONIC ATOMS WITH THE MEDIUM

V. S. EVSEEV

Laboratory of Nuclear Problems
Joint Institute for Nuclear Research
Dubna, USSR

Translated by

S. J. AMORETTY

Physics Department
Brookhaven National Laboratory
Upton, New York

I. Introduction

An interest in the depolarization of muons arose in the wake of the discovery of parity nonconservation in processes controlled by weak interactions.

The first experiment (Garwin *et al.*, 1957) to demonstrate the violation of space parity in muon decay showed a large discrepancy in the depolarization of muons of opposite charge. Since the depolarization of muons was measured in this work by observing the Larmor precession of the free muon spin after stopping the muons in a conductor with spinless nuclei (graphite), it greatly simplified the problem of developing the basic mechanism for depolarization of negative muons. As is well known, only such conductors create optimum conditions for selecting an effective depolarization mechanism for the production of a mesonic atom; the other mechanisms are either less prominent or not in evidence at all.

Shmushkevich (1959) and Dzhrbashyan (1958) were first to attribute the substantially higher depolarization of negative muons as compared with positive muons to the spin–orbit interaction of the muon in the excited states of a mesonic atom. For brevity we shall henceforth refer to this theory as the theory of cascade depolarization.

Dzhrbashyan (1958) also showed that the presence of an uncompensated magnetic moment in the electron shell of a mesonic atom must lead to additional depolarization (from two to three times, depending on the magnetic moment) because of the hyperfine interaction between the magnetic moment of the muon and the electron shell.

Absence of this effect in measuring the depolarization in graphite, according to Dzhrbashyan, favors a fast compensation of paramagnetism of the mesonic atom's electron shell (as a result of stopping, a mesonic atom possessing an electron shell similar to that of the Bohr atom is produced in carbon).

The effect of a mesonic atom's electron shell on muon depolarization, as we shall see, is central in considering the behavior of a mesonic atom in a real medium.

In the first series of experiments (Ignatenko *et al.*, 1958) measuring the depolarization of negative muons in different elements and chemical compounds whose nuclei have zero spin, no departure from the predictions of the theory of cascade depolarization was observed.

In 1968, experimental studies of the effect of a medium on the depolarization of negative muons were the initial sources of information and were concerned either with measuring the neutron asymmetry in the reaction involving nuclear absorption of a muon or with verifying the mechanism for depolarization at the moment a mesonic-atom cascade occurs in the mesonic atoms with different nuclear spins.

Überall (1959) was first to investigate the additional depolarization arising from the interaction of the magnetic moment of the muon and the nucleus. The transitions between states of the hyperfine structure in a muon–nucleus system (mesonic nucleus) and the effect of the electron shell on the probability of such transitions were investigated experimentally by a number of authors (Winston and Telegdi, 1961; Winston, 1963; Hutchinson *et al.*, 1962), and the first direct evidence for the existence of different precessional frequencies of the mesonic nucleus corresponding to the states of the hyperfine structure was obtained in a study (Babaev *et al.*, 1969) carried out at JINR and further investigated by a Belgian group (Favart *et al.*, 1970) at CERN.

The results of these experiments, in which conductors were primarily used, confirmed for the most part the predictions of the theory of cascade depolarization.

However, experimental data totally incomprehensible in terms of this theory have been gradually accumulated. Among the more important findings is the evidence for total depolarization of the negative muons (at the precessional frequency of a free muon spin) in the inert gases (Prepost *et al.*, 1960; McColm *et al.*, 1960; Kane, 1964; Buckle *et al.*, 1967, 1968) and in liquid oxygen (Buckle *et al.*, 1968) and the dependence of depolarization in water on the temperature and on the chemical binding in hydrocarbons (Buckle *et al.*, 1968; Evseev *et al.*, 1968).

All experimental data obtained by 1968 were summarized by Evseev (1968), who, on the hypothesis that the paramagnetism of mesonic atoms changes as a result of their interaction with the medium, qualitatively explained many effects associated with an isolated mesonic atom that were heretofore incomprehensible within the framework of the theory of cascade depolarization. Later this hypothesis was substantiated quantitatively by Dzhuraev and Evseev (1972, 1973), who developed a formal theory connecting the magnitude of the depolarization of the negative muons with the different parameters of the medium.

Toward the end of 1968, systematic investigation of the depolarization of negative muons in different molecular media was begun at the Laboratory of Nuclear Problems of the JINR (Evseev *et al.*, 1968; Dzhuraev and Evseev, 1973; Dzhuraev *et al.*, 1971a,b,c,d, 1972, 1973; Gurevich, 1972). The results of these and all the other investigations will be carefully analyzed in the light of the model (Evseev, 1968; Dzhuraev and Evseev, 1972, 1973) for the fast interactions of mesonic atoms with the medium.

II. *Depolarization of Negative Muons in an Isolated Mesonic Atom*

A. Spinless Nuclei

In pion decay the negative muons are completely longitudinally polarized in the pion rest frame. The muon beams from accelerators are partially polarized; when meson channels are used the polarization depends on the muon momentum and reaches an absolute value of 0.8 (Dzhuraev *et al.*, 1971a).

The muons do not seem to be depolarized noticeably in the substance as a result of slowing down (Vaisenberg, 1964). The meson loses polarization only in the mesonic atom stage. This process has been theoretically analyzed by Shmushkevich (1959), Dzhrbashyan (1958, 1959), Mann and Rose (1961), and Bukhvostov (1969) for spinless nuclei. Although the mechanisms for muon transition to the bound state have not been studied experimentally, different estimates show that the muons are found in the highly excited states of the mesonic atom with a large orbital angular momentum l.

Having undergone numerous electric dipole transitions (Auger and radiative), the muon finally reaches the K shell of the mesonic atom. Muon depolarization in the mesonic-atom stage is attributable to spin–orbit interaction of the muon in the excited states of the mesonic atom. Since the dipole–moment operator, which determines the Auger and radiative transitions in the mesonic atom, has no effect on the spin variables, further depolarization can occur only if the lifetime of a muon at the given level is longer than that of the muon spin flip due to spin–orbit interaction, i.e., if the level width of the fine structure Γ is much smaller than the fine splitting Δ:

$$\Gamma \ll \Delta \tag{1}$$

This condition holds for any excited state of the mesonic atom, where the quantity Γ is determined solely by the radiative transitions. However, for large values of n the level width is much larger than Δ because of the high Auger-transition probability, and the orbital angular momenta do not have sufficient time to reverse the muon spin. Therefore, the mesonic-atom transitions occur without additional depolarization before a certain level with principal quantum number n_0 is reached, after which the radiative transitions play a dominant role. Since near n_0 a change in the rate of Auger transitions by a factor of 10 changes n by one to two units, n_0 can be determined fairly accurately for each mesonic atom.

It should be noted that the magnitude of n_0 depends on the presence of a sufficient number of electrons in the mesonic atom to give rise to the Auger transitions. If an Auger transition is retarded because of a deficiency of electrons, depolarization will begin at a higher level $n > n_0$ and the residual polarization P in the K shell will decrease. This fact is not taken into account in the theory of cascade depolarization. The population (capture of a muon by an atom or transitions from higher levels) of the excited states with specified quantum numbers n_0, $j_0 = l_0 \pm \frac{1}{2}$, l_0, and m_0 (a projection of the total angular momentum j_0) appreciably decreases the muon polarization associated with the averaging of its spin over a large number of states. As shown by Shmushkevich (1959), the probability w_j of populating states with a given j (on the assumption that all the values of the quantum number m are equally probable) and the average value $\bar{\sigma}$ of the muon spin projection in the state with a fixed j are given by

$$w_j = \frac{2j+1}{2(2l+1)}, \qquad \bar{\sigma} = \frac{[\frac{3}{4} + j(j+1)] - l(l+1)}{3j(j+1)} \tag{2}$$

If l is large, $w_j \simeq \frac{1}{2}$ and $\bar{\sigma} = \frac{1}{3}$; i.e., the probability of populating states with $j = l + \frac{1}{2}$ and $j = l - \frac{1}{2}$ is equal and the polarization in any of these states is about one-third of that before the formation of the mesonic atom. Specific calculations show (Dzhuraev *et al.*, 1971a; see also Garwin *et al.*, 1957) that $\bar{\sigma}$ can be larger than $\frac{1}{3}$ for light mesonic atoms.

The radiative transitions from the n_0 level give rise to an additional double depolarization, since the transitions from the $j_0 = l_0 + \frac{1}{2}$ states to the K shell preserve the muon polarization and those from the $j_0 = l_0 - \frac{1}{2}$ states lead to total depolarization.

Qualitatively, this can be explained as follows. Let us consider (Shmushkevich, 1959), for example, the $5g_{9/2}$ state characterized by maximum values of l_0, j_0, and projection of the total momentum m_0 (for a given n_0). The dipole transitions from this state prior to the K orbit occur in the following manner:

$$(5g_{9/2}, m_0 = \tfrac{9}{2}) \to (4f_{7/2}, m_0 = \tfrac{7}{2}) \to (3d_{5/2}, m_0 = \tfrac{5}{2})$$
$$\to (2p_{3/2}, m_0 = \tfrac{3}{2}) \to (4s_{1/2}, m_0 = \tfrac{1}{2})$$

Since for all the states of this sequence $m_0 = j_0 = l_0 + \frac{1}{2}$, the projection of the muon spin retains a definite value equal to $\frac{1}{2}$. This means that depolarization does not occur.

If, however, l_0 is large and $j_0 = l_0 - \frac{1}{2}$, the projection of the muon spin in the ground state and in all the intermediate states of the cascade does not have a definite value and each transition is accompanied by a definite degree of depolarization. Since l_0 is large, the number of transitions in the cascade is also large and therefore the depolarization is practically total. As a result, the residual polarization must approach $\frac{1}{6}$ (with increasing l_0).

An approximate expression for $P(l_0)$ was obtained by Bukhvostov (1969):

$$P(l_0) = 0.3 \frac{l_0 + 3}{2l_0 + 1} P_0 \tag{3}$$

where P_0 is the muon polarization prior to the formation of the mesonic atom; $P(l_0)$ decreases with increasing l_0 and tends to the limit $P(l_0 \to \infty) = 0.156$.

Assuming a statistical population of the sublevels with different l_0, the residual polarization in the ground state of the mesonic atom can be written as follows (Dzhuraev *et al.*, 1972):

$$P(n_0) = 0.15 \frac{n_0 + 5}{n_0} P_0 \tag{4}$$

The population of the n_0 level depends on the initial population of the highly excited levels of the mesonic atom, from which the cascade transitions to n_0 occur. However, the initial cascade stage (in the region of the valence-electron orbits) has thus far not been studied. The cascade transitions have been theoretically investigated beginning with $n = 14$, for which the mesonic-atom orbits are located in the region of the K-electron orbits (de Borde, 1954; Burbidge and de Borde, 1933; Martin, 1963; Au-Yang and Cohen, 1968). The calculations, however, are fairly approximate, and the level population in Mann and Rose (1961) was not determined accurately (Day *et al.*, 1961); more comprehensive estimates (Martin, 1963) were made for K^- mesonic atoms, and calculations (Au-Yang and Cohen, 1968) were carried out numerically only for some heavy mesonic atoms.

To account for the experimentally observed mesonic X-ray spectra, Eisenberg and Kessler (1961, 1963) used different initial populations in l for the level with $n = 14$ in calculating the cascade transitions (de Borde, 1954; Burbidge and de Borde, 1953). The distribution in which the population is proportional to $(2l + 1) \exp(\alpha l)$ for $\alpha \simeq 0.1$ corresponds to the experimental data for light mesonic atoms (Eisenberg and Kessler, 1963; Kessler *et al.*, 1967), and the statistical distribution [the population is proportional to $(2l + 1)$] corresponds to the experimental data for intermediate mesonic atoms (Kessler *et al.*, 1967). The experimental data for mesonic X-ray spectra can also be sufficiently well described by the initial statistical distribution in l if the electron shell is strongly ionized as a result of the passage of a mesonic-atom cascade (Suzuki, 1967). For example, in the case of calcium, the K and L shells of the atom should have only two electrons each during the cascade, and in the case of magnesium, they should have only one each. As we shall see, this fact is essential for the interpretation of experimental data on depolarization of negative muons in conductors.

Muon decay in the K shell of a mesonic atom is affected by the Coulomb field of the nucleus; this field also influences the spectrum of the decay electrons, the decay probability, and the angular distribution (Vaisenberg, 1964; Johnson *et al.*, 1961). However, these effects are significant only for heavy atoms, which are not considered here.

B. Depolarization in Mesonic Atoms Whose Nuclei Have a Spin

The interaction between the magnetic moment of the muon and the nucleus, which determines the hyperfine level splitting of mesonic atoms, is strong in the mesonic atoms with nonzero nuclear spin. This interaction causes additional depolarization of muons.

Überall (1959) and Libkin (1960) calculated the additional depolarization as a function of the nuclear spin I by taking into account the hyperfine splitting of only the ground state of the mesonic atom and assuming a statistical population of the states of the hyperfine structure. The ground state of the mesonic atom in this case is split in two, with the total angular momentum $F_\pm = I \pm \frac{1}{2}$ and with the gyromagnetic ratios

$$g_+ = \frac{1}{I + \frac{1}{2}} (\mu_\mu + \mu_N), \qquad g_- = -\frac{1}{I + \frac{1}{2}} \left(\mu_\mu - \frac{I + 1}{I} \mu_N \right) \tag{5}$$

where μ_μ and μ_N are the magnetic moments of the muon and the nucleus. If $\mu_N > 0$, the F_+ state is situated above the F_- state, and vice versa.

Using this approach, the residual polarization, averaged over the two states of the hyperfine structure, can be represented in the following form, depending on the nuclear spin:

$$P_I = P_0 \frac{1}{18} \left[1 + \frac{2}{(2I + 1)^2} \right] \tag{6}$$

(P_0 is the polarization prior to the mesonic atom stage), from which it follows that the polarization can decrease by a factor of 2 to 3 because of the presence of nuclear spin.

The polarization $P_\pm$ and statistical weights $\xi_\pm$ for the states of the hyperfine structure $F_\pm$ are given by

$$\begin{aligned} P_+ &= P \frac{1}{3} \frac{2I + 3}{2I + 1}, \qquad & \xi_+ &= \frac{I + 1}{2I + 1} \\ P_- &= P \frac{1}{3} \frac{2I - 1}{2I + 1}, \qquad & \xi_- &= \frac{I}{2I + 1} \end{aligned} \tag{7}$$

where P is the residual polarization in the K shell for the spinless nucleus. The polarization in each state of the hyperfine structure must be known independently, for in comparing theory with experiment the probable fast transitions between these states must be taken into account in a number of cases (Winston and Telegdi, 1961; Winston, 1963; Hutchinson *et al.*, 1962).

The depolarization due to the hyperfine interaction and that in the excited levels of a mesonic atom was investigated by Bukhvostov *et al.* (1961; Bukhvostov, 1966) for nuclei with an arbitrary spin. Let us first consider the case in which $I = \frac{1}{2}$.

If the quantities Δ/Γ are assumed to be small for all the excited levels and the corrections due to the hyperfine splitting in the linear approximation in Δ/Γ are taken into account, we notice a peculiar characteristic: hyperfine splitting of the excited levels does not necessarily decrease the residual polarization. Physically, this can be explained by the fact that in the presence of hyperfine structure the muon polarizes the nucleus during the transitions between the excited levels; therefore, when the muon arrives in the K shell the nucleus has already been polarized to some degree.

For a strong hyperfine splitting ($\Delta/\Gamma \gg 1$), the residual muon polarization is given by

$$P_I' = \frac{5}{48} \frac{l_0 + 5}{2l_0 + 1} P_0 \tag{8}$$

A comparison with Eq. (3) shows that for the most values of l_0 the residual polarization decreases by 20 to 30% as a result of taking the hyperfine splitting of the excited levels into account. However, for small values of l_0 this decrease is negligible, and at $l_0 = 1$ the residual polarization may increase somewhat.

Since allowance for the hyperfine interaction in the excited states requires in the general case (for any I) cumbersome calculations, the investigation (Bukhvostov, 1969) was confined to the limiting cases in which the hyperfine splitting is either smaller or larger than the level width. The first case was investigated by Überall (1959) and Libkin (1960). The calculations for the second limiting case were performed on a computer (Bukhvostov, 1969).

The above holds true if the transitions between the states of the hyperfine structure are not taken into account. However, experimental data and theoretical calculations show (Winston and Telegdi, 1961; Winston, 1963; Hutchinson *et al.*, 1962) that these transitions can occur by knocking out an Auger electron from the L or the M shell of an atom whose binding energy is lower than that of the hyperfine splitting of the mesonic atom's K shell. The probability of such transitions R is much higher (at $5 \leq Z \leq 35$) than that of the decay or nuclear absorption of a muon. Therefore, the muon polarization is not constant and must change during its lifetime.

If the hyperfine splitting of the excited levels is larger than their width, the main contribution to the residual polarization comes from the F_+ state. Therefore the polarization changes little with time when the nuclear magnetic moment is negative and the F_+ level is situated below the F_- level. However, when $\mu_N > 0$ the change is considerable, especially since the muon polarization changes sign as a result of the transition between the F_+ and F_- states. When the conversion rate R is high, the average muon polarization in the K shell in both instances is of the same order of magnitude as that in the absence of conversion (with accuracy to the sign on μ_N).

A different result is obtained, however, if the hyperfine splitting of the excited levels is weak—when P_+ is approximately equal to P_- and I is large. Here the contributions from the F_+ and F_- states are compensated by changing the sign of polarization in the conversion transition, and the average muon polarization in the K orbit becomes much smaller than that in the absence of conversion. Thus, in the presence of conversion the hyperfine splitting of the excited levels increases the average residual polarization.

C. *The Effect of the Electron Shell of the Mesonic Atom on the Depolarization of the μ^- Mesons*

If a mesonic atom's electron shell has an uncompensated angular momentum J (let us for the present consider a stationary state), an additional hyperfine splitting of the levels of a mesonic atom, which may already have been split as a result of the interaction with the nuclear spin, can occur. A hyperfine splitting of the ground state of a mesonic atom, which is determined by the magnetic moment of the electron shell, is of the order of 10^9 to 10^{10} sec^{-1}; this greatly exceeds the level width, since it is determined by the probability of the decay or nuclear absorption of the muon and for light nuclei is of the order of 10^6 sec^{-1}. Therefore, the states with total angular momentum $F_+' = J + \frac{1}{2}$ and $F_-' = J - \frac{1}{2}$ produce an incoherent mixture in the mesonic atom. Since $\mu_\mu < 0$ for a μ^- meson, the F_+' state lies below the F_-' state.

Dzhrbashyan (1958, 1959) investigated the muon depolarization resulting from the interaction with the paramagnetic electron shell of a mesonic atom for spinless nuclei. Formally his result does not differ from that obtained for the interaction between the magnetic moments of the muon and the nucleus. A formula for the residual polarization,

$$P_J = PQ, \qquad Q = \frac{1}{3}\left[1 + \frac{2}{(2J+1)^2}\right] \tag{9}$$

averaged over the statistically populated states F_+' and F_-', has been obtained; P is the polarization without taking the effect of the electron shell into account.

Since the time the muon remains in the excited states of the light and intermediate mesonic atoms is much shorter (de Borde, 1954; Burbidge and de Borde, 1953) than the typical time of the hyperfine muon–electron splitting, investigation is usually confined to examining the depolarization effect of the magnetic moment of the electron shell only for the ground state of a mesonic atom, although in a number of cases this effect should be taken into account in the excited states (see Section V).

The effect of the electron shell on muon depolarization in mesonic atoms whose nuclei have $I \neq 0$ is examined in a very general way in a number of papers (Tammet, 1967, 1969a,b,c, 1970).

Since the energy of the magnetic interaction of a muon is small compared with the energy of the Coulomb interaction in a mesonic atom, the Hamiltonian of the mesonic atom can be expressed as a sum of the principal Hamiltonian H_0 and the magnetic interaction Hamiltonian H, which takes the spin–spin interaction into account. H can be considered as a perturbation. The principal Hamiltonian H_0 leaves the muon polarization constant. Since the spin–orbit interaction of the electron shell is much stronger than the other interactions involving the spin of the electron shell, its state can be characterized by the net angular momentum **J**. The following Hamiltonian of the magnetic interaction between a meson (spin $\boldsymbol{\sigma}$), an electron shell, and a nucleus is used for an isolated mesonic atom:

$$H = A\mathbf{IJ} + D\boldsymbol{\sigma}\mathbf{J} + G\boldsymbol{\sigma}\mathbf{I} \tag{10}$$

where A, D, and G are the hyperfine interaction constants. Exact values of these constants would appear to be necessary for quantitative results. It was found (Tammet, 1967, 1969a,b,c, 1970), however, that muon polarization is almost independent of these constants in a fairly broad range of their actual values in a mesonic atom. The hyperfine interaction constants A and D can be determined on the basis of the computational methods of atomic spectroscopy [see, for example, Sobel'man (1963) and Frisch (1963)]. The constant G was calculated by Winston and Telegdi (1961), Winston (1963), and Hutchinson *et al.* (1962). The D/A ratio for light and intermediate mesonic atoms may vary from 10^0 to 10^1, and the G/D ratio from 10^2 to 10^3.

In the special case $J = I = \frac{1}{2}$, the residual polarization $P_J = 0.39P_I$ for the values of A, D, and G characteristic of the actual conditions in mesonic atoms; this is approximately 25% lower than the value obtained without taking the electron-shell effect into account. For the general case (arbitrary J and I), Table 1 summarizes the results of the calculations, including the values for the residual polarization averaged over the F_+' and F_-' states,

TABLE 1

J \ I	$\frac{1}{2}$	$\frac{3}{2}$	$\frac{5}{2}$	$\frac{7}{2}$	$\frac{9}{2}$	$\frac{11}{2}$
$\frac{1}{2}$	0.39	0.34	0.33	0.33	0.33	0.33
$\frac{3}{2}$	0.21	0.29	0.27	0.30	0.31	0.32
$\frac{5}{2}$	0.19	0.16	0.20	0.24	0.27	0.29
$\frac{7}{2}$	0.18	0.14	0.15	0.18	0.22	0.25
$\frac{9}{2}$	0.17	0.14	0.14	0.15	0.18	0.21
$\frac{11}{2}$	0.17	0.13	0.13	0.14	0.15	0.18

which are assumed to be statistically populated. Numerical calculations show that the muon has a maximum polarization in the $F_+ = I + \frac{1}{2}$ sublevel, especially when I and J are small. If the spins are small, the muon in the $F_- = I - \frac{1}{2}$ sublevel is almost entirely depolarized. Polarization is appreciable in this sublevel, however, only when the spins are large. If $I \to \infty$, $P_+ \to P_0/6$, $P_- \to P_0/6$, and $P \to P_0/3$. If $J \to \infty$, the polarization approaches asymptotic values:

$$P_{I+1/2,\infty} = \frac{(I+1)(I+\frac{3}{2})}{18(I+\frac{1}{2})^2} P_0, \qquad P_{I-1/2,\infty} = \frac{I(I-\frac{1}{2})}{18(I+\frac{1}{2})^2} P_0 \tag{11}$$

Thus it is clear that the uncompensated magnetic moment of the electron shell suppresses muon polarization. This is vitally important when J is large and I is small. The electron-shell effect is not as strong, however, for nuclei with large spin.

The state of the electron shell of a mesonic atom can be determined in a Paschen–Back experiment. The muon polarization recovery in a longitudinal magnetic field was calculated in a number of papers (Tammet, 1967, 1969a,b,c, 1970), for different values of n_e and l_e of the uncompensated electron in mesonic atoms with spinless nuclei.

The muon polarization in light mesonic atoms is restored in a magnetic field of 10^2 to 10^4 G, and this can be verified experimentally. Note that a field of 10^{10} to 10^{12} G is necessary to break the spin–orbit coupling in the excited states of a mesonic atom.

III. Polarization Effects Due to the Interaction between Mesonic Atoms and the Medium

In order to evaluate experimental data on the depolarization of negative muons stopped in real media, we must consider the effect of the surrounding medium on the mesonic atom. The theory discussed in Section II ignores almost entirely the interaction between the mesonic atom and the medium. These processes, discussed in detail by Evseev (1968), Dzhuraev and Evseev (1971, 1973), Dzhuraev *et al.* (1971a,b,c,d, 1972, 1973), Gurevich (1972), can be summarized on the basis of the present level of understanding as follows:

It is clear that the medium has an effect on the mechanism for muon capture by an atom from the very outset—formation of the mesonic atom—and determines the initial population in n. If a muon is stopped in a molecular compound, it is possible for it to settle in the orbits of mesonic molecules, which are geometrically similar to the valence-electron orbits of the molecule (Gershtein *et al.*, 1969). Since the radiative transition from the states of the mesonic molecule directly to the ground state is assumed to be dominant in this model, the degree of depolarization is expected to decrease in these transitions.

However, from the analysis of experimental data on the structure of μ-mesonic X-ray spectra, obtained by using high-energy-resolution Ge(Li) spectrometers (Backenstass *et al.*, 1967; Kessler *et al.*, 1967), it follows that the probability of direct radiative transitions from the mesonic states of the molecule is less than 10^{-2} of the total strength of the K series. This means either that the probability for the appearance of the mesonic states is low, or that it is high, in which case the radiative transitions are not dominant. What probably happens, however, is that successive Auger transitions similar to those occurring in an isolated mesonic atom take place.

If this is so, then, just as in the case of an isolated mesonic atom, the transitions from the mesonic states of the molecule to the nearest split states should occur without any depolarization. Therefore the effect of the mesonic state of a molecule on muon depolarization can be neglected.

However, the surrounding medium of the mesonic atom should have a very strong effect on P because of its influence on the first part of the mesonic-atom cascade through the Auger transitions, since the number of electrons participating in this process determines the continuous development of the cascade (de Borde, 1954; Burbidge and de Borde, 1953) to the state with principal quantum number n_0, at which point the radiative transitions become dominant and the second depolarization stage begins.

Whether the muon reaches the n_0 level unimpeded also depends on the structure of the chemical bonds of the molecule's atom, which was changed to a mesonic atom. If the molecule should prove to have an insufficient number of electrons to allow the muon to reach the n_0 level, depolarization may begin from a higher level and hence be stronger, at least for the light mesonic atoms (see Section V).

It should be noted that the Auger electrons, which have a low energy in the beginning of the cascade, remain near the mesonic atom and may shortly afterward be attracted by the positively charged ion of the mesonic atom (mesonic ion) and captured into the bound state. Therefore, the charge of the mesonic ion at each cascade stage will depend on the rate of the Auger process and on the recovery rate of the electron shell, which in turn depends on the properties of the medium.

Theoretically, the mesonic ion in a metastable state with $n > n_0$ can pass to lower states on colliding with the neighboring neutral or ionized atoms and molecules. This may also lead to redistribution of the level population and a change in depolarization owing to the Stark effect.

The electron shell begins to recover after the mesonic atom cascade. Since the charge of the muon–nucleus system is one unit lower than that of the nucleus of the preceding atom, the electron shell of the mesonic atom will have a structure analogous to that of the atom of an element whose atomic number is lower by one unit (Evseev, 1968). Thus, if a muon is captured by an oxygen atom, a mesonic atom with the electron shell of atomic nitrogen is produced; we shall henceforth call this mesonic nitrogen and denote it by μN.

The recovery time of a mesonic atom's electron shell should be compared with the characteristic time for muon–electron hyperfine interaction, since the depolarizing action of the paramagnetic electron shell of the mesonic atom is equal to

$$t_{\mathrm{hf}} = 1/\nu \tag{12}$$

The hyperfine splitting in the mesonic atom is given by (Sobelman, 1963; Frisch, 1963)

$$\nu = D(\mathrm{F}_{+}' + 1) \tag{13}$$

where D is the hyperfine structure constant, which is related to the hyperfine structure constant A of an ordinary atom, a chemical analog of the mesonic atom under consideration, in the following manner:

$$D = Ag_{\mu}/g(I) \tag{14}$$

Here $g(I)$ and g_{μ} are the gyromagnetic ratios for the atomic and muonic nucleus, expressed in identical units. Thus, to determine D for mesonic boron μB, produced through atomic capture of a muon by carbon, we use $g_{\mu} = 2M_{\mathrm{p}}/m_{\mu} = 17.7$ (M_{p} and m_{μ} are the masses of the proton and the μ^{-} meson, respectively); $g(I) = 2.69/\frac{3}{2} = 1.8$ (for the nucleus of the boron isotope ^{11}B, for which A was measured); and A is different for different states of atomic boron, basic ($^{2}P_{1/2}$) boron, and metastable ($^{2}P_{3/2}$) boron.

In the first case, $A_{1/2} = 3.66 \times 10^{8}$ sec^{-1}, and in the second, $A_{3/2} = 7.34 \times 10^{7}$ sec^{-1} (Kopferman, 1958; Wessel, 1953). Thus $\nu_{\mu}\mathrm{B} = (1.4\text{–}3.6) \times 10^{9}$ sec^{-1}. Analogously (Kopferman, 1958; Lew and Wessel, 1953), for mesonic aluminum, we get $\nu_{\mu}\mathrm{Al} = (0.5\text{–}1.8) \times 10^{10}$ sec^{-1}. Mesonic nitrogen μN, which has three 2p electrons (just as atomic nitrogen, its chemical analog), can be produced in the $^{4}S_{3/2}$ state (basic), $^{2}P_{3/2}$ state, and $^{2}D_{5/2}$ state (metastable). In the pure $^{4}S_{3/2}$ state $\nu \equiv 0$; however, because of the sp hybridization, a weak hyperfine splitting ($A \simeq 10^{7}$ sec^{-1}) because of a small admixture of the s wave is observed (Heald and Beringer, 1954). A much stronger splitting ($A \simeq 10^{8}$ sec^{-1}) is evident (Goudsmit, 1931) in the metastable doublet $^2P_{3/2}$ and $^2D_{5/2}$ states. For mesonic nitrogen $\bar{\nu}_\mu N \simeq 10^{10}$ sec^{-1}.

The exact values of ν for mesonic atoms can be obtained by measuring the muon polarization in strong longitudinal or transverse magnetic field.

Total recovery of the electron shell of a mesonic atom and the recovery time can be determined by taking into account the specific conditions under which the mesonic atom is produced. These conditions, different in conductors and dielectrics, will be examined in detail for each specific case. To determine the structure of the electron shell of a mesonic atom, either the density of the free electrons near a mesonic atom must be determined or the ionization potential of the atoms of the medium must be compared with the energy of electron affinity in the mesonic atom (Dzhuraev and Evseev, 1971). If the ionization potential is higher than the energy of electron affinity, for example, the last electron of a mesonic atom, it remains a singly charged, positive mesonic ion.

Let us assume that the mesonic atom has fully recovered its paramagnetic electron shell in a time much shorter than the operating time t_{hf} of the mechanism for additional depolarization due to the hyperfine interaction of the magnetic moments of the muon and the electron shell. Any change in its paramagnetism in a time close to t_{hf} should, of course, affect the residual polarization (Evseev, 1968; Dzhuraev and Evseev, 1971, 1973). The fast interactions of a mesonic atom, which change the muon polarization, are expected to be fairly close to those occurring in atomic muonium (Gurevich, 1972; Firsov and Byakov, 1964; Shantorovich and Firsov, 1967; Gol'danskii and Firsov, 1971). These principally are: the exchange of the valence electrons between the mesonic atom and the medium; muon spin flip resulting from the scattering by a paramagnetic center; compensation of the paramagnetism in a mesonic atom (and hence cessation of muon depolarization), for example, through its entry into the chemical reaction resulting in the formation of a diamagnetic compound; change in the hyperfine structure of the ground state of a mesonic atom through its entry into the chemical reaction that produces a paramagnetic chemical compound; and muon-spin relaxation in a diamagnetic mesonic atom as a result of the interaction with the nuclear magnetic moments of the neighboring atoms.

We shall henceforth consider the fast chemical reactions of mesonic atoms with the radicals produced near a mesonic atom as the principal reason for the change in the depolarization depending on the different parameters of the condensed molecular media.

The depolarization, of course, depends on the frequency used in measuring the precession: the frequency ω of the free muon spin or the frequencies Ω of the mesonic atoms, the muon–electron shell of a mesonic atom.

IV. Procedure for Measuring Depolarization of μ^- Mesons

The depolarization of μ^- mesons is primarily studied by the method of Larmor precession of muon spin in a weak transverse magnetic field (Garwin *et al.*, 1957). The angular distribution of the electrons from the μ^- decay is given by (Vaisenberg, 1964)

$$N_e(\theta) \sim 1 + a \cos \theta \tag{15}$$

where θ is the angle between the direction of the ejected electron and the direction of muon spin, and $P/3 \equiv a$ is the asymmetry coefficient.

The facility for studying the depolarization of μ^- mesons at the JINR after 1968 was also used for studying positive muons (see Babaev *et al.*, 1966). The same procedure modified for specific cases was used in the other μ^--meson depolarization studies. Thus, for example, the asymmetry of the angular distribution of decay electrons (Ignatenko *et al.*, 1958, 1961) was determined by counting the electrons during a certain period of time when the two transverse fields were equal in absolute value and opposite in direction. Practically all the other work involved determining the time distribution (after stopping the muons in the target) of decay electrons, which, in a very general way, can be written as follows:

$$\begin{aligned} N_e(t) &= N_1(0) \exp[-t/\tau_1][1 + a_1 \cos(\omega_1 t + \phi_1)] \\ &\qquad + N_2(0) \exp[-t/\tau_2][1 + a_2 \exp[-Rt] \cos(\omega_2 t + \phi_2)] + C \end{aligned} \tag{16}$$

If the muon is stopped in a medium with two kinds of atom, then the subscripts 1 and 2 denote these atoms, and $R \equiv 0$. Thus $N(0)$ represents the number of decay electrons recorded in a time interval whose midpoint coincides with the time the muon was stopped, τ is the muon lifetime in the given mesonic atom, a the asymmetry coefficient, ω the precession frequency, ϕ the initial precession phase, which depends on the geometry of the facility and the direction of muon spin in the beam, and C the random coincidence background. If τ_1 differs greatly from τ_2, the accuracy in determining the asymmetry coefficients a_1 and a_2 will differ since the transverse magnetic field $H_\perp$ is usually chosen so that optimum conditions can be obtained for measuring the asymmetry of the long-lived mesonic atom. However, if $H_\perp$ is sufficiently large, both parameters a_1 and a_2 theoretically can be measured concurrently. The problem is more complicated if $I \neq 0$ for one of the atoms and $\tau_1 < \tau_2$, since ω_1 (if precession is dominant in one of the states of the hyperfine structure $F_\pm = I \pm \frac{1}{2}$) is much smaller than ω_2 (for the mesonic atom with $I = 0$).

If τ_1 and τ_2 are nearly similar, as is the case for the neighboring light atoms, the time distribution of decay electrons in the time of 2 to 3 μsec can be described fairly well by a single exponential curve; the contribution from two kinds of atom can be determined in this case only by comparing the average lifetime $\bar{\tau}$ with the partial lifetimes measured for each atom separately, or from the precessional frequencies if one atom has $I \neq 0$.

If the muon depolarization is measured in both states $F_{\pm} = I \pm \frac{1}{2}$, the subscripts 1 and 2 in Eq. (16) refer to these states, $\tau_1 \equiv \tau_2$, and R is the transition rate between the states of the hyperfine structure.

If the polarization is measured from the precessional frequencies (let us assume there are two) of mesonic atoms Ω_1 and Ω_2 (precession of the total angular momentum of the muon and the electron shell of the mesonic atom), $H_\perp$ must be much smaller so that the frequencies with $\Omega \gg \omega$ can fall into the appropriate recording time interval corresponding to the muon lifetime in the given substance.

Some of the precessional frequencies can be determined for different values of ω (or Ω) by analyzing the data expressed as an equation with one exponential (Eq. 16). The given values of ω, corresponding to real frequencies, have a large coefficient a and a small value of χ^2. This method is analogous to the Fourier analysis (Babaev *et al.*, 1969; Buckle *et al.*, 1968).

Small asymmetry coefficients and large magnetic fields required for heavy elements show that μ^- mesons are more difficult to work with than positive muons.

V. Depolarization of Negative Muons in Conductors and Semiconductors

Targets consisting of conductors with spinless nuclei were first used for testing the theory of cascade polarization. It was noted in one of the first theoretical studies (Dzhrbashyan, 1958, 1959) that paramagnetism of the electron shell of a mesonic atom produced, for example, on stopping a muon in graphite leads to a depolarization twice as large as that observed experimentally. Absence of the electron-shell effect was attributed to the compensation of its paramagnetism. Similar views were expressed by Hutchinson *et al.* (1963), who determined the magnetic moment of a negative muon from the precessional frequency of its free spin. However, a concrete mechanism explaining the absence of the paramagnetic effect in the mesonic atom's electron shell in the light and intermediate elements was proposed by Evseev (1968). Conductors produce mesonic atoms basically because the high-conduction electron density facilitates the recovery of the mesonic atom's electron shell destroyed by Auger transitions. In most conductors investigated the mesonic atom has unpaired valence electrons (we shall not consider now the paramagnetism resulting from incomplete inner shells). The high rate of the exchange interaction between the mesonic atom's valence electron and the conduction electrons accounts for the maximum value of P in conductors. Since the mesonic atom's valence electron does not have a definite spin direction in conductors because of exchange interaction, a mesonic atom does not have an electron magnetic moment. The mechanism for the exchange interaction between conduction electrons was first proposed (Yakovleva, 1959; Nosov and Yakovleva, 1962) to account for the large value of P_{μ^+} measured in metals at the precessional frequency of the free spin of a positive muon.

Depolarization in conductors with spinless nuclei was measured by Ignatenko *et al.* (1958, 1961), Buckle *et al.* (1968), Dzhuraev *et al.* (1971a), Telegdi (1960), Culligan *et al.* (1961), Anderson (1965), Astbury *et al.* (1961), Sundelin (1967), Sundelin *et al.* (1968), and Evseev *et al.* (1961, 1966). The results of the measurements are summarized in Table 2. If the polarization in the elements with $Z \leq 20$ were, on the average, equal to that predicted by the theory of cascade depolarization, the situation with the transition metals would be more complicated. According to the data of Ignatenko *et al.* (1958, 1961) the residual polarization of Cr, Mo, Pd, and

TABLE 2

Metal	a/a_c	References
Mg	0.82 ± 0.07	Buckle *et al.* (1968)
	1.45 ± 0.27	Ignatenko *et al.* (1958, 1961)
	0.82 ± 0.16	Telegdi (1960), Lathrop *et al.* (1961)
	0.77 ± 0.18	Astbury *et al.* (1961)
	0.70 ± 0.05	Dzhuraev *et al.* (1971a)
Ca	0.76 ± 0.04	Anderson (1965)
	0.67 ± 0.03	Sundelin (1967), Sundelin *et al.* (1968)
	1.00 ± 0.16	Evseev *et al.* (1961)
	0.89 ± 0.11	Dzhuraev *et al.* (1971a)
Ti	0.80 ± 0.08	Dzhuraev *et al.* (1971a)
Cr	0.00 ± 0.22	Ignatenko *et al.* (1958, 1961)
Zn	1.45 ± 0.32	Ignatenko *et al.* (1958, 1961)
Mo	0.12 ± 0.25	Ignatenko *et al.* (1958, 1961)
Pd	0.00 ± 0.25	Ignatenko *et al.* (1958, 1961)
Cd	1.37 ± 0.34	Ignatenko *et al.* (1958, 1961)
W	0.12 ± 0.25	Ignatenko *et al.* (1958, 1961)
Pb	1.35 ± 0.35	Ignatenko *et al.* (1958, 1961)

TABLE 3

Metal	Contents in sample (%)	I	a_+/a_0	$a(I)/a(0)$	a_-/a_c	References
^{6}Li	95.9	1	0.452 ± 0.018	—	$<\|0.01\|$	Favart *et al.* (1970)
^{7}Li	99.7	$\frac{3}{2}$	0.455 ± 0.018	0.31	0.047 ± 0.013	Favart *et al.* (1970)
^{9}B	100	$\frac{3}{2}$	—	0.31	0.360 ± 0.021	Babaev *et al.* (1969)
		$(\mu_N < 0)$			0.376 ± 0.016	Favart *et al.* (1970)
^{27}A1	100	$\frac{5}{2}$	0.16 ± 0.16	—	—	Ignatenko *et al.* (1961)

W is zero, whereas the polarization of Ti (Dzhuraev *et al.*, 1971a), related to transition metals, is about the same as that of the neighboring metals with diamagnetic shells. The values of the first four transition metals can be explained (Ignatenko *et al.*, 1958, 1961) by the paramagnetism's depolarizing effect due to the incomplete inner shells of the mesonic atom whose electrons do not participate in the conduction and hence do not change the spin state because of the exchange interaction. However, this simple explanation does not hold for Ti. The mesonic atom μSc has an unpaired electron in the 3d inner shell and two electrons in the 4s diamagnetic outer valence shell, and the polarization at the frequency ω is large. This probably means that the atomic paramagnetism in the titanium lattice of magnetic scandium μSc is absent, i.e., s as well as d electrons participate in the exchange interaction between the conduction electrons, and that bonds with neighboring lattice atoms are established. However, additional experiments with transition metals must be performed to remove all ambiguity.

The depolarization of metals whose nuclei have $I \neq 0$ was investigated by Dzhrbashyan (1958, 1959), Winston and Telegdi (1961), Winston (1963), Hutchinson *et al.* (1962), Babaev *et al.* (1969), and Favart *et al.* (1970). The experimental data are given in Table 3. According to the theory (Überall, 1959; Bukhvostov, 1969; Libkin, 1960; Bukhvostov and Shmushkevich, 1961; Bukhvostov, 1966), the decrease of the residual polarization in ^{6}Li, ^{7}Li, and ^{9}Be is 2–3 times larger than in graphite at the precessional frequencies of the F_+ state, and the theory (Bukhvostov, 1966; Bukhvostov and Shmushkevich, 1961) predicting fairly low residual polarization in ^{7}Li and ^{9}Be in the F_- states and a change in direction of the muon spin in these mesonic atoms was confirmed (Favart *et al.*, 1970).

The results of an experimental study of the depolarization of μ^- mesons in

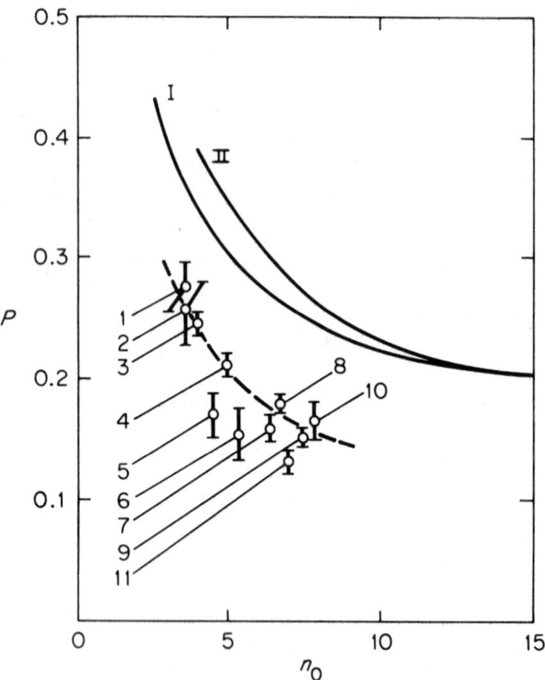

Fig. 1. Dependence of the residual polarization on the quantum number n_0 of the mesonic atom's level below which depolarization is considerable. Curve I was calculated by using the statistical population in l_0 of the level n_0, and curve II was calculated by using the population deduced from the higher cascade transitions in carbon. The arabic numerals correspond to the experimental data for the following substances: 1, average for ^6Li and ^4Li; 2, LiH; 3, Be; 4, graphite; 5, average for ^{10}B and ^{11}B; 6, N_2; 7, Mg; 8, Si; 9, Ca; 10, Cr; and 11, S.

semiconductors are given in Table 4. Since the crystal lattice of boron is diamagnetic and mesonic beryllium has a diamagnetic electron shell, we include in Fig. 1 the polarization in boron (averaged over two isotopes).

Dzhuraev *et al.* (1971a) used pure silicon (n type, 200 ohm/cm) and silicon with an admixture of 3×10^{18} cm^{-3} antimony atoms (n type, 0.01 ohm/cm) at different temperatures to determine the role of the exchange between the mesonic atom's valence electrons and the conduction electrons. Although the number of current carriers varied several orders of magnitude, the residual polarization remained large (average value relative to graphite 0.91 ± 0.03) and constant within $\simeq 10\%$ limit of experimental error. Since the depolarization of positive muons in semiconductors

TABLE 4

Semiconductor	I	a_+/a_c or a/a_c for $I = 0$	$a(I)/a(0)$	a_-/a_c	$R \times 10^5\ sec^{-1}$	References
^{10}B	3	0.193 ± 0.027	0.27	0.018 ± 0.013	2.1 ± 0.5	Favart *et al.* (1970)
^{11}B	$\frac{3}{2}$	0.284 ± 0.034	0.37	0.022 ± 0.014	3.3 ± 0.5	Favart *et al.* (1970)
	0	0.67 ± 0.17				Astbury *et al.* (1961)
^{28}Si		0.83 ± 0.03				Sundelin (1967), Sundelin *et al.* (1968)
Pure Si						Dzhuraev *et al.* (1971a)
$T = 300°K$	0	0.85 ± 0.08				
$T = 80°K$		0.90 ± 0.10				
Si with impurities						
$T = 300°K$	0	0.98 ± 0.06				
$T = 80°K$		0.90 ± 0.06				
		0.55 ± 0.12				Ignatenko *et al.* (1958, 1961)
^{31}P	$\frac{1}{2}$	0				Winston and Telegdi (1961), Winston (1963), Hutchinson *et al.* (1962)
		-0.035 ± 0.0133				Babaev *et al.* (1969)
CuO (in oxygen)						
$T = 300°K$	0	0.65 ± 0.10				Dzhuraev *et al.* (1971a)
$T = 80°K$		0.15 ± 0.07				
PbS (in sulfur)	0	1.27 ± 0.40				Dzhuraev *et al.* (1971a)

(Andrianov *et al.*, 1969) greatly depends on the amount of impurity and on the temperature, the exchange-interaction mechanism cannot be dominant in stopping the negative muons in silicon. Dzhuraev *et al.* (1971a) proposed another mechanism showing specific properties of the mesonic atoms in this medium. Stopping in silicon produces mesonic aluminum μAl, an acceptor atom for the target lattice. If the mesonic aluminum occupies the acceptor level, the paramagnetism of the electron shell is compensated immediately. A large residual polarization in Si can be explained by assuming that μAl occupies the level a time of $\leq 3 \times 10^{-9}$ sec, since the residual polarization has the characteristic operating time of the mechanism for paramagnetic depolarization due to the interaction of the muon spin with the magnetic moment of the unpaired electron; this leads to additional double (as compared with graphite) decrease of polarization.

Total depolarization in phosphorus (red and black) is determined (Winston and Telgedi, 1961; Winston, 1963; Hutchinson *et al.*, 1962; Babaev *et al.*, 1969) by the fast $F_+ \rightarrow F_-$ transitions.

Paramagnetic mesonic atoms (μN and μP) with similar structure of the valence zone are produced in two other semiconductors, CuO and PbS, in which the residual polarization in the light elements was measured. The mechanism for Si apparently can also be used for these semiconductors; however, the chemical reactions of mesonic atoms must be taken into account.

Large polarization in CuO at room temperature and the temperature dependence (see Table 4) are attributed to the presence of fast chemical reactions producing, for example, a diamagnetic (Selwood, 1966) compound CuμNO in a time much shorter than 10^{-10} sec (for μN).

Knowing the muon polarization in the beam (Dzhuraev *et al.*, 1971a), we can determine the absolute residual polarization in graphite ($P = 0.208 \pm 0.011$) and hence in all the other conductors, and compare more closely the experimental data with the theory of cascade depolarization. The results of such a comparison by Dzhuraev *et al.* (1971a) are shown in a slightly modified form in Fig. 1. Curve I was constructed according to Eq. (4) on the basis of the statistical population of l in all the levels, and curve II corresponds to the level population determined by calculating the mesonic-atom cascade in a carbon atom after the initial settling of the muon in the level $n = 14$ (Eisenberg and Kessler, 1961). The value of n_0 corresponds to the level of mesonic atom in which (n_0 is different for different Z) the Auger transition rate is 10^2 times higher than that of the radiative transitions (de Borde, 1954; Burbidge and de Borde, 1953) in circular orbits. The averaged data for metals and ^{7}LiH, B, and N_2 are shown in this figure. We shall consider later the reason for including the data on lithium hydride and nitrogen (dielectrics). The polarization of elements whose nuclei have $I \neq 0$ was increased according to the correction (Babaev *et al.*, 1969) for additional depolarization due to hyperfine interaction between the magnetic moments of the muon and the nucleus in the ground state of a mesonic atom (pertinent correction factors are listed in column 5 of Table 3 and column 4 of Table 4). The polarization is somewhat smaller if the hyperfine splitting of the excited states (Favart *et al.*, 1970; Bukhvostov and Shmushkevich, 1961; Bukhvostov, 1966) is taken into account. It follows from the data of Fig. 1 that if the dependence of the measured polarization on n_0 is taken into account, the theory of cascade depolarization clearly does not agree with the experiment: additional depolarization occurs at the given n_0. This depolarization can be accounted for by assuming that the electron shell of the mesonic atoms of all the metals is strongly ionized during mesonic-atom cascade.

This is indirectly confirmed by the initial statistical distribution in l, which satisfactorily describes (Suzuki, 1967) the structure of the K series of the light and intermediate mesonic atoms; however, the strongly ionized shells, for example, in calcium should have only two electrons (during the decay) in the s and p shells, and only one electron in each shell in magnesium.

If only one electron is present in the K electron shell of a mesonic atom during the cascade (Suzuki, 1967), then, assuming that the Auger transition rate decreases by two orders of magnitude, additional depolarization through the hyperfine interaction between the muon spin and the 1s electron would be possible in the levels near n_0. Such suppression of Auger transitions in metals is possible because of fairly long relaxation time of electrons, of the order of 10^{-13} sec (Wert and Thompson, 1964) (Auger transition probability is of the order of 10^{16} sec^{-1}) (de Borde, 1954; Burbidge and de Borde, 1953). The rate of an Auger transition accompanied by direct knocking out of a conduction electron is very low (Primakoff, 1959). This complementary mechanism, however, is not considered unique, since P would increase several dozen percent with decreasing temperature (the relaxation time in metals decreases with decreasing temperature). The experiment of Dzhuraev *et al.* (1971a) shows no evidence of a temperature dependence in graphite.

Since P (Fig. 1) is noticeably smaller for boron and molecular nitrogen than the neighboring substances with spinless nuclei, additional depolarization because of the complex interaction between $\mathbf{I}$, $\boldsymbol{\sigma}$, and $\mathbf{J}$ occurs in the semicollapsed electron shell during the mesonic-atom cascade (Tammet, 1967, 1969a,b,c, 1970). The depolarization effect of the spin–spin relaxation should also be taken into account.

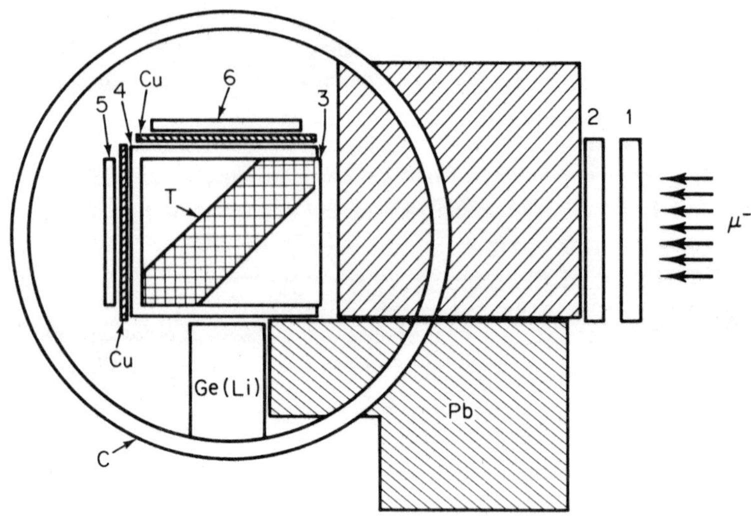

FIG. 2. Arrangement of equipment for the muon beam experiment. 1–6, plastic scintillators; 1234, telescope for separating stopped muons; 45 and 46, telescopes for recording μ-decay electrons; T, target; C, Helmholtz coils; Cu, 4-mm-thick copper filters; Ge(Li), mesonic X-ray detector with a 27-mm³ sensitive volume; dimensions of counters 1 and 2, $10 \times 10 \times 1$ cm³.

The depolarization of the negatively charged muons stopped in a substance and the structure of μ-mesonic X-ray spectra have heretofore been studied independently. There is, however, a connection between these phenomena, since the main muon depolarization occurs in the part of the mesonic-atom cascade in which the radiative transitions are dominant.

A correlation between the residual muon polarization in the K shell of a mesonic atom and the structure of the energy spectrum of mesonic X rays can be obtained, for example, by measuring the degree of depolarization of muons traveling different paths in the mesonic-atom cascade; the theory of cascade depolarization can be examined in greater detail in this experiment.

Arlt *et al.* (1973) measured depolarization by the method of muon spin precession in weak ($H \simeq 50$ G) transverse magnetic field and isolated the stopped negative muons accompanying the emission of the discrete lines of the K series of mesonic X-ray radiation of a carbon atom in graphite and paraffin. The experimental arrangement is shown in Fig. 2. The mesonic X rays were recorded by Ge(Li) spectrometer with a 27-cm³ crystal and an energy resolution of nearly 2.2 keV at $E_X = 100$ keV. The electrons from the μ decay were recorded by two telescopes. The time distribution of signals from the electron telescopes was measured relative to the μ stopping signal, which accompanied the recording of the mesonic X-ray quantum. The detector separated four energy regions in the spectrum corresponding to the np → 1s transitions: 2p → 1s, 3p → 1s, 4p → 1s, and (≥5p) → 1s; the dimensions and shape of the graphite target are shown in Fig. 2 (the dimension in the direction perpendicular to the plane of the figure is 8 cm), and the paraffin target occupied the entire space between counters 3 and 4.

By using the absolute residual polarization of muons in graphite measured elsewhere (Dzhuraev *et al.*, 1971a), we can express in absolute scale the data for different lines of the K series, since the asymmetry coefficient in the angular distribution of the electrons from the μ decay was measured under the same conditions as in the coincidence of mesonic X-ray quanta and the integral value for all the μ stops.

Experimental dependence of the residual polarization P on the mesonic X-ray transition in graphite and paraffin is compared in Fig. 3 with the results of a calculation based on the theory of cascade depolarization. It was shown elsewhere (Dzhuraev *et al.*, 1971a) (see Fig. 1) that the residual polarization P in all the conductors, including graphite, is approximately one-and-a-half times smaller than that predicted by theory. The data in Fig. 3 show that additional depolarization in graphite because of the

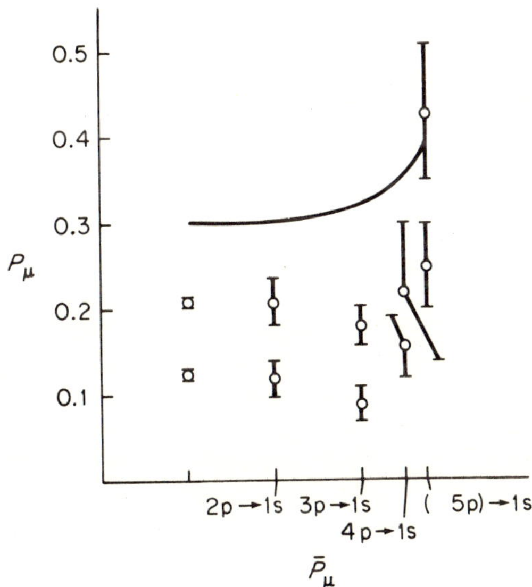

Fig. 3. Dependence of the residual polarization of muons on the mesonic-atom transition (○, graphite; ●, paraffin; solid curve, calculation based on the theory of depolarization in a mesonic-atom cascade).

paramagnetism of the mesonic atom's electron shell partially destroyed by Auger transitions during the cascade is present in the mesonic-atom level with $n = 4$, since the $(\geq 5p) \rightarrow 1s$ transition is accompanied by a depolarization nearly as large as the calculated. The behavior of P vs n determined experimentally for the transitions from the levels with $n \leq 4$ is consistent with the predictions of the theory of cascade depolarization.

The dependence of P on the energy of the mesonic-atom transition in paraffin is generally the same as in graphite. This means that in the first approximation the difference in the values of P for graphite and paraffin should be attributed to factors other than the difference in the muon cascades (the structure of the spectrum for these two substances is different). According to the model (Evseev, 1968; Dzhuraev and Evseev, 1972, 1973), which takes into account the chemical interaction of mesonic atoms after the cascade, this difference is attributed to additional depolarization in paraffin because of the paramagnetism of the valence zone of the mesonic atom's electron shell and the competition between this depolarization and the rate of entry of a mesonic atom into the chemical reaction resulting in the formation of diamagnetic compounds.

By increasing the accuracy of the experiment, we can estimate more accurately the increase of P in the region $n \simeq 4$ to 5 and determine whether the additional depolarization in graphite caused by the paramagnetism of the mesonic atom's electron shell in the muon cascade is different from that in paraffin.

VI. Depolarization of Negative Muons in Inert Gases

Experiments show that total depolarization in liquid helium occurs at the precessional frequency ω of free muon spin (Prepost *et al.*, 1960; Kane, 1964) and at the precessional frequency Ω of the triplet state of mesonic hydrogen (μH) (McColm *et al.*, 1960; Buckle *et al.*, 1968). Zero residual polarization in liquid argon (Prepost *et al.*, 1960) was measured at the frequency ω [for liquid helium $a/a_c = -0.02 \pm 0.08$ (Kane, 1964), 0.04 ± 0.11 (Buckle *et al.*, 1968), and 0.44 ± 0.32 (Prepost *et al.*, 1960), and for liquid argon $a/a_c = 0.17 \pm 0.12$ (Kane, 1964), where a and a_c are the asymmetry coefficients of the substance and graphite].

Total depolarization of completely pure inert gas should occur at the frequency ω if the mesonic atom has a paramagnetic electron shell (Evseev, 1968) (mesonic hydrogen or mesonic halogen). Because of the chemical inertness of the medium surrounding the mesonic atom, this state remains until the muon decays, i.e., precession occurs only at the frequency Ω_i.

Absence of polarization in the triplet state of mesonic hydrogen at the frequency Ω is attributed (Evseev, 1968) theoretically to the conversion of the triplet state to the singlet state (with the consequent fast depolarization) through the collision between a mesonic atom and the molecules of the paramagnetic impurities. If the cross section of the triplet–singlet conversion for the hydrogen atom (and hence the mesonic hydrogen) in O_2 and NO is (Mokrushin and Gol'danskii, 1967) $\sigma_c \simeq 2 \times 10^{-15}$ cm^2 and the average relative collision rate in liquid helium is $v_{He} = 3.5 \times 10^4$ cm/sec, then the conversion rate γ_c is given by

$$\gamma_c = n_c \sigma_c v_{He} \tag{17}$$

where n_c is the number of molecules per cm^3 of the converter. Since a small amount of O_2 or NO impurity ($n_c \simeq 2 \times 10^{17}$, which is 10^{-5} of the number of helium atoms) in liquid helium can lead to a total muon depolarization in 10^{-7} sec, the concentration of the paramagnetic molecules in this type of experiment should be measured directly in the working volume of the target.

The discovery of the mesonic atom's precessional frequencies in highly purified gaseous neon was reported at the Fourth International Conference on High Energy Physics and Nuclear Structure (Gurevich, 1972; Varlamov *et al.*, 1971). These frequencies correspond to the closed electron shell of neutral mesonic fluorine μF. Gaseous He and Ne in a 60-G field have zero polarization.

VII. Formal Theory of Depolarization of μ^- Mesons in Condensed Molecular Media

A. Specific Properties of Molecular Media

Dzhuraev *et al.* (1973) used water to illustrate depolarization of μ^- mesons in molecular media. Since the fundamental process here is the formation of a free radical zone around the mesonic atom, it should be studied carefully beginning with the development of the mesonic X-ray cascade.

Numerous authors proposed different mechanisms for the transition of a μ^- meson to the bound states (Mann and Rose, 1961; Au-Yang and Cohen, 1968; Gershtein *et al.*, 1969).

According to the hypothesis of Mann and Rose (1961), atomic capture of a muon occurs through an Auger process in the region of maximum electron density, i.e., the mesonic atom level of $n \simeq$ 14–15, on the average. However, this occurrence in higher or lower lying levels is also possible.

If only thermalized muons are captured, as in the model for large mesonic molecules (Gershtein *et al.*, 1969; Varlamov, 1971), the probability of finding a molecule in any region is proportional to the electron density in the region. It follows that, for example, 20% of all the muons in a water molecule populate the valence electron region, i.e., the level with $n > n'$ (Zinov *et al.*, 1967):

$$n' = \left[\frac{Z_1 L}{2(Z_2/Z_1)^{1/2} + 1}\right]^{1/2} \tag{18}$$

where L is the internuclear spacing in units of meson Bohr radius; for water, $Z_1 = 8$, $Z_2 = 1$, $L \simeq 400$, and $n' = 28$. In the other cases, the muon settles immediately in the separated oxygen orbits for which $n \leq 28$. We shall henceforth consider only these cases.

If the initial population occurs in $n \leq 14$, then nine electrons (the tenth is replaced by a muon) in the H_2O molecule participate in the first part of the cascade; the muon reaches the level n_0 determined by the theory of cascade depolarization. Since the radiative dipole transition probability decreases with increasing n ($\Gamma_X \sim n^{-3}$), and the Auger transition probability Γ_a increases with increasing n, $\Gamma_a \gg \Gamma_X$ if n of the mesonic atom is large. According to the estimates of de Borde and Burbidge (1953; de Borde, 1954), $\Gamma_a/\Gamma_X > 10^5$ for $Z \leq 10$ and $n \geq 8$, and the Auger transition rate is of the order of 10^{16} sec^{-1}.

If the levels with $n > 14$ are populated, the transitions will be delayed significantly, since the neighboring levels in this region, after one or two Auger transitions, no longer seem to have sufficient energy to extract additional electrons. As indicated by de Borde and Burbidge (1953; de Borde, 1954), the recombination processes of a mesonic atom ionized by Auger transitions in the collisions with atoms or molecules of the surrounding medium are dominant. The recombination processes and Auger transitions can be compared with the collision time of the molecules or atoms at the specified temperature. Specifically, for water this time (of the order of 10^{-13} sec) is much shorter than Γ_X^{-1}, and depolarization does not occur in the region n under consideration. Experimental proof of the validity of this assumption is the absence in the μ-mesonic X-ray spectrum of high-energy transitions in water. An analysis of the data (Backenstoss *et al.*, 1967; Kessler *et al.*, 1967) shows that the intensity of the np $\rightarrow$ 1s transitions for $n > 7$ does not exceed 0.5%.

The average charge of the mesonic ion μN is small before the meson reaches the level $n \simeq 15$. However, the transitions from this level to $n = n_0$ occur so fast [approximately 10^{-15} sec (Vaisenberg, 1964; de Borde, 1954; Burbidge and de Borde, 1953)] that the emitted Auger electrons do not have time to return. This should produce in the place occupied by the water molecule a system of three positively charged centers that experience strong Coulomb repulsion.

This occurs, for example, when a vacancy is created in the K shell of iodine in the CH_3I molecule; experimental evidence (Karlson and White, 1965, 1970) of C, H, and I ions with charge and kinetic energy can be accounted for by relatively fast and complete destruction (because of autoionization) of the valence zone of all the atoms of a molecule. The average energy of the ions corresponds to the Coulomb repulsion energy (the "Coulomb explosion" of a molecule). The destruction time of the valence zone is assumed to be much shorter than that of separation of the ions, i.e., 10^{-15} sec, which means that the "mobility" of the valence electrons in the molecule is very high. The "Coulomb explosion" effect has been observed in many molecules (see Karlson and White, 1965, 1970).

On the basis of the standard model for a water molecule (see, for example, Bloch, 1969), the energy of "Coulomb explosion" can be easily calculated by removing all the electrons from the molecule. The maximum value of this energy is 220 eV, of which the meson receives only 7 eV and the rest is divided between two protons that fly apart at an angle close to π. As a result of slowing down, the protons establish a zone of radius ~10–20 Å, which is saturated by H and OH radicals. A rough estimate based on Libby's mechanism (Libby, 1947) shows that about 20 pairs of H and OH radicals are produced by slowing down the protons. The Coulomb repulsion in the large molecules diminishes as a result of the distribution of the electric charge over the entire molecule.

The emitted Auger electrons produce "spurs" isotropically distributed around the mesonic atom (Dzhuraev and Evseev, 1973; Dienes and Vineyard, 1957); each spur has up to six pairs of H and OH radicals (Vereshchinskii and Pikaev, 1963). If the main contribution comes from the electrons emitted following the transitions with $n < 14$ and having a *total energy* of about 5 keV and *mean free path* (Cole, 1969a,b) of the order of 150 Å, roughly 10^2 pairs of radicals will be produced in an area of this radius. The radiolysis of water produced by a muon track prior to atomic capture is difficult to determine, since the muon's average kinetic energy at which capture occurs is not known exactly. The absorption of the soft photons of the mesonic X rays near a mesonic atom can also contribute to radical formation.

The free radical zone is produced in about 3×10^{-13} to 3×10^{-12} sec, determined primarily by the slowing-down time of protons. We are basically interested in the radical density of the chemical reaction region of a mesonic atom. We shall discuss this problem below.

Thermalization of a mesonic atom, which acquires an electron shell by slowing down through the collisions with the surrounding molecules, occurs concurrently with the radical formation. The recombination energy is released by gamma radiation through the impact mechanism in the time of 10^{-13} sec, determined by the collision frequency between the mesonic atom and the water molecules. Total thermalization of the 7-eV mesonic atom occurs in less than 7 to 8 collisions, i.e., $\lesssim 3 \times 10^{-12}$ sec. If mesonic nitrogen is produced in the ground state $1s^2 2s^2 2p^3$, its term is $^4S_{3/2}$. However, mesonic nitrogen can also be produced in the $^2S_{3/2}$ and $^2D_{5/2}$ metastable states. To account for the initial paramagnetism in the mesonic nitrogen, we must assume that the probability of its occurrence in these states is high, since there is no hyperfine splitting in the $^4S_{3/2}$ state (Heald and Beringer, 1954). The experiments (Dzhuraev *et al.*, 1971b) determining the depolarization in water as a function of $H_\perp$ (Dzhuraev *et al.*, 1971b) (see Section IX,D) favor this assumption.

There are two basic depolarization mechanisms for the precession at the frequency ω:

(a) If we ignore further interaction of the mesonic nitrogen with the surrounding medium (mainly with the produced radicals), which compensates the paramagnetism in its shell, then further muon depolarization through the interaction with the magnetic moment of the electron shell is possible. The residual polarization P can be determined, with due allowance for this mechanism, by using Eq. (9). Q varies numerically from $\frac{1}{2}$ to $\frac{1}{3}$. The transition frequency between the states of the hyperfine structure can be determined for the metastable states of the mesonic nitrogen from the data of Sobel'man (1963) and Heald and Beringer (1954); $\nu_{\mu N} \simeq 10^{+10}$ sec^{-1} [see Eqs. (13) and (14)].

(b) Since the residual polarization is usually determined from the precession of free muon spin in the transverse magnetic field, the apparent depolarization determined by the precession of the total magnetic moment of the electron shell and the muon should also be taken into account. The frequency Ω_i of this precession is approximately two orders of magnitude higher than the precessional frequency ω of the free muon spin. In the field $H_\perp = 50$ G, $\omega \simeq 5 \times 10^6$ rad sec^{-1} and the mesonic nitrogen in the $^2P_{3/2}$ and $^2D_{5/2}$ states has a number of frequencies with the average value $\bar{\Omega} \simeq 1.3 \times 10^9$ rad sec^{-1}.

On the basis of the physical meaning of the parameter Q, we can assume that the depolarization, determined by the interaction of the total magnetic moment of the muon and the paramagnetic electron shell with an external magnetic field, occurs in Q cases, whereas in the other cases $(1 - Q)$ the depolarization is determined by the concurrent effect of both depolarization mechanisms mentioned above.

If the mesonic nitrogen is free when the measurements are taken ($\simeq 10^{-7}$ sec), the apparatus tuned to the frequency ω will show total depolarization.

B. Formal Theory of Depolarization in Water and in Other Molecular Media

The reaction of a mesonic atom with the radicals accounts (Dzhuraev and Evseev, 1971) for the fact that the residual polarization in water measured at the frequency ω is nonvanishing (Buckle *et al.*, 1968; Evseev *et al.*, 1968).

The reaction of mesonic nitrogen with the radicals occurs under the condition of recombination and diffusion of radicals in the reaction volume. A quantitative analysis of these effects by means of diffusion kinetics is rather complex. As is well known (Vereshchinskii and Pikaev, 1963), because of a large concentration of radicals in the track of a heavy particle (this example is ideally suited for simulating conditions of chemical interaction between μN and the radicals), the reaction between H and OH, which produces molecular products, is fairly short. For example, the recombination processes in water (Dyne and Kennedy, 1960) are significant only in the range of 10^{-11} to 10^{-10} sec, whereas the concentration of free radicals and molecular products is practically constant in the range of 10^{-10} to 10^{-8} sec. This is because the diffusion of the radicals and the reduction of their density occurs at $\sim 10^{-10}$ sec. We shall therefore assume that the concentration of the radicals and molecular products is independent of time when the reaction of thermalized mesonic nitrogen is dominant. We shall further assume that in the first approximation the H and OH concentrations are also independent of temperature. It is important to note that the formalism discussed below is valid only if the first collision between a mesonic atom and the radical produces a chemical compound. Otherwise special account must be taken of depolarization due to scattering by a paramagnetic center (see Section VI).

The rate of entry of free mesonic nitrogen into chemical reaction can be expressed as follows:

$$-\frac{\partial}{\partial t}[\mu\text{N}] = [\mu\text{N}]_0\lambda_0 e^{-\lambda_0 t} \tag{19}$$

where $[\mu\text{N}]_0$ and $[\mu\text{N}]$ are the initial mesonic nitrogen concentration and the concentration after time t, respectively, and λ_0 is the rate of any chemical reaction.

The reaction rate in this case is determined by the reaction probability W_N. For normalization we shall consider the reactions occurring only in the reaction volume V. Assuming that this volume contains only one μN atom, i.e.,

$$[\mu N]_0 = 1/V \tag{20}$$

the probability that the mesonic nitrogen will enter into the chemical reaction at time t can be written in the form:

$$W_N(t) = \lambda_0 e^{-\lambda_0 t} \tag{21}$$

As noted earlier, the precession at the frequency ω occurs only when the mesonic nitrogen produces a diamagnetic compound. The probability $W_N{}^d$ for the production of a diamagnetic product at time t is given by

$$W_N{}^d(t) = \lambda e^{-\lambda_0 t} \tag{22}$$

where λ is the reciprocal of the time the mesonic nitrogen enters into the chemical reaction resulting in the formation of a diamagnetic compound.

Let us now consider the effect of the chemical reaction of a mesonic atom on muon depolarization. The time distribution of decay electrons [see Eq. (16)] was determined by the precession of the muon spin. The second term in each parenthesis [denoted by $x(t)$] represents the precession of the vector whose direction is determined by the magnetic moment of the muon, and the absolute value is equal to the asymmetry coefficient of the decay electrons. We obtained B by averaging (over the time of entry of the mesonic nitrogen into the chemical reaction) the asymmetry coefficient B_0 corresponding to the polarization P_0, and the expression for $x(t)$ for the apparent paramagnetic depolarization has the form

$$\chi_1(t) = B_0 \overline{\cos(\omega t + \Omega t_1)} \tag{23}$$

where t_1 is the time the mesonic nitrogen enters into the chemical reaction that produces a diamagnetic product. Thus the term Ωt_1 represents the initial phase of the muon precession for each particular event of the chemical reaction of mesonic nitrogen. The average value is given by

$$\chi_1(t) = \frac{\lambda}{\lambda_0} \int_0^{T_\mu} \cos(\omega t + \Omega t_1) W_N(t_1)\, dt_1 \Big/ \int_0^{T_\mu} W_N(t_1)\, dt_1 \tag{24}$$

where T_μ is the initial observation time of the decay electrons and t the current observation time. Assuming that there is no free mesonic nitrogen when observations begin, i.e., $\lambda_0 T_\mu \gg 1$, we obtain

$$\chi_1(t) = B_1 \cos(\omega t + \phi_0) \tag{25}$$

where

$$B_1 = B_0 \frac{\lambda}{\lambda_0}\left[1 + \left(\frac{\Omega}{\lambda_0}\right)^2\right]^{-1/2}, \qquad \text{and} \qquad \phi_0 = \arctan \frac{\Omega}{\lambda_0} \tag{26}$$

respectively, are the asymmetry coefficient and the initial precession of the free (freed by chemical reaction) meson spin with the frequency ω, which correspond to the transverse magnetic field $H_\perp$. As shown above, this mechanism for the apparent paramagnetic depolarization is effective in Q cases (for metastable mesonic nitrogen $Q \approx 0.35$–0.50). The analysis of the experimental data (see Sections VIII and IX) is practically independent of the choice of Q.

For the remaining $1 - Q$ cases in which the depolarization is determined by the interaction between the total magnetic moment and the external magnetic field and by the hyperfine interaction between the magnetic moment of a muon and the electron shell of a mesonic radical, the precession of the magnetic moment of a muon can be written as follows:

$$\chi_2(t) = B_0 \cos[\omega t + f(t_1/\nu)\pi + \Omega t_1] \tag{27}$$

where $f(t_1/\nu)$ is the function for the frequency of transitions between the levels of the hyperfine structure ν and the time t_1 of entry into the chemical reaction producing a diamagnetic product. The function $f(t_1/\nu)$ is such that if the integer part of t_1/ν is even, $f(t_1/\nu) = +1$ and if it is odd, $f(t_1/\nu) = -1$. Thus the precession of free muon spin can be written as follows:

$$\chi_2(t) = B_2 \cos \omega t \tag{28}$$

where

$$B_2 = B_0 \frac{\lambda}{\lambda_0} \frac{\cosh(\lambda_0/\nu) - \cos(\Omega/\nu)}{\cosh(\lambda_0/\nu) + \cos(\Omega/\nu)} \tag{29}$$

is the asymmetry coefficient of the decay electrons in the depolarization resulting from hyperfine interaction occurring in $1 - Q$ cases.

Because of the participation of both mechanisms, we observe the total precession with amplitude B:

$$\begin{aligned} B &= \{(QB_1)^2 + [(1 - Q)B_2]^2 + 2[Q(1 - Q)B_1B_2] \cos B_1 \cdot B_2\}^{1/2} \\ &= B_0 \frac{\lambda}{\lambda_0}\left\{\left[\frac{\lambda_0^2}{\lambda_0^2 + \Omega^2}\right]\left[Q^2 + (1 - Q)^2 \frac{\cosh(\lambda_0/\nu) - \cos(\Omega/\nu)}{\cosh(\lambda_0/\nu) + \cos(\Omega/\nu)}\right.\right. \\ &\qquad \left.\left. + 2Q(1 - Q) \frac{\tanh(\lambda_0/\nu)}{\cosh(\lambda_0/\nu) + \cos(\Omega/\nu)}\right]\right\} \end{aligned} \tag{30}$$

and initial phase given by

$$\tan \phi = \frac{QB_1 \sin \phi_0}{(1 - Q)B_2 + QB_1 \cos \phi_0} \tag{31}$$

For $\Omega \ll \nu$ and $\lambda_0 \gg \Omega$, we obtain from Eq. (30) the following relation taken from Dzhuraev and Evseev (1971):

$$B = B_0 \frac{\lambda}{\lambda_0} \left[Q + (1 - Q) \tanh \frac{\lambda_0}{2\nu} \right] \tag{32}$$

Expression (30) can be used in a quantitative analysis of the experimental data on depolarization of negative muons. Examples of this are given in the following sections.

The dependence [Eq. (30)] for mesonic nitrogen (oxygen-containing substances) and mesonic boron (carbon-containing substances) for different precession fields $H_\perp$ is shown in Fig. 4.

VIII. Depolarization of μ^- Mesons in Organic Compounds

Although one of the first papers on depolarization of μ^- mesons (Ignatenko *et al.*, 1958, 1961) found no evidence of a difference in residual polarization in metals on the one hand, and hydrocarbons and water on the other, other papers (Buckle *et al.*, 1968; Evseev *et al.*, 1968) published concurrently firmly rejected this conclusion (see Table 5). The residual polarization in hydrocarbons proved to be about half of that in graphite. This problem was analyzed by Evseev (1968) on the basis of the interaction of mesonic boron (μB) with the medium. However, more recent findings (Dzhuraev and Evseev, 1971, 1973) show that the behavior of mesonic boron is very similar to that of mesonic nitrogen.

A. Connection between the Residual Polarization of Muons and the Radiation Resistance of a Substance

One of the physicochemical characteristics of a substance, which has, as we shall see, a bearing on the depolarization of negative muons, is its radiation stability. In radiation chemistry a quantitative analysis of this characteristic uses the value $G_{\mathrm{R}}\frac{1}{100}$ eV, which represents the number of stabilized radicals formed by absorbing a 100-eV dose of ionizing radiation (Pshezhetskii, 1962; Truby *et al.*, 1965). To suppress the recombination

TABLE 5

Substance	a/a_c	References
Styrene	0.54 ± 0.03	Evseev *et al.* (1968)
Polystyrene	0.32 ± 0.04	Evseev *et al.* (1968)
	1.05 ± 0.25	Ignatenko *et al.* (1958)
Polyethylene	0.45 ± 0.09	Buckle *et al.* (1968)
	0.52 ± 0.04	Evseev *et al.* (1968)
	0.95 ± 0.24	Ignatenko *et al.* (1958)
Paraffin	0.56 ± 0.09	Buckle *et al.* (1968)
	0.50 ± 0.05	Evseev *et al.* (1968)
Cyclohexane	0.51 ± 0.04	Evseev *et al.* (1968)
	0.45 ± 0.05	Buckle *et al.* (1968)
Toluene	0.30 ± 0.06	Evseev *et al.* (1968)
Benzene	0.41 ± 0.05	Evseev *et al.* (1968)
Phenyl cyclohexane	0.43 ± 0.04	Evseev *et al.* (1968)
Polyvinyl	0.33 ± 0.09	Buckle *et al.* (1968)

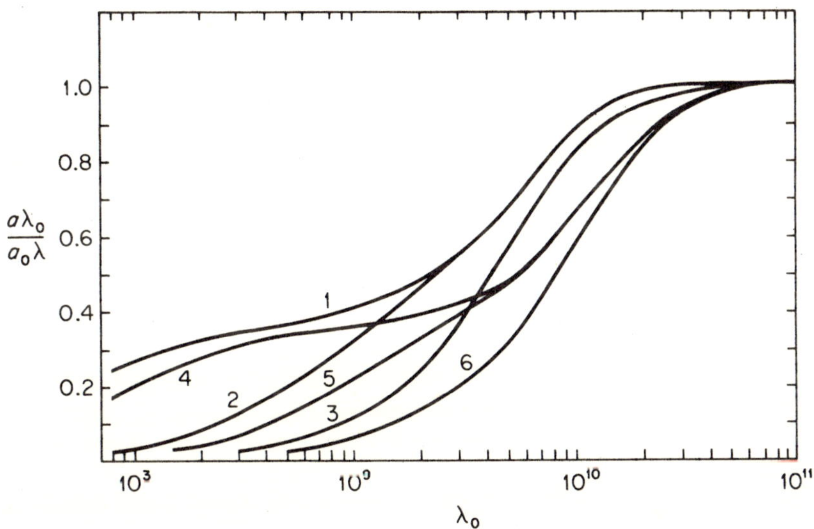

FIG. 4. Dependence of $a\lambda_0/a_0\lambda$ on λ_0. For mesonic boron (curves 1–3) $\bar{\Omega} = 7.3 \times 10^8$ rad sec^{-1}, $\bar{\nu} = 4.6 \times 10^9$ sec^{-1} (1, $H_\perp = 5$ G; 2, $H_\perp = 50$ G; 3, $H_\perp = 500$ G); for mesonic nitrogen $\bar{\Omega} = 1.3 \times 10^9$ rad sec^{-1}, $\bar{\nu} \simeq 10^{10}$ sec^{-1} (4, $H_\perp = 5$ G; 5, $H_\perp = 50$ G; 6, $H_\perp = 250$ G).

of radicals, experiments to determine G_R at low temperature should be performed.

In the radiolysis, for example, of hydrocarbons in the liquid and solid phases, primary radicals are known (Pshezhetskii, 1962) to be formed mainly as a result of breaking the C—H bond, since the hydrogen atom can easily leave the Franck–Rabinowitsch cage (Franck and Rabinowitsch, 1934). The hydrogen atoms can enter into chemical reactions even at liquid nitrogen temperature—recombine, tear away from the molecules of the medium, or combine with them in the case of compounds with multiple bonds. Thus, in the radiolysis of hydrocarbons radicals are formed in the following channels (Nauka, 1972):

$$\begin{aligned} &RH \rightsquigarrow (RH)^* \rightarrow R^\circ + H^\circ && I' \\ &RH + H^\circ \rightarrow R_1^\circ + H_2 && I'' \\ &H^\circ + H^\circ \rightarrow H_2 && I''' \\ &RH + H^\circ \rightarrow RH_2^\circ && I''' \end{aligned} \qquad (33)$$

As shown in Section VII,A, production of a mesonic atom in the ground state must be accompanied by radiolysis of the surrounding medium. It seems that the radiolysis of all hydrocarbons is the same, since muons settle principally on carbon atoms. The number of radicals formed near a mesonic atom determines the rate of its entry into the chemical reaction and hence the degree of depolarization.

Numerous attempts were made to determine G_R; references to the main literature and values of G_R for a large number of organic compounds are given in Nauka (1972). The values of G_R in this paper were obtained by radiolysis of substances by an ~1.6-MeV electron beam at a temperature of 180–150°C. Only fairly heavy radicals with relatively small diffusion coefficients can be stabilized at this temperature (and hence contribute to G_R).

Table 6 gives the experimental data for substances with known G_R and relative (to graphite) asymmetry coefficients in the angular distribution of μ-decay electrons a/a_c, which are proportional to the relative residual polarization.

The compounds listed in Table 6 are broken down into two groups according to G_R: aromatic compounds with $G_R < 1$, and compounds with saturated bonds having $G_R \geq 1$. Molin *et al.* (1962) attribute this difference to the difference in the energy state of the first level of electron excitation of the molecule. Since in the compounds with multiple bonds this level lies below the C—H binding energy, the probability of dissociation of the longest-lived first excited level diminishes markedly compared with that of molecules having saturated bonds, for which the first excited state lies above the dissociation energy of the C—H bond.

The data of Table 6 show direct correlation between G_R and a/a_c: residual muon polarization depends on the magnitude of G_R. We can choose from the cited compounds those that, because of radiolysis, produce radicals reacting similarly with mesonic boron (μB). These compounds are aromatic hydrocarbons and their derivatives, except for anisole [whose probability for atomic capture of a muon into the OCH_3 group is anomalously high ($\geq 30\%$)] (Dzhuraev and Evseev, 1973), cyclohexane, and polyethylene. Phenol also belongs to this group, since for this compound the probability of capture by an oxygen atom is $\leq 7\%$. In addition to atomic hydrogen, all these compounds produce radicals large enough to prevent migration from the zone of chemical activity of mesonic boron (Dzhuraev *et al.*, 1972).

The experimental data on a/a_c and G_R for these compounds are compared empirically in Fig. 5 (the errors for G_R, equal to 20%, are not shown in the figure). The smooth curve was obtained by analyzing the experimental data by the least-squares method and with use of Eq. (30), assuming that

$$\lambda_0 = \Lambda G_R \tag{34}$$

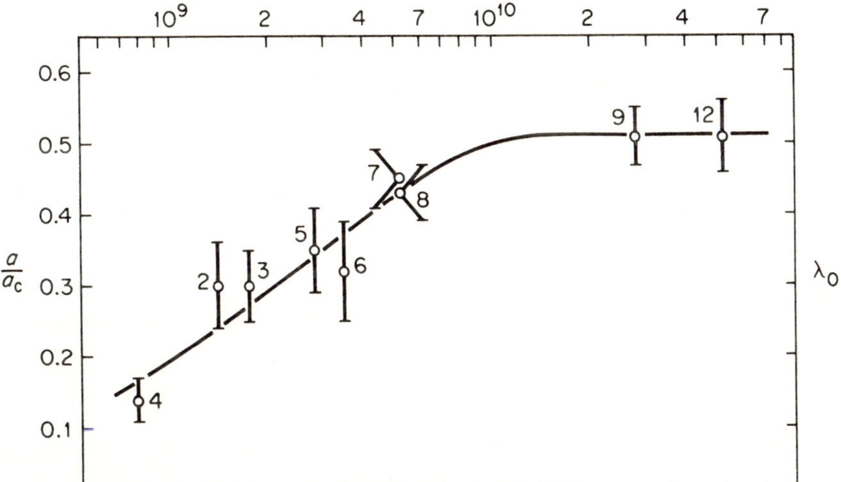

FIG. 5. Dependence of the degree of depolarization of negative muons on G_R. The numbers denote the compounds from Table 6.

TABLE 6

Number	Substance	Structure	a/a_c	References	G_R	References
1	Diphenyl	Ph—Ph	0.14 ± 0.03	Dzhuraev and Evseev (1972)	0.045 ± 0.009	Nauka (1972)
2	Toluene	Ph—CH_3	0.30 ± 0.06	Evseev *et al.* (1968)	0.08 ± 0.02	Nauka (1972)
3	Phenol	Ph—OH	0.30 ± 0.05	Dzhuraev and Evseev (1972)	0.10 ± 0.02	Nauka (1972)
4	Anisole	Ph—O—CH_3	0.58 ± 0.02	Dzhuraev and Evseev (1972)	0.10 ± 0.02	Nauka (1972)
5	Benzene	Ph	0.35 ± 0.07	Dzhuraev *et al.* (1972)	0.16 ± 0.03	Nauka (1972)
6	Chlorobenzene	Ph—Cl	0.32 ± 0.07	Dzhuraev and Evseev (1972)	0.20 ± 0.04	Nauka (1972)
7	Styrene	Ph—CH=CH_2	0.45 ± 0.04	Dzhuraev *et al.* (1972)	0.30 ± 0.06	Nauka (1972)

8	Phenyl cyclohexane	Ph—C_6H_{11}	0.43 ± 0.04	Evseev *et al.* (1968)	0.30 ± 0.06	Nauka (1972)
9	Cyclohexane	C_6H_{12}	0.51 ± 0.04	Evseev *et al.* (1968)	1.6 ± 0.03	Nauka (1972)
10	Tetrahydrofuran	CH_2—CH_2—O—CH_2—CH_2 (ring)	0.64 ± 0.05	Dzhuraev and Evseev (1972)	1.7 ± 0.3	Nauka (1972)
11	Acetone	CH_3—C(=O)—CH_3	0.63 ± 0.05	Dzhuraev and Evseev (1972)	1.8 ± 0.9	Alger *et al.* (1959)
12	Polyethylene	$(C_2H_4)_n$	0.51 ± 0.05	Evseev *et al.* (1968)	3.0 ± 1.5	Lawton *et al.* (1960)
13	Methanol	CH_3OH	0.75 ± 0.05	Dzhuraev *et al.* (1972)	5.5 ± 2.7	Alger *et al.* (1959)
14	Ethanol	C_2H_5OH	0.67 ± 0.05	Dzhuraev *et al.* (1972)	6.0 ± 3.0	Alger *et al.* (1959)

The following parameters were obtained: $a_0/a_c \times \lambda/\lambda_0 = 0.51 \pm 0.03$ and $\Lambda = (1.7 \pm 0.3) \times 10^{10}$ sec^{-1} for $G_R = 1$ (without taking the errors of G_R into account). Using the data of Fig. 5, we can determine the rate of entry of μB into the chemical reaction with radicals produced by stopping a muon in different hydrocarbons at room temperature. The systematic error connected with inaccurate determination of G_R produces a $\pm 27\%$ total error in Λ. The chemical reaction rates of μB at room temperature obtained in this manner for benzene and styrene are $(2.7 \pm 0.7) \times 10^9$ sec^{-1} and $(5.1 \pm 1.4) \times 10^9$ sec^{-1}, and within the error limits are consistent with the corresponding values obtained from the temperature dependences a/a_c of these substances [see Dzhuraev *et al.* (1971c) and Section VIII,D].

The correlation between a/a_c and G_R confirms the assumption that radiolysis of the medium occurs at the place where a mesonic atom is produced, and the chemical reaction rate of a mesonic radical μB with radicals formed by radiolysis is comparable to the frequency of hyperfine splitting.

Since the average time of entry of μB into the chemical reaction in condensed media is of the order of $2\text{–}3 \times 10^{-10}$ sec (at room temperature), a/a_c, in contrast to G_R, should be sensitive to (a) the radicals produced in the channel (VIII–I′), and (b) radicals formed by breaking the C—C bond, since mesonic boron can be an acceptor of these radicals in the Franck–Rabinowitsch cage (Franck and Rabinowitsch, 1934). The probability of chemical reaction between mesonic boron and atomic hydrogen is low because of hydrogen's high mobility; this reduces its concentration in the chemical reaction region of mesonic boron. Because of the correlation between a/a_c and G_R, the radicals produced in the channels (VIII–I″) and (VIII–I″″) are assumed to be situated in the chemical interaction radius of μB.

It can be seen from Table 6 that a/a_c in oxygen-containing compounds (except for phenol, for the reason stated above) is much larger than in hydrocarbons. This can also be accounted for by the hypothesis of fast chemical reactions of mesonic atoms. The point is that the mesonic radical of nitrogen (μN) produced by capture of a muon by an oxygen atom enters very rapidly into the chemical reaction with radicals and molecules of organic compounds resulting in the formation primarily of diamagnetic cyanogen H—C $\equiv \mu$N (Fomin, 1967).

The fact that Buckle *et al.* (1968) did not observe precessional frequencies of mesonic atoms in some hydrocarbons indicates [just as in the case of water and aqueous solutions of hydrogen peroxide (Dzhuraev *et al.*, 1971b)] that the mesonic atom does not exist in the free state in condensed media before measurements are made ($T_\mu \simeq 10^{-7}$ sec).

B. Change in Depolarization of μ Mesons on Going from a Monomer to a Polymer

This effect was initially observed by Evseev *et al.* (1968) and studied in more detail by Dzhuraev *et al.* (1971d). Methyl methacrylate, butyl mcthacrylate, isoprene, highly purified vinyl acetate in monomer and polymer form, and three types of polystyrene [emulsion-type ($M = 1.3 \times 10^5$), block-type ($M = 3.0 \times 10^5$), and suspension-type ($M = 1.5 \times 10^5$)] were used as targets. Structural formulas for monomers and polymers are given in Table 7. Monomers were distilled twice: inert medium was used and a stabilizer was introduced to prevent polymerization and formation of oxide and peroxide compounds.

The asymmetry coefficients a, adjusted for different target thicknesses,

TABLE 7

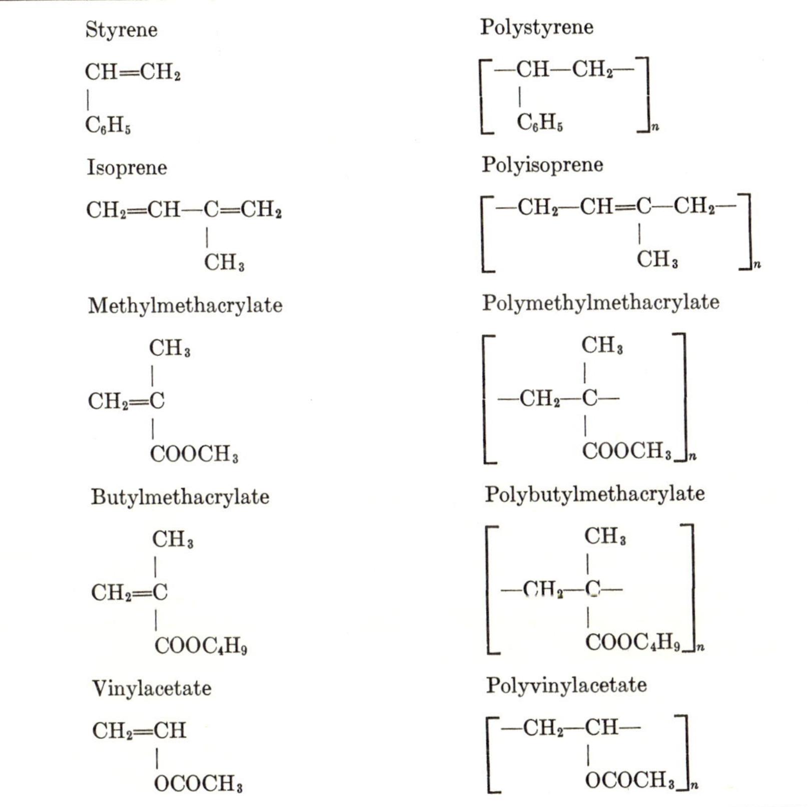

TABLE 8

Substance	a/a_c	$\frac{a(\text{monomer})}{a(\text{polymer})}$	$\bar{\tau}$ (μsec)	b_0	b_0
Vinylacetate	0.859 ± 0.035	1.51 ± 0.12	1.93 ± 0.01	0.35 ± 0.08	0.33
Polyvinylacetate	0.570 ± 0.033				
Butylmethacrylate	0.830 ± 0.037	1.28 ± 0.09	1.97 ± 0.01	0.35 ± 0.08	0.21
Polybutylmethacrylate	0.650 ± 0.037				
Methylmethacrylate	0.754 ± 0.026	1.29 ± 0.09	1.98 ± 0.01	0.30 ± 0.07	0.30
Polymethylmethacrylate	0.583 ± 0.035				
Styrene[a]	0.535 ± 0.029				
Polystyrene					
block-type[a]	0.322 ± 0.041				
block-type	0.309 ± 0.059				
suspended	0.414 ± 0.060				
emulsion	0.403 ± 0.059				
average	0.357 ± 0.027	1.50 ± 0.14	2.04 ± 0.01		
Isoprene	0.619 ± 0.063	0.96 ± 0.13	2.03 ± 0.01		
Polyisoprene	0.646 ± 0.057				

[a] Taken from Evseev *et al.* (1968).

are compared with a_c for graphite (a/a_c) in column 2 of Table 8. We notice first that the ratio a/a_c decreases (by a factor of 1.4 on the average) in all monomer–polymer pairs being investigated (except for the isoprene-polyisoprene pair) on going from a monomer to a polymer, in spite of considerable differences in molecule weight and molecular structure. This effect can be explained in the light of the physical processes associated with formation of a mesonic atom in molecular compounds (Evseev, 1968; Dzhuraev and Evseev, 1971, 1973).

Polymers differ from monomers in their greater ability to transfer excitation energy along the chain of carbon (in the instance) atoms (Nauka, 1966); this decreases the free radical density near the mesonic atom and hence the probability for the formation of dimagnetic compounds.

Absence of this effect in an isoprene–polyisoprene pair may be attributed to the presence of a double bond in the polymer chain. In all the other pairs investigated by us, the double bonds become single bonds as a result of polymerization. The result for polyisoprene can be understood if we assume that double bonds inhibit energy transfer along the polymer chain. Since the process of intramolecular energy migration in polyisoprene is indeed highly suppressed (Nauka, 1965), high radical density is maintained (in contrast to a monomer) at the place where the mesonic atom is formed.

Comparison of the values of a/a_c for all high-molecular polymers investigated (see Tables 5 and 8) shows that polystyrene and polyvinyl have the lowest polarization. This indicates their high stability against radiation damage, typical of all aromatic compounds (see Section VIII,A); this property diminishes the number of radicals near a mesonic atom and hence reduces the value of λ_0.

On the other hand, since ionizing radiation has a destructive effect on polybutyl methacrylate, polyisoprene, polymethyl methacrylate, and polyvinyl acetate (Bovey, 1958), their values of λ_0 and λ/λ_0 are large. Moreover, as indicated above, the mesonic radical μN produced in oxygen-containing substances actively interacts with molecules of the medium to produce diamagnetic cyanogen.

The average lifetime of a muon $\bar{\tau}$ is much shorter in oxygen-containing compounds than in hydrocarbons (see column 4 of Table 8). This is attributed to capture of a fraction of the muons by oxygen atoms. The probability b_0 that a muon in a compound containing hydrogen, carbon, and oxygen atoms will decay in the K orbit of oxygen can be determined by using the formula

$$b_0 = \frac{\bar{\tau} - \tau_c}{\tau_o - \tau_c} \times \frac{\tau_o}{\bar{\tau}}$$

where τ_c and τ_o are the muon lifetimes in graphite and oxygen, respectively. The hydrogen contribution was neglected, since within the limits of experimental error τ_c does not differ from the muon lifetime in hydrocarbons. By assuming that $\tau_c = 2.04 \pm 0.01$ μsec and $\tau_o = 1.84 \pm 0.1$ μsec (Dzhuraev *et al.*, 1971c) we obtain for three oxygen containing substances the values of b_0 listed in column 5 of Table 8; the values of b_0 shown in column 6 correspond to our estimate, based on the Z law, if all the muons captured by hydrogen are intercepted by binding atoms of the corresponding hydrogen-containing groups. Comparison of the data in the last two columns of Table 8 shows that some of the substances investigated have a tendency to draw the muons away from hydrogen-containing groups and toward oxygen.

C. Dependence of the Degree of Depolarization on the Length of the Hydrocarbon Chain

To determine the connection between μ-meson depolarization and the process of intramolecular energy migration, Dzhuraev *et al.* (1972) used the method of free muon spin precession to measure the relative (to graphite) residual polarization a/a_c in a number of aliphatic alcohols $C_nH_{2n+1}OH$ and alkyl chlorides $C_nH_{2n+1}Cl$.

Experimental data are shown in Fig. 6. In both types of compound, a/a_c first increases with increasing number n_c of carbon atoms in the chain, reaches a peak, and falls off to the limiting value of ≈ 0.5, which is determined by the asymmetry coefficient of the saturated compound (polyethylene). The values of a/a_c were determined for alcohols. These values, averaged over the atomic capture probability, are the same as those for mesonic boron and mesonic nitrogen produced on the settling of a muon in carbon and oxygen atoms, respectively. In the case of alkyl chlorides the values of a/a_c pertain only to mesonic boron. Since the points for alcohols are situated on the same curve as those for alkyl chlorides, they must depend on the length of the hydrocarbon chain.

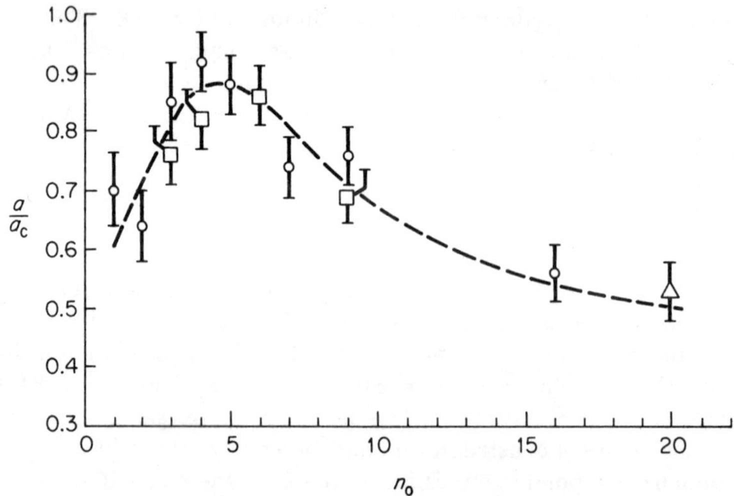

FIG. 6. Dependence of the relative residual polarization of negative muons in alcohol (○) and alkyl chlorides (□) on the number n_c of carbon atoms in the chain. The data for alcohol and alkyl chlorides were taken from Dzhuraev *et al.* (1972) and those for paraffin (△) from Dzhuraev *et al.* (1971c).)

The data may be interpreted as follows. At present the size of molecules does not exceed that of the zone in which a mesonic atom reacts chemically, and the intramolecular migration of a fraction of the energy released in the production of a mesonic atom does not appreciably reduce the radical density within the limits of this zone, since it is here that all the energy is finally released. The increase of a/a_c with increasing n_c for alcohols with $n_c = 1$ to 5 is attributed to several factors. First, an increase in the length of the chain is accompanied by an increase in the total number of valence electrons in the molecule. Because of high electron mobility, the negative charge will migrate (Naukà, 1965; Molin *et al.*, 1961) along the chain of carbon atoms to a mesonic atom in which Auger transitions lead to ionization of the surrounding parts of the molecule. Increasing the number of valence electrons in the molecule improves the conditions for Auger transitions and increases the number of emitted Auger electrons and of radicals formed near the mesonic atom. Second, since the radiation resistance of an alcohol molecule with small n_c can be reduced by increasing the length of its chain (Bovey, 1958), the radical density, and hence the contribution of radicals producing diamagnetic compounds with mesonic atoms, will increase. Increasing the molecular weight of alcohol will increase its viscosity, and this will further increase the density of large radicals in the interaction zone by decreasing their diffusion rate.

As n_c increases, part of the energy is removed from the chemical reaction zone of the mesonic atom. This decreases the radical density and a/a_c. The tendency of a/a_c to approach the limit when n_c is large can be explained by assuming that the energy migrates a distance of more than 5 to 10 chain links, which is consistent with current ideas (Gusynin and Tal'roze, 1960); the chemical interaction radius of a mesonic atom is smaller in size than a molecule with this chain length. Because the chain molecules have unique conformational properties, they are much smaller than a chain extended in one direction, which at $n_c = 5$ to 10 has a length of 12 to 25 Å (Flory, 1969).

If the average radius r of chemical reaction of a mesonic atom is compared with the average distance s between two ends of a molecule, we can see that for $n_c = 10$, $s \simeq 10$ Å (Flory, 1969) and hence $r \leq 10$ Å. The conformational properties of aliphatic alcohols and alkyl chlorides are assumed to be the same as those of paraffins. On the other hand, the value of r can be determined by using the expression $r \approx (2D')^{1/2}t$ [where D' is the diffusion coefficient of the mesonic atom and $t = 1/\Lambda$ is the average time of entry of the mesonic atom into the chemical reaction, determined by analyzing the temperature dependence of a/a_c; see Dzhuraev *et al.* (1971c) and Section VIII,G], assuming that mesonic atom reactions in condensed media can be described by the theory of chemical reactions limited by diffusion (Koldin, 1966; Iguerabide *et al.*, 1964; Elkana *et al.*, 1968).

The value of D' for mesonic boron (when $n_c = 10$, the mesonic nitrogen contribution can be ignored) in alcohols with $n_c \simeq 10$ can be determined as follows. Since $D' \sim 1/\eta$ (Koldin, 1966; Iguerabide *et al.*, 1964; Elkana *et al.*, 1968) (η is the viscosity of the medium), and at the room temperature of water the diffusion coefficient for mesonic boron is the same as that for atomic hydrogen, i.e., $D'_{\mu B} \leq 10^{-4}$ cm sec^{-1} (Pshezhetskii, 1962), the viscosity of alcohol with $n_c \simeq 10$ is about an order of magnitude larger than that of water, and therefore $D'_{\mu B} \leq 10^{-5}$ cm^2 sec^{-1}. The average time of entry of mesonic boron into the chemical reaction is of the order of 10^{-10} sec (see Section VIII,B), and $v \leq 5$ Å. This value is consistent with that obtained above if we take into account the initial path length (5–10 Å) of the "hot" mesonic atom, which receives energy from "Coulomb explosion" or from the recoil in the cascade transitions.

D. Temperature Dependence of μ^- Meson Depolarization in Organic Compounds

Dzhuraev *et al.* (1971c, 1973) showed that a/a_c increases with increasing temperature of styrene, benzene, methyl alcohol, *N*-butyl alcohol, polystyrene, and polybutyl methacrylate. Experimental data are shown in Figs. 7 and 8 and Table 9. The data for the first two substances can be analyzed (Dzhuraev and Evseev, 1973) by using Eq. (30) on condition that

$$\lambda_0 = \Lambda(T/300) \exp(-E_a/R'T) \tag{35}$$

(R' is the Boltzmann constant, equal to 1.987 cal/deg) and by assuming that mesonic atom reactions can be described by the theory of chemical reactions in condensed phases limited by diffusion (Koldin, 1966;

TABLE 9

Substance	$a_{300°K}/a_{80°K}$	References
Styrene	3.6 ± 0.8	Dzhuraev *et al.* (1971c)
Benzene	2.7 ± 0.7	Dzhuraev *et al.* (1971c)
Butanol	2.7 ± 0.4	Dzhuraev *et al.* (1973)
Polybutylmethacrylate	1.57 ± 0.13	Dzhuraev *et al.* (1973)
Polystyrene	1.33 ± 0.31	Dzhuraev *et al.* (1973)

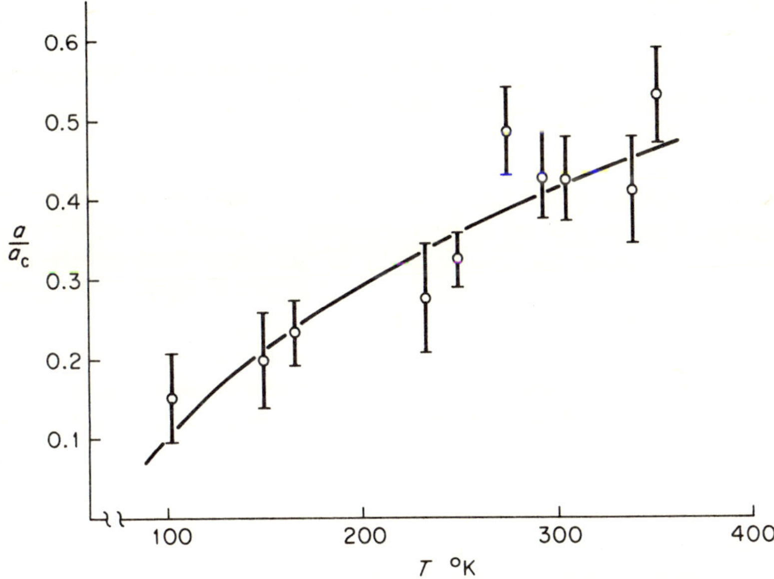

FIG. 7. Temperature dependence of depolarization in styrene.

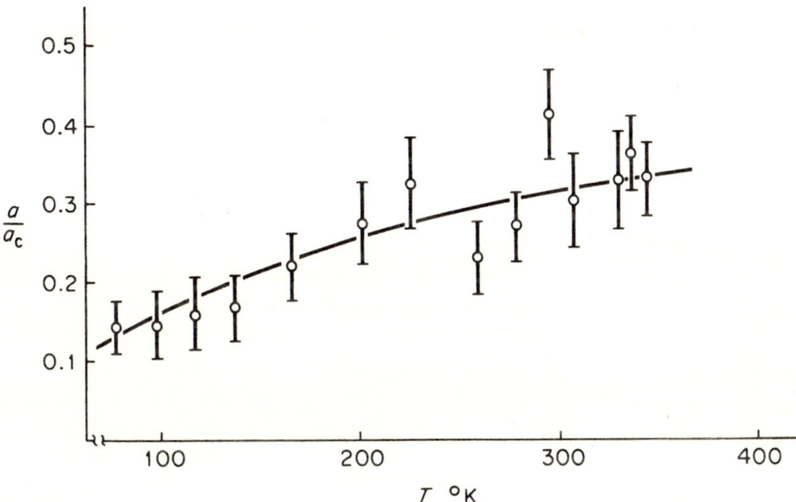

FIG. 8. Temperature dependence of depolarization in benzene.

Iguerabide *et al.*, 1964; Elkana *et al.*, 1968); i.e., the constant of the chemical reaction rate is equal to unity.

Parameters in Table 10 (first column for each substance) are compared with those from Dzhuraev *et al.* (1971c), where

$$\lambda_0 = \Lambda(T/300)^i \tag{36}$$

that is, the activation energy was not taken into account. Because of this, the parameter i turned out to be unjustifiably large. However, the values of λ_0 obtained within the error limits by using both methods are the same. It should be noted that the values of λ_0 obtained in approximation (36) were corrected [unlike those of Dzhuraev *et al.* (1971c)], which is consistent with the corrected value $\bar{\nu}_{\mu B} = 2.5 \times 10^9\ \text{sec}^{-1}$ of Dzhuraev and Evseev (1973) (see Section III). The fairly small values of E_a are typical of radical–radical reactions (Koldin, 1966; Iguerabide *et al.*, 1964; Elkana *et al.*, 1968).

Of great importance is the agreement between the parameters $a_0/a_c \times \lambda/\lambda_0$ and λ_0 obtained by analyzing the temperature dependence of depolarization in styrene and benzene, and those obtained by analyzing the correlation between a/a_c and G_R (see Section VIII,A).

The increase of a/a_c with increasing temperature is direct proof of the existence of paramagnetism in μB. If μB is produced, for example, as a diamagnetic ion μB^+, then a/a_c would decrease with increasing temperature, since the reactions producing paramagnetic compounds of μB would become dominant (those producing diamagnetic compounds do not change the muon polarization).

The temperature dependence of a/a_c is weaker in polymers than in other

TABLE 10

Substance	$a_0\lambda/a_c\lambda_0$	E_a (cal)	i	$\Lambda \times 10^{-10}$ (sec^{-1})	$\lambda_0 \times 10^{-10}$ (sec^{-1})
Water	0.40±0.02	232±176	—	5.2 ±4.0	3.5 ±2.7
	0.39±0.02	—	2.3±1.1	—	3.2 ±2.4
7.5 (by weight) aqueous solution of hydrogen peroxide	0.60±0.02	197±29	—	5.3 ±4.0	3.8 ±2.8
	0.59±0.07	—	1.4±0.4	—	2.4 ±1.5
Styrene	0.60±0.34	345±135	—	0.55±0.16	0.31±0.09
	0.49±0.05	—	3.7±1.3	—	0.60±0.25
Benzene	0.54±0.16	0±275	—	0.18±0.09	0.18±0.09
	0.36±0.03	—	1.9±0.6	—	0.60±0.40

organic compounds (see Table 9). This is consistent with the general relationship between a/a_c and λ_0, since radiolysis of polymers decreases the radical density and shifts the variation range of a/a_c (with changing T) toward lower λ_0, where a/a_c changes little with changing λ_0 (see Fig. 4).

The temperature dependence of muon depolarization in styrene and benzene was evaluated without taking the effect of the phase transition into account. A search for an abrupt change of a/a_c for styrene at the phase transition point ($T \simeq -30°C$) was conducted by Dzhuraev *et al.* (1973). The ratio of the asymmetry coefficients for solid styrene to liquid styrene near the phase transition point is $a_{liq}/a_{sol} = 0.93 \pm 0.08$; this eliminates with sufficient accuracy the effect of the phase transition on muon depolarization in hydrocarbons.

IX. Depolarization of Negative Muons in Water and in Aqueous Solutions

A. *Temperature Dependence of Depolarization*

The experimental data for water (Babaev *et al.*, 1969) and a 7.5% aqueous solution of hydrogen peroxide (Dzhuraev *et al.*, 1971b, 1973) evaluated according to Eqs. (30), (35), and (36) are given in Table 10 and Figs. 9 and 10. The increase of the temperature dependence and λ_0 by an order of magnitude over that characteristic of mesonic boron indi-

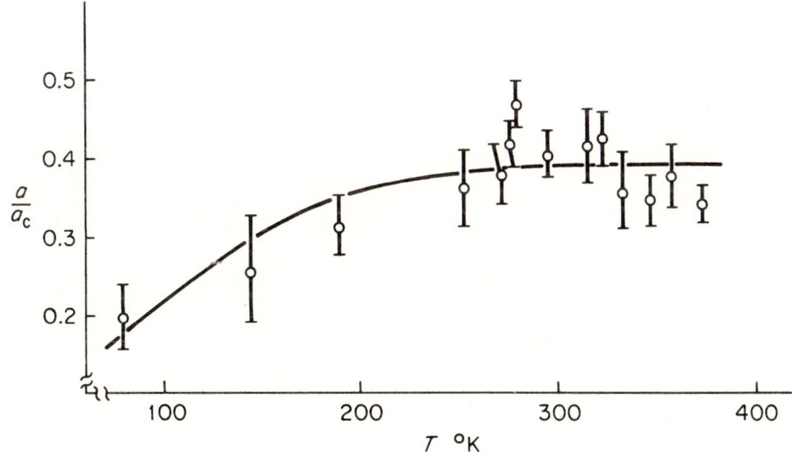

FIG. 9. Temperature dependence of residual polarization in water.

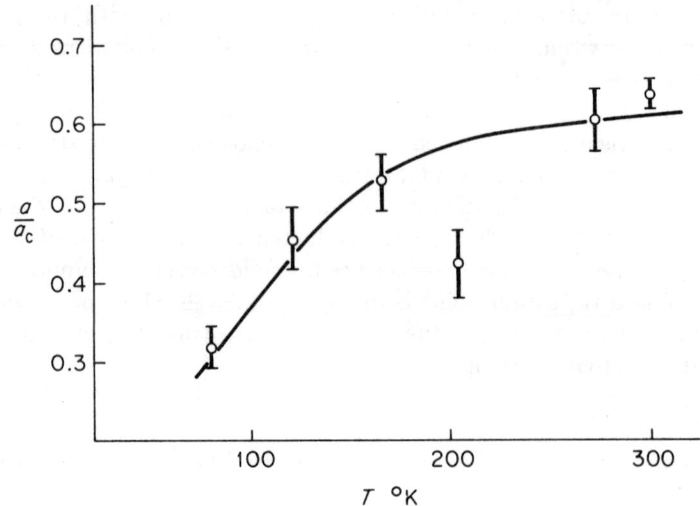

FIG. 10. Temperature dependence of the residual polarization of muons in aqueous solution (7.5% by weight) of hydrogen peroxide.

cates that mesonic nitrogen reactions are more active in water. It should be noted that the temperature dependence observed experimentally is described by the part of the theoretical curve (at low temperature) having the maximum slope. This also holds true for hydrocarbons. We conclude from this that the contribution of the reaction of "hot" mesonic atoms is negligible, for otherwise the temperature dependence would have a flatter slope.

Some authors [see, for example, Seitz and Kocher (1956)] assume that strong local heating of the medium may occur as a result of radiolysis produced by highly ionized particles. The strong temperature dependence of negative muon depolarization indicates that the hot zone cools (assuming there was heating) at the location of the "Coulomb explosion" in less than 10^{-11} to 10^{-10} sec. These findings are in agreement with those of Seitz and Kocher (1956).

B. Depolarization in Aqueous Electrolytic Solutions

To verify the general behavior of a mesonic atom in a medium, Dzhuraev *et al.* (1971b) measured a/a_c in numerous aqueous solutions and in heavy water. Aqueous solutions of ammonia [ammonium hydroxide (NH_4OH), ammonium chloride (NH_4Cl), hydrogen chloride (hydrochloric acid HCl), potassium hydroxide (KOH), hydrazine (N_2H_4), and hydrogen peroxide (H_2O_2)] were used as targets.

Table 11 gives the values of a, taken from Dzhuraev *et al.* (1971b), relative to a_c for graphite reduced to the same recording threshold of decay electrons. The concentrations of dissolved substances based on the number of molecules are given in parentheses. All the values were obtained at room temperature. Table 11 compares the value of a for KOH (Dzhuraev *et al.*, 1971a) with that for water (Ignatenko *et al.*, 1958, 1961; Buckle *et al.*, 1968).

All the solutions investigated can be broken down into two categories. KOH and HCl solutions in water are strong electrolytes, since the dissociation constant of these compounds in water ("Handbook of Chemistry," 1964) is $K_D \simeq 10^7$. Depolarization was measured in these solutions in order to determine the manner in which the mesonic atom μN interacts with the ionic radicals K^+, H^-, OH^-, and Cl^-.

According to current views (Denisov, 1970, 1971), the reactions of neutral H and OH radicals with H^+, OH^-, and Cl^- ionic radicals in aqueous solutions are greatly suppressed compared with those between neutral radicals. This is attributable to the presence of hydration shells in ionic radicals. Although the reaction mechanism is not clearly defined, it is assumed that the neutral radical reacts with the ion through substitution of a ligand in the inner coordination sphere (shell of hydration) or electron transfer across the medium.

The reaction between mesonic nitrogen and ions in aqueous electrolytic

TABLE 11

Substance	a/a_c
H_2O	0.403 ± 0.018
NH_4OH (0.24)	0.427 ± 0.028
NH_4Cl (0.11)	0.396 ± 0.064
N_2H_4 (0.15)	0.343 ± 0.041
H_2O_2 (0.14)	0.691 ± 0.029
KOH (0.15)	0.328 ± 0.034
KOH (granules)	0.170 ± 0.050
HCl (0.07)	0.429 ± 0.078
HCl (0.23)	0.443 ± 0.056
D_2O	0.424 ± 0.044
H_2O	0.38 ± 0.11[a]
H_2O	1.07 ± 0.18[b]

[a] From Buckle *et al.* (1968).
[b] From Ignatenko *et al.* (1958).

solutions can be observed by reducing the value of a compared with that of pure water, since this increases the fraction of paramagnetic interaction products of μN.

The fact that the value of a for aqueous solutions of KOH and HCl within a 10% error limit is close to that for pure water indicates that the contribution (not larger than the statistical error) of the reaction between μN and ionic radicals is negligible.

Since NH_4OH, NH_4Cl, and N_2H_4 solutions are fairly weak electrolytes (the dissociation constant $K_D \simeq 10^{-5}$ to 10^{-6}) ("Handbook of Chemistry," 1964), they can be regarded as molecular solutions, similar to the aqueous solution of H_2O_2. The hydrogen bond lattice can be strengthened (see, for example, Naberukhin and Rogov, 1971) in any nonelectrolytic aqueous solutions of 0.10–0.15 mole fraction concentration by increasing the energy and number of hydrogen bonds. The molecules of the nonelectrolyte occupy the free space in water. The equivalence (within statistical error limits) of the values for water and aqueous solutions of NH_4OH, NH_4Cl, and N_2H_4 means that for a nonelectrolyte concentration of 0.1–0.2 mole fraction, a is insensitive within 10% to the change in ordering of the structure of water. Structural effects in aqueous solutions of KOH and HCl electrolytes are unimportant at the concentrations used by us (Bloch, 1969).

The sharp increase of a (by a factor of 1.8) in weak aqueous solutions of H_2O_2 is attributed to the appearance of new (compared with pure water) direct reaction channels between μN and H_2O_2 molecules to produce diamagnetic compounds (see Section IX,C). The interaction between μN and H, OH, and HO_2 neutral radicals formed in an H_2O_2 aqueous solution upon destruction of the medium near a mesonic atom, may apparently be neglected, since the contribution of these radicals is proportional to the mole fraction of H_2O_2; i.e., it does not exceed 15%. This apparently also holds true for other electrolytic and nonelectrolytic solutions.

The value of a for heavy water is the same as that for ordinary water. This seems reasonable from the viewpoint of the general behavior of a mesonic atom, since the isotopic effects associated with destruction of the medium surrounding a mesonic atom, which affect the radical density, or those of the constants for the reaction rate between radicals and mesonic nitrogen, must be small (Acad. Sci. USSR., 1962; Melander, 1964).

C. Concentration Dependence of Depolarization in Aqueous Solutions of Hydrogen Peroxide

The experimental data (Dzhuraev *et al.*, 1971b, 1973) are shown in Fig. 11. Since the conditions $\nu \gg \Omega$ and $\lambda_0 \gg \Omega$ for water and aqueous solutions of H_2O_2 (which form new reaction channels for mesonic nitrogen) are

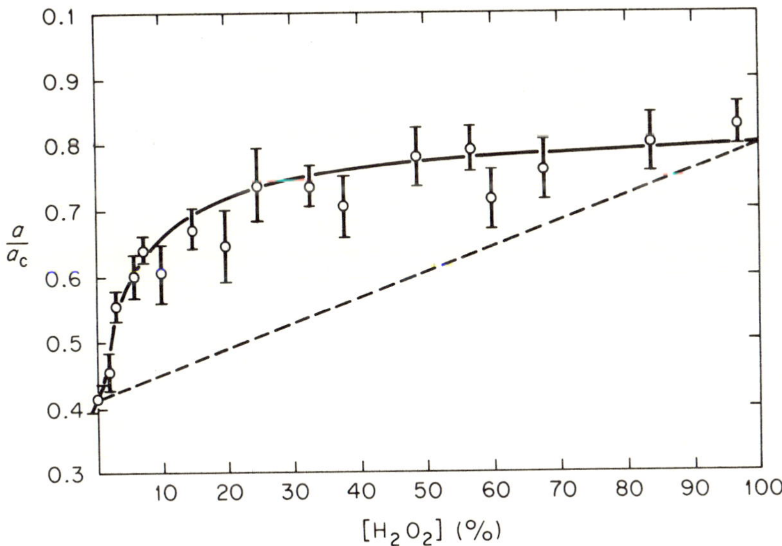

FIG. 11. Dependence of the residual polarization in aqueous solutions of hydrogen peroxide on the weight concentration of H_2O_2.

satisfied at room temperature, and a/a_c remains constant with increasing temperature, we can assume that

$$a/a_c = a_0\lambda/a_c\lambda_0 \tag{37}$$

For a two-component solution, this ratio can be written as follows:

$$\frac{a}{a_c} = \frac{a_0}{a_c}\frac{\lambda'(1 - U) + pU}{\lambda_0'(1 - U) + p_0U} \tag{38}$$

Here λ_0' and p_0 are chemical reaction rates of mesonic nitrogen in pure water and pure hydrogen peroxide, respectively (λ and P are reactions producing diamagnetic compounds),

$$U = \frac{[H_2O_2]}{[H_2O_2] + [H_2O]} = \frac{[H_2O_2]}{(1 - M_{H_2O_2}/M_{H_2O}d)[H_2O_2] + (A'/M_{H_2O_2})} \tag{39}$$

where M is the molecular weight, A' Avogadro's number, and $d = 1.46$ (density of H_2O_2). Using numerical values, we can write

$$\frac{a}{a_c} = \frac{a_0}{a_c}\frac{\lambda}{\lambda_0}\frac{1 + (0.30(p/\lambda') - 0.39)[H_2O_2]}{1 + (0.30(p_0/\lambda_0') - 0.39)[H_2O_2]} \tag{40}$$

Here the H_2O_2 concentration is given in 10^{22} cm^{-3}. The solid line in Fig. 11 shows the data evaluated by the least-squares method with use of Eq. (40); using $\bar{\lambda}_0 = (2.8 \pm 1.5) \times 10^{10}$ sec^{-1} (average value for water and a 7.5% solution of H_2O—see Table 10), we obtain $p_0 = (6.9 \pm 5.0) \times 10^{11}$ sec^{-1}. The error of p_0 is due to the large error of $\bar{\lambda}_0$. The broken line in Fig. 11 denotes the case in which μN, produced in one of the components of the solution, does not react with molecules or radicals of the other component.

Since p_0 is the reaction rate of μN in pure H_2O_2, we can determine the reaction rate constant $K = (2.7 \pm 1.9) \times 10^{-11}$ sec^{-1} cm^2, which coincides within an order of magnitude with the constants for chemical reactions limited by diffusion (Vereshchinskii and Pikaev, 1963; Pikaev, 1965; Denisov, 1970, 1971; Koldin, 1966; Iguerabide *et al.*, 1964; Elkana *et al.*, 1968).

It is important to note that a/a_c changes markedly in aqueous solution of H_2O_2 when the H_2O_2 concentration is of the order of 10^{21} cm^{-3}. This means that the absolute concentration of H and OH radicals must be approximately the same in order to account for the value of a/a_c measured in water.

If we take the concentration of H_2O molecules (3.34×10^{22} cm^{-3}) into account, we must assume that the destruction of the substance in the neighborhood of the mesonic atom is fairly small ($\sim$10%) in order to account for the short lifetime of the mesonic atom in water, 10^{-11} sec (at room temperature). Since destruction of the medium at the end of a muon track and on slowing-down fragments from "Coulomb explosion" is similar to that produced by recoil atoms in nuclear transformations, the data for small destruction of the medium near a mesonic atom and chemical stabilization time should be consistent with those for recoil atoms obtained by other methods (Bondarevskii *et al.*, 1971).

To determine the constants for the rate of chemical reaction of μN with H and OH in water, we must calculate the radical concentration near a mesonic atom as a function of space coordinates and time.

The higher value of a/a_c (as compared with water) in concentrated H_2O_2 indicates that a large fraction of the chemical reaction between μN and H_2O_2 is responsible for the production of diamagnetic compounds, for example,

$$\mu\text{N} + \text{H}_2\text{O}_2 \to \begin{cases} \text{H}\mu\text{NO} + \text{OH} \\ \text{OH}\mu\text{NO} + \text{H} \end{cases} \tag{41}$$

Dzhuraev *et al.* (1973) determined the ratio of residual polarization at room temperature to that at liquid nitrogen temperature for 97% H_2O_2:

$P_{300}/P_{80} = 1.39 \pm 0.08$. The ratio $P_{300}/P_{80} = 1.85 \pm 0.15$ for a 7.5% aqueous solution of H_2O_2 (Dzhuraev *et al.*, 1971b).

Since $\lambda_0 \sim [H_2O_2]$, increasing λ_0 in concentrated hydrogen peroxide will reduce the temperature dependence of depolarization. Since in a 7.5% (by weight) aqueous solution of H_2O_2 the reaction of mesonic nitrogen with H_2O_2 molecules contributes to λ_0 about as much as that with the radicals from radiolysis of water, λ_0 should increase by a factor of 5 to 10 if the H_2O_2 concentration is increased to 97%. Thus a/a_c should plateau at 30–60°K, if λ_0 is proportional to the first power of the temperature and the activation energy E_a in the $\mu N + H_2O_2$ reaction is ignored.

The activation energy must be taken into account because of the temperature dependence of depolarization in H_2O_2 in the range of 80 to 300°K. Dzhuraev *et al.* (1973) obtained a value of $E_a = 197 \pm 29$ cal. Addition of an exponential factor in the expression for λ_0 at this value of E_a accounts for the experimental result.

D. Muon Depolarization in Water and in Aqueous Solutions of H_2O_2 in Transverse and Longitudinal Magnetic Fields

It follows from Eq. (30) and Fig. 4 that the average time of entry of a mesonic atom into chemical reaction can be determined independently by measuring a/a_c in different transverse fields.

The hyperfine interaction constant $\nu_{\mu N}$ for the electron shell of mesonic nitrogen can be determined independently by measuring the depolarization in a longitudinal magnetic field in water.

Assuming that at $H_{||} = 0$, $a/a_c = 0.4$, and that the longitudinal magnetic field restores polarization prior to $a/a_c = 1$, we can write the expression for residual polarization in the same form as that for muonium (see Breit and Hughes, 1957):

$$P = P_0 \left[0.4 \pm 0.6 \frac{y^2}{1 + y^2} \right] \tag{42}$$

where $y = H_{||}/H_0$ and H_0 is the strength of the longitudinal magnetic field which restores 50% of the polarization.

To clarify all the problems considered above, Dzhuraev *et al.* (1971b) determined a experimentally for different longitudinal $H_{||}$ and transverse $H_{\perp}$ magnetic fields in water and aqueous solutions of hydrogen peroxide. The experimental arrangement for measuring a as a function of H is shown in Fig. 12. A solenoid (5) produced the longitudinal (in the direction of meson spin in the beam) field up to $H_{||} = 1250$ G. A lead slug (8) pre-

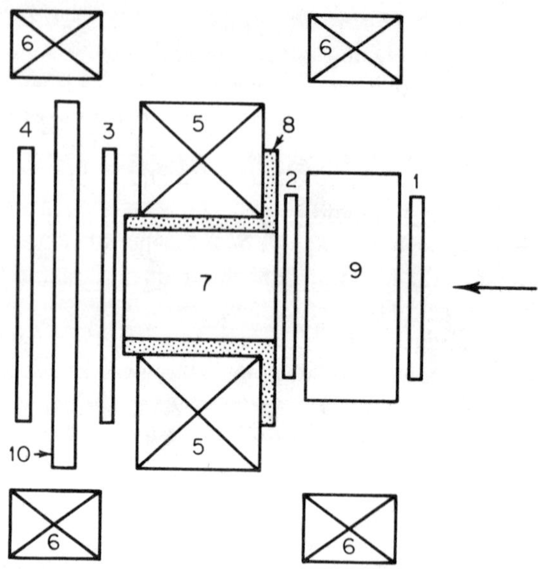

FIG. 12. Apparatus for measuring depolarization in transverse and longitudinal magnetic fields (side view). 1, 2: scintillation counters of the muon telescope; 3, 4: scintillation counters for recording μ-decay electrons; 5: solenoid for the longitudinal field; 6: Helmholtz coil for the transverse field; 7: target enclosed in a lead shield; 8, 9: muon moderators; 10: electron telescope filter.

vented muons from stopping and decaying in the aluminum frame of the solenoid and eliminated the second component in the time-dependent distribution of decay electrons; this component is of nearly the same magnitude as the oxygen component. The component produced by stopping muons in lead can be easily eliminated by delaying electron recording several fractions of a microsecond. We used a water target 80 mm in diameter and 70 mm thick.

There was no evidence within the limits of experimental accuracy ($\pm 5\%$) of a focusing effect of decay electrons produced by the strong field $H_{||}$ (μ^+ mesons with a graphite target were used in the experiment).

Helmholtz coils (6) produced a transverse magnetic field up to $H_{\perp} \simeq 200$ G, measured with an accuracy of a fraction of 1%. The longitudinal field was measured with an accuracy of $\pm 5\%$. The large error is due to the highly inhomogeneous field $H_{||}$ in the target's solenoid.

Time-dependent distributions of electrons obtained experimentally were evaluated on a computer by the least-squares method according to Eq. (16).

For measurements in a transverse field the asymmetry coefficient was defined as parameter a in Eq. (16), and for measurements in a longitudinal field the asymmetry coefficient a' was determined by comparing the count N' at zero time in the longitudinal field with N_0 at zero time in the transverse field according to the equation $a' = (N' - N_0)/N'$.

Table 12 gives the experimental data for water and an aqueous solution of H_2O_2. The values of a were corrected (for $H_\perp$) by taking into account a number of points in a single precession period.

As noted above, by measuring a as a function of the transverse magnetic field intensity, we can uniquely determine the time a mesonic atom enters into the first chemical reaction. If this time is of the order of $t \approx \pi/2$, then a will decrease by a factor of ≈ 2 as $H_\perp$ increases from 20 to 200 G (see Fig. 4); if, however, there is no field dependence, then the time the mesonic atom enters into the chemical reaction will be of the order of 10^{-11} sec (at room temperature) (Dzhuraev and Evseev, 1971, 1973). It was indicated (Dzhuraev and Evseev, 1971, 1973) that under certain conditions the

TABLE 12

Substance	T (°K)	$H_{\parallel}$ (G)	$H_\perp$ (G)	a, a'
	300	1265	—	0.0171 ± 0.0035
H_2O	300	1000	—	0.0241 ± 0.0058
	300	800	—	0.0146 ± 0.0058
		Average: 0.0180 ± 0.0026		
	300	—	50	0.0202 ± 0.0018
H_2O	300	—	200	0.0207 ± 0.0025
H_2O	77	200	—	0.0110 ± 0.0019
	77	—	50	0.0105 ± 0.0013
	300	—	43	0.0295 ± 0.0022
	300	—	190	0.0268 ± 0.017
$H_2O + H_2O_2$ (15%)[a]				
	300	—	24	0.0308 ± 0.0020
	300	—	—	
	300	—	47	0.0292 ± 0.0016
	300	—	70	0.0285 ± 0.0022
$H_2O + H_2O_2$ (7.5%)[a]				
	300	—	106	0.0274 ± 0.0021
	300	—	197	0.0290 ± 0.0020

[a] By weight.

initial precessional phase ϕ will depend on $H_\perp$. However, we could not determine ϕ precisely under our experimental conditions.

We obtained the following values for H_2O_2 by the least-squares method according to the equation $a = a^* + bH_\perp$: $a^* = 2.982 \pm 0.15$ and $b = -(1.28 \pm 1.45) \times 10^{-3}\ G^{-1}$. Thus a is independent of $H_\perp$ within the error limits. A similar result was obtained for pure water (see Table 12).

We conclude from this that the time of entry of the mesonic atom into the chemical reaction in water and aqueous solutions of H_2O_2 is of the order of 10^{-11} sec (at room temperature); this is consistent with the result obtained by studying the temperature dependence of residual polarization in water and aqueous solutions of H_2O_2 (Section IX,A).

It was possible to determine the lower limit of ν in mesonic nitrogen because a is independent of $H_{||}$ within the 20% error. The field H_0, which is about 3000 G according to Eq. (42), corresponds to a 20% change of P/P_0. Since for muonium $H_0 = 1588$ G and $\nu = 4.46 \times 10^9\ \text{sec}^{-1}$ (Vaisenberg, 1964), for mesonic nitrogen $\nu_{\mu N} \geq 10^{10}\ \text{sec}^{-1}$, consistent with the value given in Section III.

The limiting value of $\nu_{\mu N}$ favors the assumption that the electron shell of mesonic nitrogen is not formed in the ground state for the most part ($\nu \simeq 10^9\ \text{sec}^{-1}$).

X. Depolarization of μ^- Mesons in Inorganic Dielectrics

The data for dielectrics whose atoms have nuclei with $I \neq 0$ are given in Table 13.

Total depolarization in liquid hydrogen is consistent with the theory (Gershtein, 1959) based on the mechanism for muon transfer from one proton to another. The value of a/a_c for ^7LiH, measured at the frequency of the F_+ state, coincides within the limits of error with that obtained for metallic ^7Li (see Table 3). This can be illustrated by noting that mesonic helium (μHe) can quickly acquire a diamagnetic electron shell and is not involved in the chemical reaction during the muon lifetime. Because of this, the value of a/a_c for LiH, corrected with allowance for the hyperfine muon-nuclear interaction, is shown in Fig. 1 along with the data for metals.

In contrast to crystal boron, polarization in B_4C at the frequency of the F_+ state is nearly zero. Since at $Z \leq 5$ the transition probability R between states of the hyperfine structure is lower than the muon decay probability (Winston and Telgedi, 1961; Winston, 1963; Hutchinson *et al.*, 1962) and R may depend on the structure of the valence shell, this divergence can be attributed to the higher transition probability R in B_4C because of the

TABLE 13

Substance	a_+/a_c or a/a_c, for $I = 0$	References
H_2 (liquid)	0.005 ± 0.005	Ignatenko *et al.* (1958)
$^7Li\ H$	0.388 ± 0.003	Babaev *et al.* (1969)
N_2 (liquid)	0.270 ± 0.036	Babaev *et al.* (1969)
O_2 (liquid)	0.09 ± 0.11	Buckle *et al.* (1968)
Al_2O_3[a]	0.05 ± 0.06	Dzhuraev *et al.* (1971a)
Cr_2O_3[a]	0.12 ± 0.06	Dzhuraev *et al.* (1971a)
B_2O_3[a]	0.09 ± 0.02	Dzhuraev *et al.* (1971a)
C_aO[a]	0.05 ± 0.03	Dzhuraev *et al.* (1971a)
SiO_2 (fused)[a]	0.12 ± 0.06	Dzhuraev *et al.* (1971a)
SiO_2 (crystal)[a]	0.09 ± 0.06	Dzhuraev *et al.* (1971a)
PbO[a]	0.28 ± 0.06	Dzhuraev *et al.* (1971a)
KOH[a]	0.17 ± 0.05	Dzhuraev *et al.* (1971a)
NaOH[a]	0.25 ± 0.08	Dzhuraev *et al.* (1971a)
	0.70 ± 0.17	Dzhuraev *et al.* (1971a)
	1.05 ± 0.20	Ignatenko *et al.* (1958)
	0.96 ± 0.34	Telegdi (1960), Culligan *et al.* (1961)
S_{2n}	0.58 ± 0.13	Astbury *et al.* (1961)
	0.55 ± 0.03	Sundelin (1967), Sundelin *et al.* (1968)
	0.99 ± 0.15	Evseev *et al.* (1966)
B_4C	< 0.02	Dzhuraev *et al.* (1971a)

[a] In oxygen.

eligibility of a larger number of electrons for the Auger process, which facilitates the $F_+ \rightarrow F_-$ transition.

The case of N_2 is fairly straightforward. The residual polarization at the frequency of the F_+ state proved to be markedly smaller than the nuclear spin effect would indicate. This is attributed to several factors. First, according to estimates of Winston and Telgedi (1961; Winston, 1963; Hutchinson *et al.*, 1962) $R \approx \Lambda_{\text{distr}}$ for nitrogen, and because the frequencies of the F_+ and F_- states differ by only 15%, the decrease of a/a_c may be connected in some imperceptible way with the $F_+ \rightarrow F_-$ transitions (a/a_c is equal to about 0.087 in the F_- state) and insufficient statistical accuracy (Babaev *et al.*, 1969). Second, since mesonic carbon (μC) has a diamagnetic electron shell, a paramagnetic compound, say NμC, can produce additional muon depolarization, since all the processes peculiar to water and other molecular media should accompany the formation of a mesonic atom of μC in N_2, i.e., formation of neutral and charged nitrogen atoms. Depolarization can also occur because of a collision between μC and the molecules of paramagnetic impurities. The value of $a(I)/a(0)$ for N_2 is 0.37 (Babaev *et al.*, 1969).

Total depolarization in O_2 is determined by the scattering of μN by paramagnetic O_2 molecules (Buckle *et al.*, 1968; Evseev, 1968).

The value of a/a_c for oxides measured in oxygen is about 0.1, on the average (Dzhuraev *et al.* (1971a). If mesonic nitrogen atom will not react with molecules of the medium or radicals produced by destruction of the medium, then a/a_c will not differ from zero. The average value of $a/a_c \simeq 0.1$ is attributed to the compensated paramagnetism of μN due to fast chemical reactions ($t_1 \leq 1/\nu$).

Although an S_2 molecule is paramagnetic (Selwood, 1966), ordinary fused sulfur used in depolarization experiments has ring-shaped diamagnetic molecules. The slightly lower polarization in sulfur than in conductors with the nearest Z is attributed to fast chemical reactions in mesonic phosphorus (μP). Polarization at frequency ω can be preserved by adding μP to one end of the chain of sulfur atoms, which was broken as a result of the radiation effect of Auger electrons and fragments of the Coulomb explosion. In this case, however, the other end of the chain, which has an uncompensated electron shell, may cause additional depolarization because of spin–spin relaxation.

XI. Conclusion

Many problems of depolarization of negative muons can now be explained either qualitatively or quantitatively. The basic, fast depolarization in the cascade clearly must be attributed to spin–orbit interaction of a muon in the excited states of a mesonic atom. The dominant role (for understanding slower depolarization) of the interaction of the muon spin with the electron shell of a mesonic atom was demonstrated. A large number of experimental data indicates that interaction between the electron shell of a mesonic atom and the medium, which accounts for the diversity of polarization effects, must be taken into account.

Chemical reactions between a mesonic atom and the atoms and molecules surrounding it, in which the mesonic atom behaves as a typical atomic radical, were demonstrated.

It is important to determine the connection between the degree of depolarization and the radiation resistance of a substance and to account for this effect with allowance for radiolysis of the medium surrounding the mesonic atom.

Many problems remain to be solved. Not everything is clear, for example, about the mechanism of additional depolarization (as compared with that predicted by the theory of cascade depolarization) in metals. It is not absolutely clear why polarization at the precessional frequency of a mesonic atom has not been observed in pure helium gas. It would be desirable to understand why mesonic nitrogen is produced mostly (as follows from the experimental data for depolarization) with a metastable electron shell.

The concept of interaction between mesonic atoms and the medium was developed parallel to and in close connection with the investigation of the physics and chemistry of atomic muonium, for which the experimental data are interpreted along similar principles. Atomic muonium and mesonic atoms are rare, "radioactive" isotopes of ordinary atoms and are very useful in studying the behavior of these atoms under different conditions. The qualitative diversity and abundance of information that further studies involving polarized effects in mesonic atoms could yield may be the basis for development of new and unique methods of investigating the electron structure of matter and the kinetics of chemical reactions.

Acknowledgments

I should like to thank A. P. Bukhvostov, A. A. Dzhuraev, V. S. Roganov, and V. G. Firsov for useful discussions.

References

Acad. Sci. USSR (1962). "Ionization Potentials and Electron Affinity" (Handbook). Moscow.

Alger, R. S., Anderson, T. H., and Webb, L. A. (1959). *J. Chem. Phys.* **30,** 695.

Arl't, R., Evseev, V. S., Ortlepp, G. Kh., Roganov, V. S., Sabirov, B. M., and Khaupt, Kh. (1973). J.I.N.R. P15-7202, Dubna. Paper presented at the *Int. Conf. High-Energy Phys. Nucl. Structure, 5th, Upsala, Sweden* June 18–22.

Anderson, E. W. (1965). Thesis, Nevis-136.

Andrianov, D. A., Myasishcheva, G. G., Obukhov, Yu. V., Roganov, V. S., Firsov, V. G., and Fistul', V. I. (1969). *J.E.T.P.* **56,** 1195.

Astbury, A., Hutterby, P. M., Hussain, M., Kemp, M. A., Muirhead, H., and Woodhead, T. (1961). *Proc. Phys. Soc.* **78,** 1145.

Au-Yang, M. J., and Cohen, M. L. (1968). *Phys. Rev.* **174,** 568.

Babaev, A. I., Balats, M. Ya., Myasishcheva, G. G., Obukhov, Yu. V., Roganov, V. G., and Firsov, V. G. (1966). *J.E.T.P.* **50,** 877.

Babaev, A. I., Evseev, V. S., Myasishcheva, G. G., Obukhov, Yu. V., Roganov, V. S., and Chernogorova, V. A. (1969). *Yad. Fiz.* **10,** 964.

Backenstoss, G., *et al.* (1967). *Phys. Lett.* **25B,** 365.

Bloch, A. M. (1969). "Structure of Water and Geological Processes." "Nedra," Moscow.

Bondarevskii, S. I., Murin, A. N., and Seregin, P. P. (1971). *Advan. Chem.* **40,** 95.

Bovey, F. A. (1958). "The Effects of Ionizing Radiation on Natural and Synthetic High Polymers." Wiley (Interscience), New York.

Breit, G., and Hughes, V. W. (1957). *Phys. Rev.* **106,** 1293.

Buckle, D. C., Kane, J. R., Siegel, R. T., and Wetmore, R. J. (1967). *Bull. Amer. Phys. Soc. Ser. 11* **12,** 7, 1050.

Buckle, D. C., Kane, J. R., Siegel, R. T., and Wetmore, R. J. (1968). *Phys. Rev. Lett.* **20,** 14, 705.

Bukhvostov, A. P. (1966). *Yad. Fiz.* **4,** 83.

Bukhvostov, A. P. (1969). *Yad. Fiz.* **9,** 107.

Bukhvostov, A. P., and Shmushkevich, I. M. (1961). *J.E.T.P.* **41,** 1895.

Burbidge, C. R., and de Borde, A. H. (1953). *Phys. Rev.* **89,** 189.

Cole, A. (1969a). *Radiat. Res.* **38,** 1.

Cole, A. (1969b). *Advan. Radiat. Chem.* **1,** 33.

Culligan, G., Lathrop, J. E., Telegdi, V. L., and Winston, R. W. (1961). *Phys. Rev. Lett.* **7,** 458.

Day, T. B., Rodberg, L. S., Snow, G. A., and Sucher, J. (1961). *Phys. Rev.* **123,** 1051.

de Borde, A. H. (1954). *Proc. Phys. Soc.* **A67,** 57.

Denisov, E. T. (1970). *Usp. Khim.* **39,** 1.

Denisov, E. T. (1971). *Usp. Khim.* **40,** 1.

Dienes, G. J., and Vineyard, G. H. (1957). "Radiation Effects in Solids." Wiley (Interscience), New York [Translation published by IIL, Moscow, 1960].

Dyne, P. J., and Kennedy, J. M. (1958). *Can. J. Chem.* **36,** 1518.

Dyne, P. J., and Kennedy, J. M. (1960). *Can. J. Chem.* **38,** 61.

Dzhrbashyan, V. A. (1958). *J.E.T.P.* **35,** 307.

Dzhrbashyan, V. A. (1959). *J.E.T.P.* **36,** 277.

Dzhuraev, A. A., and Evseev, V. S. (1971). J.I.N.R. P14-6023, Dubna; *J.E.T.P.* **62,** 1167 (1972).

Dzhuraev, A. A., and Evseev, V. S. (1973). J.I.N.R. P14-7212, Dubna (to be published in *J.E.T.P.*).

Dzhuraev, A. A., Evseev, V. S., Myasishcheva, G. G., Obukhov, Yu. V., and Roganov, V. S. (1971a). J.I.N.R. P1-6020, Dubna; *J.E.T.P.* **62,** 1424 (1972).

Dzhuraev, A. A., Evseev, V. S., Obukhov, Yu. V., and Roganov, V. S. (1971b). J.I.N.R. P14-6203, Dubna; *J.E.T.P.* **62,** 2210 (1972).

Dzhuraev, A. A., Evseev, V. S., Obukhov, Yu. V., and Roganov, V. S. (1971c). J.I.N.R. P14-6203, Dubna; *Yad. Fiz.* **16,** 114 (1972).

Dzhuraev, A. A., Evseev, V. S., Obukhov, Yu. V., Roganov, V. S., and Kholodov, N. I. (1971d). J.I.N.R. P14-6206, Dubna; *Yad. Fiz.* **16,** 121 (1972).

Dzhuraev, A. A., Evseev, V. S., Obukhov, Yu. V., Roganov, V. S., Frontas'eva, M. V., and Kholodov, N. I. (1972). J.I.N.R. P6-6822, Dubna; *J.E.T.P.* **64,** 1930 (1973).

Dzhuraev, A. A., Evseev, V. S., Obukhov, Yu. V., and Roganov, V. S. (1974). J.I.N.R. P14-7213, Dubna; *J.E.T.P.* **66,** 2, 1433.

Eisenberg, J., and Kessler, D. (1961). *Nuovo Cimento* **19,** 1195.

Eisenberg, J., and Kessler, D. (1963). *Phys. Rev.* **130,** 2349.

Elkana, J., Feitelson, J., and Katchalski, E. (1968). *J. Chem. Phys.* **48,** 2399.

Evseev, V. S. (1968). J.I.N.R. P14-4052, Dubna.

Evseev, V. S., Komarov, V. I., Kush, V. Z., Roganov, V. S., Chernogorova, V. A., and Shimchak, M. M. (1961). *J.E.T.P.* **41,** 306.

Evseev, V. S., Kil'binger, F., Roganov, V. S., Chernogorova, V. A., and Shimohak, M. M. (1966). *Yad. Fiz.* **4,** 545.

Evseev, V. S., Roganov, V. S., Chernogorova, V. A., Myasishcheva, G. G., and Obukhov, Yu. V. (1968). *Yad. Fiz.* **8,** 742.

Favart, D., Brouillard, F., Grenacs, L., Igo-Kemenes, P., Lipnik, P., and Macq, P. C. (1970). *Phys. Rev. Lett.* **25,** 1348.

Firsov, V. G., and Byakov, V. M. (1964). *J.E.T.P.* **47,** 1074.

Flory, P. J. (1969). "Statistical Mechanics of Chain Molecules." Wiley (Interscience), New York.

Fomin, O. K. (1967). *Usp. Khim.* **36,** 1701.

Franck, J., and Rabinowitsch, E. (1934). *Trans. Faraday Soc.* **30,** 120.

Frisch, S. E. (1963). "Optical Atomic Spectra." Fizmatgiz, Moscow and Leningrad.

Garwin, R. L., Lederman, L. M., and Weinrich, M. (1957). *Phys. Rev.* **105,** 1415.

Gershtein, S. S. (1959). *J.E.T.P.* **34,** 463.

Gershtein, S. S., Petrukhin, V. I., Ponomarev, L. I., and Prokoshkin, Yu. D. (1969). *Usp. Fiz. Nauk* **97,** 3.

Gol'danskii, V. I., and Firsov, V. G. (1971). *Ann. Rev. Phys. Chem.* **22,** 209.

Goudsmit, S. (1931). *Phys. Rev.* **37,** 663.

Gurevich, I. I. (1972). *Proc. Int. Conf. High Energy Phys. Nucl. Structure, 4th* J.I.N.R. D1-6349, p. 411, Dubna.

Gusynin, V. I., and Tal'roze, V. L. (1960). *Dokl. Akad. Nauk SSSR* **135,** 1160.

"Handbook of Chemistry" (1964). Vol. 3. Nauka, Moscow and Leningrad.

Heald, M. A., and Beringer, R. (1954). *Phys. Rev.* **96,** 645.

Hutchinson, D. P., Menes, Y., and Shapiro, G. (1962). *Phys. Rev. Lett.* **9,** 516.

Hutchinson, D. P., Mones, J., Shapiro, G., and Patlach, A. M. (1963). *Phys. Rev.* **131,** 1362.

Ignatenko, A. E., Egorov, L. B., Khalupa, V., and Chultem, D. (1958). *J.E.T.P.* **35,** 1131.

Ignatenko, A. E., Egorov, L. B., Khalupa, V., and Chultem, D. (1961). *Nucl. Phys.* **23,** 75.

Iguerabide, J., Dillon, M., and Burton, M. (1964). *J. Chem. Phys.* **40,** 3040.

Johnson, W. R., O'Connell, R. F., and Mullin, C. J. (1961). *Phys. Rev.* **124,** 904.

Kane, J. R. (1964). Thesis CAR-882-9.

Karlson, T. A., and White, R. M. (1965). *Chem. Effects Transformations, Vienna* **1,** 23.

Karlson, T. A., and White, R. M. (1970). *Int. Discuss. Progr. Probl. Mod. Radiat. Chem., Prague* **1,** 91.

Kessler, D., *et al.* (1967). *Phys. Rev. Lett.* **18,** 1179.

Koldin, E. (1966). "Fast Reactions in Solutions." Mir, Moscow.

Kopferman, H. (1958). "Nuclear Moments." Academic Press, New York.

Lawton, E. J., Balwit, J. S., and Powell, R. S. (1960). *J. Chem. Phys.* **33,** 395.

Low, H., and Wessel, G. (1953). *Phys. Rev.* **90,** 1.

Libby, W. (1947). *J. Amer. Chem. Soc.* **69,** 2523.

Libkin, E. (1960). *Phys. Rev.* **119,** 815.

Mann, R. A., and Rose, M. E. (1961). *Phys. Rev.* **121,** 239.

Martin, A. D. (1963). *Nuovo Cimento* **27,** 1358.

McColm, D., Ziock, K., Hughes, V. W., Penman, S., and Prepost, R. (1960). *Bull. Amer. Phys. Soc. Ser. 11* **5,** 1, 75.

Melander, L. (1964). "Isotopic Effects in Chemical Reactions." Mir, Moscow.

Mokrushin, A. D., and Gol'danskii, V. I. (1967). *J.E.T.P.* **53,** 2.

Molin, Yu. P., Chkheidze, I. I., Buben, N. Ya., and Voevodskii, V. V. (1961). *Kinet. Catal.* **2,** 192.

Molin, Yu. N., Chkheidze, I. I., Kaplan, E. P., Buber, N. Ya., and Voevodskii, V. V. (1962). *Kinet. Catal.* **3,** 674.

Naberukhin, Yu. I., and Rogov, B. A. (1971). *Advan. Chem.* **40,** 3, 369.

Nauka (1965). *Proc. Symp. Elementary Processes High-Energy Chem.* p. 194. Nauka, Moscow.

Nauka (1966). "Polymer Radiation Chemistry," p. 202. Nauka, Moscow.

Nauka (1972). "Free Radical States in Chemistry." Nauke, Siberian Div. Sov. Acad. Sci., Novosibirsk.

Nosov, V. G., and Yakovleva, N. V. (1962). *J.E.T.P.* **43,** 1750.

Pikaev, A. K. (1965). "Pulsed Radiolysis in Water and in Aqueous Solutions." "Nauka," Moscow.

Primakoff, H. (1959). *Rev. Mod. Phys.* **31,** 802.

Prepost, R., Hughes, V. W., Penman, S., McColm, D., and Ziock, K. (1960). *Bull. Amer. Phys. Soc. Ser. 11* **5,** 175, 75.

Pshezhetskii, S. D. (1962). "Radiation-Chemical Reaction Mechanisms." G'oskhimizdat, Moscow.

Seitz, F., and Kocher, J. S. (1956). *Solid State Phys.* **2,** 307.

Selwood, P. (1956). "Magnetochemistry." Wiley (Interscience), New York.

Shantorovich, V. P., and Firsov, V. G. (1967). *Vestnik Acad. Nauk SSSR* **5,** 47.

Shmushkevich, I. M. (1959). *Nucl. Phys.* **11,** 419.

Sobel'man, I. I. (1963). "Introduction to the Theory of Atomic Spectra." Nauka, Moscow.

Sundelin, R. M. (1967). Thesis, CAR-882-22.

Sundelin, R. M., Edelstein, R. M., Suzuki, A., and Tokahashi, K. (1968). *Phys. Rev. Lett.* **20,** 1201.

Suzuki, A. (1967). *Phys. Rev. Lett.* **19,** 1005.

Tammet, E. (1967). *Izv. Akad. Nauk Estonian SSR* **16,** No. 2, 135.

Tammet, E. (1969a). *Izv. Acad. Nauk Estonian SSR* **18,** No. 1, 120.

Tammet, E. (1969b). *Izv. Acad. Nauk Estonian SSR* **18,** No. 2, 193.

Tammet, E. (1969c). *Izv. Acad. Nauk Estonian SSR* **18,** No. 3, 350.

Tammet, E. (1970). *Yad. Fiz.* **11,** 4, 840.

Telegdi, V. L. (1960). *Proc. Rochester Conf., 10th,* 713.

Truby, F. K., Wallace, D. C., and Hesse, J. E. (1965). *J. Chem. Phys.* **42,** 3845.

Überall, H. (1959). *Phys. Rev.* **114,** 1640.

Vaisenberg, A. O. (1964). "μ Meson." Nauka, Moscow.

Varlamov, V., Dobretsov, Yu., Dolgoshein, B., Kirillov-Ugryumov, V., and Rogozhin, A. (1971). *Proc. Int. Conf. High-Energy Phys. Nucl. Structure, 4th* D1-5983, Dubna.

Vereshchinskii, I. V., and Pikaev, A. K. (1963). "Introduction to Radiation Chemistry." Acad. Sci. USSR, Moscow.

Wert, C. A., and Thompson, R. M. (1964). "Physics of Solids." McGraw-Hill, New York.

Wessel, G. (1953). *Phys. Rev.* **92,** 1581.

Winston, R. (1963). *Phys. Rev.* **129,** 2766.

Winston, R., and Telegdi, V. (1961). *Phys. Rev. Lett.* **7,** 104.

Yakovleva, N. V. (1959). *J.E.T.P.* **35,** 970.

Zinov, V. G., Konin, A. D., and Mukhin, A. I. (1965). *Yad. Fiz.* **2,** 859.

Zinov, V. G., Konin, A. D., and Mukhin, A. I. (1967). *Yad. Fiz.* **5,** 591.

INDEX

R

S

T

A 5
B 6
C 7
D 8
E 9
F 0
G 1
H 2
I 3
J 4